National Center for Construction Education and Research

Carpentry Level One

Annotated Instructor's Guide

Prentice
Hall

Upper Saddle River, New Jersey
Columbus, Ohio

This information is general in nature and intended for training purposes only. Actual performance of activities described in this manual requires compliance with all applicable operating, service, maintenance, and safety procedures under the direction of qualified personnel. References in this manual to patented or proprietary devices do not constitute a recommendation of their use.

Prentice Hall

nccer

10 9 8 7 6 5 4 3 2
ISBN 0-13-060477-1

Preface

This volume was developed by the National Center for Construction Education and Research (NCCER) in response to the training needs of the construction and maintenance industries. It is one of many in the NCCER's standardized craft training program. The program, covering more than 30 craft areas and including all major construction skills, was developed over a period of years by industry and education specialists. Sixteen of the largest construction and maintenance firms in the U.S. committed financial and human resources to the teams that wrote the curricula and planned the nationally-accredited training process. These materials are industry-proven and consist of competency-based textbooks and instructor's guides.

The NCCER is a not-for-profit educational entity affiliated with the University of Florida and supported by the following industry and craft associations:

PARTNERING ASSOCIATIONS

- American Fire Sprinkler Association
- American Society for Training & Development
- American Welding Society
- Associated Builders and Contractors, Inc.
- Associated General Contractors of America
- Association for Career and Technical Education
- Carolinas AGC, Inc.
- Carolinas Electrical Contractors Association
- Citizens Democracy Corps
- Construction Industry Institute
- Construction Users Roundtable
- Design-Build Institute of America
- Merit Contractors Association of Canada
- Metal Building Manufacturers Association

- National Association of Minority Contractors
- National Association of State Supervisors for Trade and Industrial Education
- National Association of Women in Construction
- National Insulation Association
- National Ready Mixed Concrete Association
- National Utility Contractors Association
- National Vocational Technical Honor Society
- North American Crane Bureau
- Painting & Decorating Contractors of America
- Portland Cement Association
- SkillsUSA–VICA
- Steel Erectors Association of America
- Texas Gulf Coast Chapter ABC
- U.S. Army Corps of Engineers
- University of Florida
- Women Construction Owners & Executives, USA

Some of the features of the NCCER's standardized craft training program include:

- A proven record of success over many years of use by industry companies.
- National standardization providing portability of learned job skills and educational credits that will be of tremendous value to trainees.
- Recognition: upon successful completion of training with an accredited sponsor, trainees receive an industry-recognized certificate and transcript from the NCCER.
- Compliance with Apprenticeship, Training, Employer, and Labor Services (ATELS) requirements (formerly BAT) for related classroom training (CFR 29:29).
- Well-illustrated, up-to-date, and practical information.

FEATURES OF THIS BOOK

Capitalizing on the well-received campaign last year, NCCER and Prentice Hall are continuing to publish select textbooks in color. *Carpentry Level One* incorporates the new design and layout, along with color photos and illustrations, to present the material in an easy-to-use format. Special pedagogical features augment the technical material to maintain the trainees' interest and foster a deeper appreciation of the trade.

Inside Track provides a head start for those entering the field by presenting tricks of the trade from master carpenters.

Think About It uses "What If" questions to help trainees apply theory to real-world experiences and put ideas into action.

Case Histories emphasize the importance of safety by citing examples of the costly (and often devastating) consequences of ignoring safe work practices and OSHA regulations.

Profiles in Success share the experiences of and advice from successful tradespersons in the carpentry field.

We're excited to be able to offer you these improvements and hope they lead to a more rewarding learning experience. As always, your feedback is welcome! Please let us know how we are doing by visiting NCCER at www.nccer.org or e-mail us at info@nccer.org.

ANNOTATED INSTRUCTOR'S GUIDES—DESIGNED TO MAKE YOUR JOB EASIER!

We understand that the course material you choose may have a profound effect on your success. The best way we can help to ensure your success is to provide you with the finest tools we have available. That's why NCCER and Prentice Hall are releasing select titles as Annotated Instructor's Guides. Each Annotated Instructor's Guide is actually the Trainee Module enhanced with specific directions to the instructor, space for the instructor's notes, suggestions for session break-outs, a comprehensive Materials and Equipment list, and teaching tips to coincide with the performance examinations, laboratories, and demonstrations.

The *Carpentry Level One Annotated Instructor's Guide* is packaged with an accompanying test booklet, which includes the performance tests, written module exams, and answer keys. Also included in this package are task checklists (job sheets) for each module. Please note that the Transparency Masters package and Test Scrambling Software are sold separately for an additional charge. PowerPoint® transparency masters on CD-ROM also are now available.

For more information on the Annotated Instructor's Guides or to find out which titles are planned for annotations in the future, please contact your Prentice Hall Sales Specialist or NCCER Customer Service at info@nccer.org.

Acknowledgments

This curriculum was revised as a result of the farsightedness and leadership of the following sponsors:

Anderson Construction Company of Ft. Gaines

Allied Construction Industries

Carolinas Associated General Contractors

AGC Oregon-Columbia Chapter

Camden County High School, Kingsland, GA

Construction Education Foundation of Georgia

Guilford Technical Community College, Jamestown, NC

Jones County Comprehensive High School, Gray, GA

Loften High School, Gainesville, FL

This curriculum would not exist were it not for the dedication and unselfish energy of those volunteers who served on the Authoring Team. A sincere thanks is extended to:

Mick Anderson

Tommy Caldwell

Richard Davis

John Hoerlein

R. P. Hughes

Carlos Jones

Jonathan D. Liston

Ed Prevatt

We would also like to thank the following reviewers for contributing their time and expertise to this endeavor:

Brian Mate

David Scalf

Jerry Schwengels

Duane Sellers

Pamela Stacy

Johnny Trotter

Van Yates

A final note: This book is the result of a collaborative effort involving the production, editorial, and development staff at Prentice Hall and the National Center for Construction Education and Research. Thanks to all of the dedicated people involved in the many stages of this project.

Contents

27101-01   Orientation to the Trade 1.i

27102-01   Wood Building Materials, Fasteners,
and Adhesives 2.i

27103-01   Hand and Power Tools 3.i

27104-01   Floor Systems 4.i

27105-01   Wall and Ceiling Framing 5.i

27106-01   Roof Framing 6.i

27107-01   Windows and Exterior Doors 7.i

Orientation to the Trade

27101-01

MODULE OVERVIEW

This module introduces the carpentry trainee to the carpentry trade, including the apprenticeship process and the opportunities within the trade.

PREREQUISITES

Please refer to the Course Map in the Trainee Module. Prior to training with this module, it is suggested that the trainee shall have successfully completed the following modules:

Core Curriculum

LEARNING OBJECTIVES

Upon completion of this module, the trainee will be able to:

1. Describe the history of the carpentry trade.
2. Identify the stages of progress within the carpentry trade.
3. Identify the responsibilities of a person working in the construction industry.
4. State the personal characteristics of a professional.
5. Explain the importance of safety in the construction industry.

PERFORMANCE OBJECTIVES

This is a knowledge-based module—there is no performance profile examination.

NCCER STANDARDIZED CRAFT TRAINING PROGRAM

The National Center for Construction Education and Research (NCCER) provides a standardized national program of accredited craft training. Key features of the program include instructor certification, competency-based training, and performance testing. The program provides trainees, instructors, and companies with a standard form of recognition through a National Craft Training Registry. The program is described in full in the *Guidelines for Accreditation*, published by the NCCER. For more information on standardized craft training, contact the NCCER at P.O. Box 141104, Gainesville, FL 32614-1104, 352-334-0911, visit our Web site at www.NCCER.org, or e-mail info@NCCER.org.

HOW TO USE THIS ANNOTATED INSTRUCTOR'S GUIDE

Each page presents two sections of information. The larger section displays each page exactly as it appears in the Trainee Module. The narrow column ties suggested trainee and instructor actions to each page and provides icons to call your attention to material, safety, audiovisual, or testing requirements. The bottom of each page includes space for your notes.

If you see the Teaching Tip icon, that means there is a teaching tip associated with this section. Also refer to any suggested teaching tips at the end of the module.

SAFETY CONSIDERATIONS

Ensure that the trainees are equipped with appropriate personal protective equipment.

PREPARATION

Before teaching this module, you should review the Module Outline, the Learning Objectives, and the Materials and Equipment List. Be sure to allow ample time to prepare your own training or lesson plan and gather all required equipment and materials.

MATERIALS AND EQUIPMENT LIST

Materials:

Transparencies
Markers/chalk
Module Examinations*
Videotape (optional), *Careers in Construction: Carpentry*
Exploring Careers in Construction (optional)

Equipment:

Overhead projector and screen
Whiteboard/chalkboard
Appropriate personal protective equipment
Television and videocassette recorder (optional)

*Located in the Test Booklet packaged with this Annotated Instructor's Guide.

ADDITIONAL RESOURCES–

This module is intended to present thorough resources for task training. The following reference is suggested for both instructors and motivated trainees interested in further study. This is optional material for continued education rather than for task training.

> *Careers in Construction: Carpentry*, videotape. Gainesville, FL: The National Center for Construction Education and Research.

TEACHING TIME FOR THIS MODULE

An outline for use in developing your lesson plan is presented below. Note that each Roman numeral in the outline equates to one session of instruction. Each session has a suggested time period of 2½ hours. This includes 10 minutes at the beginning of each session for administrative tasks and one 10-minute break during the session. Approximately 2½ hours are suggested to cover *Orientation to the Trade*. You will need to adjust the time required for hands-on activity and testing based on your class size and resources.

Topic	Planned Time
Session I. Orientation to the Trade	
A. Introduction	_____
B. History of Carpentry	_____
C. Modern Carpentry	_____
D. Opportunities in the Construction Industry	_____
1. Formal Construction Training	_____
2. Apprenticeship Program	_____
a. Youth Apprenticeship Program	_____
b. Apprenticeship Standards	_____
3. Responsibilities of the Employee	_____
a. Professionalism	_____
b. Honesty	_____
c. Loyalty	_____
d. Willingness to Learn	_____
e. Willingness to Take Responsibility	_____
f. Willingness to Cooperate	_____
g. Rules and Regulations	_____
h. Tardiness and Absenteeism	_____
4. What You Should Expect From Your Employer	_____
5. What You Should Expect From a Training Program	_____
6. What You Should Expect From the Apprenticeship Committee	_____

E. Human Relations   _____
  1. Making Human Relations Work   _____
  2. Human Relations and Productivity   _____
  3. Attitude   _____
  4. Maintaining a Positive Attitude   _____
F. Employer and Employee Safety Obligations   _____
G. Summary   _____
  1. Summarize module
  2. Answer questions
H. Module Examination   _____
  1. Trainees must score 70% or higher to receive recognition from the NCCER.
  2. Record the testing results on Craft Training Report Form 200 and submit the results to the Training Program Sponsor.

Module 27101-01

Orientation
to the Trade

Course Map

This course map shows all of the modules in the first level of the Carpentry curriculum. The suggested training order begins at the bottom and proceeds up. Skill levels increase as a trainee advances on the course map. The training order may be adjusted by the local Training Program Sponsor.

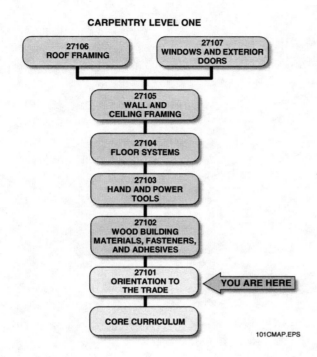

CARPENTRY LEVEL ONE

27106
ROOF FRAMING

27107
WINDOWS AND EXTERIOR DOORS

27105
WALL AND CEILING FRAMING

27104
FLOOR SYSTEMS

27103
HAND AND POWER TOOLS

27102
WOOD BUILDING MATERIALS, FASTENERS, AND ADHESIVES

27101
ORIENTATION TO THE TRADE

YOU ARE HERE

CORE CURRICULUM

101CMAP.EPS

Assign reading of Module 27101.

MODULE 27101 CONTENTS

1.0.0 INTRODUCTION 1.1

2.0.0 HISTORY OF CARPENTRY 1.1

3.0.0 MODERN CARPENTRY 1.6

4.0.0 OPPORTUNITIES IN THE CONSTRUCTION INDUSTRY 1.8
    4.1.0 Formal Construction Training 1.10
    4.2.0 Apprenticeship Program 1.10
    4.2.1 *Youth Apprenticeship Program* 1.10
    4.2.2 *Apprenticeship Standards* 1.11
    4.3.0 Responsibilities of the Employee 1.12
    4.3.1 *Professionalism* 1.12
    4.3.2 *Honesty* 1.12
    4.3.3 *Loyalty* 1.13
    4.3.4 *Willingness to Learn* 1.13
    4.3.5 *Willingness to Take Responsibility* 1.13
    4.3.6 *Willingness to Cooperate* 1.13
    4.3.7 *Rules and Regulations* 1.13
    4.3.8 *Tardiness and Absenteeism* 1.14
    4.4.0 What You Should Expect from Your Employer 1.15
    4.5.0 What You Should Expect from a Training Program 1.15
    4.6.0 What You Should Expect from the Apprenticeship
            Committee 1.15

5.0.0 HUMAN RELATIONS 1.15
    5.1.0 Making Human Relations Work 1.15
    5.2.0 Human Relations and Productivity 1.16
    5.3.0 Attitude 1.16
    5.4.0 Maintaining a Positive Attitude 1.16

6.0.0 EMPLOYER AND EMPLOYEE SAFETY OBLIGATIONS 1.17

SUMMARY 1.19

REVIEW QUESTIONS 1.19

PROFILE IN SUCCESS 1.20

GLOSSARY 1.21

APPENDIX A 1.22

ANSWERS TO REVIEW QUESTIONS 1.25

Figures

Figure 1 Balloon framing 1.3
Figure 2 Western platform framing 1.4
Figure 3 Typical commercial construction 1.6
Figure 4 Typical residential construction 1.7
Figure 5 Example of finish carpentry 1.7
Figure 6 Opportunities in the construction industry 1.8
Figure 7 Safety is everyone's responsibility 1.17
Figure 8 OSHA standards govern workplace safety 1.18

Instructor's Notes:

_____
_____
_____
_____
_____
_____
_____
_____

Orientation to the Trade

Ensure you have everything required to teach the course. Check the Materials and Equipment List at the front of this Instructor's Guide.

Objectives

Upon completion of this module, the trainee will be able to:

1. Describe the history of the carpentry trade.
2. Identify the stages of progress within the carpentry trade.
3. Identify the responsibilities of a person working in the construction industry.
4. State the personal characteristics of a professional.
5. Explain the importance of safety in the construction industry.

Prerequisites

Successful completion of the following Task Modules is recommended before beginning study of this Task Module: Core Curriculum.

Required Trainee Materials

1. Trainee Task Module
2. Appropriate personal protective equipment

1.0.0 ◆ INTRODUCTION

Opportunity is driven by knowledge and ability, which are in turn driven by education and training. This program of the National Center for Construction Education and Research (NCCER) was designed and developed by the construction industry for the construction industry. It is the only nationally-accredited, competency-based construction training program in the United States. A *competency-based* program requires that the trainee demonstrate the ability to safely perform specific job-related tasks in order to receive credit. This approach is unlike other apprentice programs that merely require a trainee to put in the required number of hours in the classroom and on the job.

The primary goal of the NCCER is to standardize construction craft training throughout the country so that both employers and employees will benefit from the training, no matter where they are located. As a trainee in an NCCER program, you will become part of a national registry. You will receive a certificate for each level of training you complete. If you apply for a job with any participating contractor in the country, a transcript of your training will be available. If your training is incomplete when you make a job transfer, you can pick up where you left off because every participating contractor is using the same training program. In addition, many technical schools and colleges are using the program.

2.0.0 ◆ HISTORY OF CARPENTRY

Primitive carpentry developed in forest regions during the latter years of the Stone Age, when early humans improved stone tools so they could be used to shape wood for shelters, animal traps, and dugout boats. Between 4000 and 2000 BCE, Egyptians developed copper tools, which they used to build vaults, bed frames, and furniture. Later in that period, they developed bronze tools and bow drills. An example of the Egyptians' skill in mitering, mortising, dovetailing, and paneling is the intricate furniture found in the tomb of Tutankhamen ("King Tut"). European carpenters did not produce such furniture until the Renaissance (1300 to 1500 CE), although they used timber to construct dwellings, bridges, and industrial equipment. In Denmark and ancient Germany, Neolithic people (around 5000 BCE) built rectangular houses from timbers that were nearly 100

Show Transparency 1, Course Objectives.

Discuss the NCCER apprenticeship program. Explain that each trainee's academic record is maintained in the National Registry, and describe how the registry works. Refer the trainees to *Appendix A* in the Trainee Module.

Discuss the types of framing.

Show Transparencies 2 and 3 (Figures 1 and 2).

Discuss the history of carpentry.

INSIDE TRACK

Tools

Tools are an essential part of carpentry. Although modern tools have advanced beyond the primitive stone tools used by our ancestors, they serve the carpenter's needs in much the same way. Carpentry, like other trades, relies on tools to make difficult tasks easier. If you take the time to learn the proper way to handle and use your tools safely, you'll be able to work much more effectively and produce a high-quality product.

101P0101.TIF

feet long. In England, the mortised and fishtailed joints of the stone structures at Stonehenge indicate that advanced carpentry techniques were known in ancient Britain. Before the Roman conquest of Britain (100 CE), its carpenters had already developed iron tools such as saws, hatchets, rasps, and knives. They even had turned-wood objects made on primitive pole lathes.

In the Middle Ages, carpenters began a movement toward specialization, such as shipwrights, wheelwrights, turners, and millwrights. However, general-purpose carpenters were still found in most villages and on large private estates. These carpenters could travel with their tools to outlying areas that had no carpenters or to a major building project that required temporary labor. During this period, European carpenters invented the carpenter's brace (a tool for holding and turn-

ing a drill bit). The plane, which the Romans had used centuries earlier, reappeared about 1200 CE The progress of steelmaking also provided for advancements in the use of steel-edged tools and the advent of crude iron nails. Wooden pegs were used to hold wooden members together before the use of nails. Screws were invented in the 1500s.

The first castles and churches in northern Europe were constructed of timber. When the great stone buildings replaced those made of timber, skilled carpenters built the floors, paneling, doors, and roofs. The erection of large stone buildings also led to the inventions of scaffolding for walls, framework for arch assembly, and pilings to strengthen foundations. Houses and other smaller buildings were still made of timber and thinner wood. Clay was used to fill the gaps between the beams.

The art of carpentry contributed significantly to the grandeur of the great buildings of the Renaissance. Two noted masterpieces of timber construction are the outer dome of St. Paul's Cathedral in London and the 68-foot roof of the Sheldonian Theater in Oxford. After the Renaissance, other examples of architecture requiring skilled carpentry appeared, including the mansard roof with its double slope, providing loftier attics, broad staircases, and sashed windows. These architectural features were incorporated in homes constructed in Colonial America. George W. Snow introduced balloon-frame construction (*Figure 1*) in Chicago in 1840, which proved to be a much cheaper and quicker method because it used machine-made studs and nails. In balloon framing, the studs run from the bottom floor to the uppermost rafters. This method gives the structure exceptional ability to handle strong winds, but requires very long studs that are difficult to manufacture, transport, and store. Because of these problems, balloon framing has almost disappeared. It is used to some extent in Florida to frame the gable ends of buildings in order to provide protection from hurricanes. Today, platform (western) framing (*Figure 2*) has almost completely replaced balloon-frame construction.

This brief history illustrates that carpentry has a long and rich heritage. It also shows that carpentry is an ever-changing trade. You will inevitably discover that learning never ends as you practice the carpentry trade because new and better ways of construction will continue to emerge.

CARPENTRY LEVEL ONE—TRAINEE MODULE 27101

Instructor's Notes:

_____

_____

_____

_____

_____

_____

_____

_____

CARPENTRY LEVEL ONE—INSTRUCTOR'S GUIDE MODULE 27101

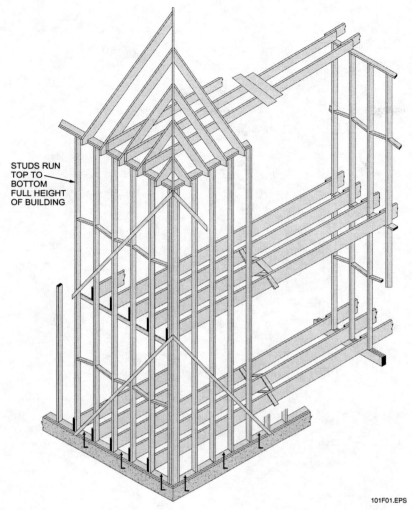

STUDS RUN
TOP TO
BOTTOM
FULL HEIGHT
OF BUILDING

101F01.EPS

Figure 1 ◆ Balloon framing.

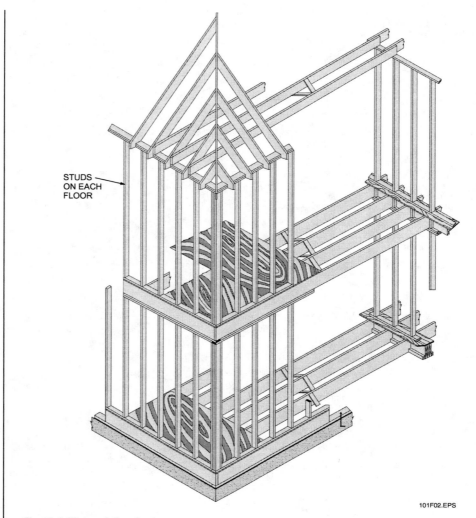

STUDS ON EACH FLOOR

101F02.EPS

Figure 2 ◆ Western platform framing.

Instructor's Notes:

Ancient Construction

Modern carpentry is a continuum from ancient times. Carpenters and other craftsmen were responsible for construction of the world's most celebrated examples of historical architecture. From the lavish interiors of ancient Egypt's great pyramids to the outer dome of London's St. Paul's Cathedral (shown here), the rich heritage of carpentry can be found across the world.

101P0102.TIF

3.0.0 ◆ MODERN CARPENTRY

The scope of carpentry has expanded in modern times with the use of synthetic building materials and ever-improving tools. Today's carpenters must not only know about wood, but also about materials such as particleboard, gypsum wallboard, suspended ceiling tiles, plastics, and laminates. They must also know how to use many modern tools, fasteners, construction techniques, and safety procedures.

The duties of carpenters can vary significantly from one job to another. A carpenter who works for a commercial contractor may work primarily with concrete, steel, and preformed building materials (*Figure 3*). A carpenter who does residential work is more likely to work with wood-frame construction and wood finish materials, but will also encounter an increasing variety of preformed and prefabricated building materials (*Figure 4*).

In the construction industry, carpentry is commonly divided into two categories: rough and finish. Examples of rough carpentry include erecting frameworks, scaffolds, and wooden forms for

101F03.TIF

Figure 3 ◆ Typical commercial construction.

concrete, as well as building docks, bridges, and supports for tunnels and sewers. Finish carpentry (*Figure 5*) includes building stairs; installing doors, cabinets, wood paneling, and molding; and putting up acoustical tiles. Skilled carpenters do both rough and finish work.

The duties of carpenters vary even within the broad categories of rough and finish carpentry. The type of construction, size of the company, skill of the carpenter, community size, and other factors affect the carpenter's work. Carpenters who are employed by a large contractor, for example, may specialize in one area, such as laying hardwood floors, while others who are employed by a small firm may build wall frames, put in insulation, and install paneling. They may even perform concrete finishing, welding, and painting. The duties of carpenters also vary because each job is unique.

Carpenters often have great freedom in planning and performing their work. However, carpentry techniques are standard, and most jobs involve the following steps to some extent:

- Using the construction drawings, receiving instructions from the supervisor, or both, carpenters first do the layout, measure the area, and prepare a takeoff of the materials.
- Local building codes are consulted to determine the materials to be used.
- Wood or other material is cut or shaped with various hand and power tools such as saws and drills.
- Carpenters then join the materials with nails, screws, or glue.
- Finally, the accuracy of the work is checked using levels, rulers, and framing squares.

Carpenters also use powder-actuated and pneumatic tools and operate power equipment such as personnel lifts, equipment and material lifts, and small earth-moving machines. You will learn more about the various tools and construction methods used by carpenters in other modules of this program.

As in other building trades, the carpenter's work is active and sometimes strenuous. Prolonged standing, climbing, and squatting are often necessary. Many carpenters work outside under adverse weather conditions. Carpenters risk injury from slips or falls, from contact with sharp or rough materials, and from the use of sharp tools and power equipment. Being new to the trade increases the chance of being injured. Therefore, it is essential that you rely on the knowledge of more experienced workers, learn applicable safety procedures, and wear appropriate personal protective equipment.

CARPENTRY LEVEL ONE—TRAINEE MODULE 27101

Instructor's Notes:

_____
_____
_____
_____
_____
_____
_____
_____

101F04.TIF

Figure 4 ◆ Typical residential construction.

101F05.TIF

Figure 5 ◆ Example of finish carpentry.

Show Transparency 4 (Figure 6).

Explain the professional levels through which a carpenter can progress.

Obtain a copy of the NCCER publication *Exploring Careers in Construction* and allow the trainees to examine it. Emphasize the wide variety of opportunities available in the carpentry trade.

Show the trainees the video *Careers in Construction: Carpentry,* available from the NCCER.

4.0.0 ◆ OPPORTUNITIES IN THE CONSTRUCTION INDUSTRY

The construction industry employs more people and contributes more to the nation's economy than any other industry. Our society will always need new homes, roads, airports, hospitals, schools, factories, and office buildings. This means that there will always be a source of well-paying jobs and career opportunities for carpenters and other construction trade professionals. As shown in *Figure 6*, the opportunities are not limited to work on construction projects. A skilled, knowledgeable carpenter can work in a number of areas.

As a construction worker, a carpenter can progress from apprentice through several levels:

- Journeyman carpenter
- Master carpenter
- Foreman/lead carpenter
- Supervisor
- Safety manager
- Project manager/administrator
- Estimator
- Architect
- General contractor
- Construction manager
- Contractor/owner

Journeyman carpenter – After successfully completing an apprenticeship, a trainee becomes a journeyman. The term *journeyman* originally meant to journey away from the master and work alone. A person can remain a journeyman or advance in the trade. Journeymen may have additional duties such as supervisor or estimator. With larger companies and on larger jobs, journeymen often become specialists.

Master carpenter – A master craftsperson is one who has achieved and continuously demonstrates the highest skill levels in the trade. The master is a mentor and teacher of those to follow. Master carpenters often start their own businesses and become contractors/owners.

Foreman/lead carpenter – This individual is a front-line leader who directs the work of a crew of craft workers and laborers.

Supervisor – Large construction projects require supervisors who oversee the work of crews made up of foremen, apprentices, and journeymen. They are responsible for assigning, directing, and inspecting the work of construction crew members.

Safety manager – An individual responsible for project safety and health-related issues, including development of the safety plan and procedures, safety training for workers, and regulatory compliance.

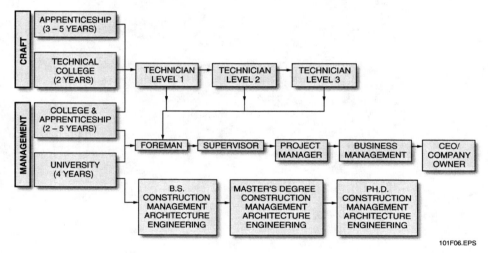

Figure 6 ◆ Opportunities in the construction industry.

Instructor's Notes:

_____

_____

_____

_____

_____

_____

_____

_____

Careers in Carpentry

Project manager/administrator – Business management and administration deal with controlling the scope and direction of the business and dealing with such concerns as payroll, taxes, and employee benefits. Larger contracting firms may have one or several managers/ administrators. This person is responsible for worker output and must determine the best methods to use and the way to apply workers to accomplish the job. A project administrator is one who is responsible for a contractor's support operations, such as accounting, finance, and secretarial work.

Estimator – Estimators work for contractors and building supply companies. They make careful estimates of the materials and labor required for a job. Based on these estimates, the contractor submits bids for jobs. Estimating requires a complete understanding of construction methods as well as the materials and supplies required. Only experienced carpenters who possess good math skills and the patience to prepare detailed, accurate estimates are employed to do this work. This is a highly responsible position since errors in estimates can result in financial losses to the contractor. Depending on the size and type of the business, the job of estimating may be done by the owner, manager, administrator, or an estimating specialist. Today's estimators need solid computer skills because advances in computer software have revolutionized the field of estimating.

Architect – An architect is a person who is licensed to design buildings and oversee their construction. A person normally needs a specialized degree in architecture to qualify as an architect.

General contractor – A general contractor is an individual or company that manages an entire construction project. The general contractor plans and schedules the project, buys the materials, and usually contracts with carpentry, plumbing, electrical, and other trade contractors to perform the work. The general contractor usually works with architects, engineers, and clients in planning and implementing a project. General contracting is a natural career path for a master carpenter because, of the many trades involved in a construction project, the carpenter generally plays the largest role and is more likely to have knowledge of the other trades. The general (prime) contractor is also responsible for safety on site.

Construction manager – The role of the construction manager (CM) is different from that of the general contractor. The CM is usually hired by the building owner to represent the owner's interests on the project. The CM is the individual who works with the general contractor and architect to ensure that the building meets the owner's requirements.

Contractor/owner – Construction contractors/ owners are those who have established a contracting business. Generally, they hire apprentices, journeymen, and master carpenters to work for them. Depending upon the size of the business, contractors may work with the crew or they may manage the business full-time. Very small contractors may have only one or two people do everything, including managing the business, preparing estimates, obtaining supplies, and doing the work on the job. This group includes specialty subcontractors who perform specialized tasks such as framing, interior trim work, and cabinet installation.

More than any other construction worker, the carpenter is likely to become a construction specialist, with knowledge of many trades. This makes carpentry work interesting and challenging and creates a great variety of career opportunities.

The important thing to learn is that, regardless of the path you choose, a career is a lifelong learning process. To be an effective carpenter, you need to keep up-to-date with new tools, materials, and

As time permits, invite guests who hold positions at different levels and occupations in the carpentry trade to speak to the trainees.

methods. If you choose to work your way into management or to someday start your own construction business, you need to learn management and administrative skills on top of keeping your carpentry skills honed. Every successful manager and business owner started the same way you are starting, and they all have one thing in common: a desire and willingness to keep on learning. The learning process begins with apprentice training.

As you develop your carpentry skills and gain experience, you will have the opportunity to earn greater pay for your services. There is great financial incentive for learning and growing within the trade. You can't get to the top, however, without learning the basics.

4.1.0 Formal Construction Training

Over the past twenty years, the rate of formal training within the construction industry has been declining. Until the establishment of the NCCER, the only opportunity for formal construction training was through the Federal Department of Labor, Bureau of Apprenticeship and Training (BAT). The National Apprenticeship Act of 1937, commonly referred to as *the Fitzgerald Act,* officially established BAT. The federal government recently created the Office of Apprenticeship, Training, Employer and Labor Services (ATELS) that consolidated both BAT and new employer–labor relations responsibilities.

The federal government established registered apprenticeship training via the Code of Federal Regulations (CFR) 29:29, which dictates specific requirements for apprenticeship, and CFR 29:30, which dictates specific guidelines for recruitment, outreach, and registration into BAT-approved apprenticeship programs.

Compared to the overall employment in the construction industry, the percentage of enrollment in BAT-style programs has been less than 5 percent for the past decade. BAT programs rely upon mandatory classroom instruction and on-the-job training (OJT). The classroom instruction required is 144 hours per year while the OJT requirement is 2,000 hours per year. A typical BAT program requires 8,000 hours of OJT and 576 hours of related classroom training prior to getting the journeyman certificate dispensed by the BAT.

Craft training via the BAT has not been changed for 30 years, which is believed to be one reason for the lack of use of this program in the construction industry today. Education and training throughout the country is undergoing significant change. As education, political, financial, and student factions argue over the direction and future of edu-

cation, educators and researchers have been learning and applying new techniques to adjust to how today's students learn and apply their education.

NCCER is an independent, private educational foundation founded and funded by the construction industry to solve the training problem plaguing the industry today. The basic idea of the NCCER is to supplant governmental control and credentialing of the construction workforce with *industry-driven* training and education programs. NCCER departs from traditional classroom learning and has adopted a pure *competency-based* training regimen. Competency-based training means that instead of requiring specific hours of classroom training and set hours of on-the-job training, you simply have to prove that you know what is required and can demonstrate that you can perform the specific skill. NCCER also uses the latest technology, interactive computer-based training, to deliver the classroom portions of the training. All completion information for every trainee is sent to the NCCER and kept within the National Registry. The National Registry can then confirm training and skills for workers as they move from company to company, state to state, or even within their own company (see *Appendix A*).

The dramatic shortage of skills within the construction workforce, combined with the shortage of new workers coming into the industry, is forcing the industry to design and implement new training initiatives to combat the problem. Whether you enroll in a BAT program, an NCCER program, or both, it is critical that you work for an employer who supports a national, standardized training program that includes credentials to confirm your skill development.

4.2.0 Apprenticeship Program

Apprentice training goes back thousands of years; its basic principles have not changed in that time. First, it is a means for individuals entering the craft to learn from those who have mastered the craft. Second, it focuses on learning by doing; real skills versus theory. Although some theory is presented in the classroom, it is always presented in a way that helps the trainee understand the purpose behind the skill that is to be learned.

4.2.1 Youth Apprenticeship Program

A Youth Apprenticeship Program is also available that allows students to begin their apprentice training while still in high school. A student entering the carpentry program in eleventh grade may complete as much as one year of the NCCER Stan-

Explain the regulations and requirements governing formal construction training.

Emphasize the benefits of apprenticeship programs.

As time permits, introduce the trainees to members of a youth apprenticeship program, and allow them to observe the apprentices in training.

Instructor's Notes:

_____

_____

_____

_____

_____

_____

_____

_____

dardized Craft Training four-year program by high school graduation. In addition, the program, in cooperation with local craft employers, allows students to work in the trade and earn money while still in school. Upon graduation, the student can enter the industry at a higher level and with more pay than someone just starting the apprenticeship program.

This training program is similar to the one used by National Center for Construction Education and Research learning centers, contractors, and colleges across the country. Students are recognized through official transcripts and can enter the second year of the program wherever it is offered. They may also have the option of applying the credits at a two-year or four-year college that offers degree or certificate programs in the construction trades.

4.2.2 Apprenticeship Standards

All apprenticeship standards prescribe certain work-related or on-the-job training. This on-the-job training is broken down into specific tasks in which the apprentice receives hands-on training during the period of the apprenticeship. In addition, a specified number of hours is required in each task. The total number of hours for the carpentry apprenticeship program is traditionally 8,000, which amounts to about four years of training. In a competency-based program, it may be possible to shorten this time by testing out of specific tasks through a series of performance exams.

In a traditional program, the required on-the-job training may be acquired in increments of 2,000 hours per year. Layoffs or illness may affect the duration.

The apprentice must log all work time and turn it in to the Apprenticeship Committee (discussed later) so that accurate time control can be maintained. Another important aspect of keeping work records up-to-date is that after each 1,000 hours of related work, the apprentice will receive a pay increase as prescribed by the apprenticeship standards.

The classroom-related instruction and work-related training will not always run concurrently due to such reasons as layoffs, type of work needed to be done in the field, etc. Furthermore, apprentices with special job experience or coursework may obtain credit toward their classroom requirements. This reduces the total time required in the classroom while maintaining the total 8,000-hour on-the-job training requirement. These special cases will depend on the type of program and the regulations and standards under which it operates.

Informal on-the-job training provided by employers is usually less thorough than that provided through a formal apprenticeship program. The degree of training and supervision in this type of program often depends on the size of the employing firm. A small contractor who specializes in home building may provide training in only one area, such as rough framing. In contrast, a large general contractor may be able to provide training in several areas.

For those entering an apprenticeship program, a high school or technical school education is desirable, as are courses in carpentry, shop, mechanical drawing, and general mathematics. Manual dexterity, good physical condition, a good sense of balance, and a lack of fear of working in high places are important. The ability to solve arithmetic problems quickly and accurately and to work closely with others is essential. You must have a high concern for safety.

The prospective apprentice must submit to the apprenticeship committee certain information. This may include the following:

- Aptitude test (General Aptitude Test Battery or GATB Form Test) results (usually administered by the local Employment Security Commission)
- Proof of educational background (candidate should have school(s) send transcripts to the committee)
- Letters of reference from past employers and friends
- Results of a physical examination
- Proof of age
- If the candidate is a veteran, a copy of Form DD214
- A record of technical training received that relates to the construction industry and/or a record of any pre-apprenticeship training
- High school diploma or General Equivalency Diploma (GED)

The apprentice must:

- Wear proper safety equipment on the job
- Purchase and maintain tools of the trade as needed and required by the contractor
- Submit a monthly on-the-job training report to the committee
- Report to the committee if a change in employment status occurs
- Attend classroom-related instruction and adhere to all classroom regulations such as that for attendance

Emphasize the standards with which an apprenticeship must conform.

Explain the requirements for apprenticeship application.

4.3.0 Responsibilities of the Employee

In order to be successful, the professional must have the skills to use current trade materials, tools, and equipment to produce a finished product of high quality in a minimum period of time. A carpenter must be adept at adjusting methods to meet each situation. The successful carpenter must continuously train to remain knowledgeable about the technical advancements in trade materials and equipment and to gain the skills to use them. A professional carpenter never takes chances with regard to personal safety or the safety of others.

4.3.1 Professionalism

The word *professionalism* is a broad term that describes the desired overall behavior and attitude expected in the workplace. Professionalism is too often absent from the construction site and the various trades. Most people would argue that it must start at the top in order to be successful. It is true that management support of professionalism is important to its success in the workplace, but it is more important that individuals recognize their own responsibility for professionalism.

Professionalism includes honesty, productivity, safety, civility, cooperation, teamwork, clear and concise communication, being on time and prepared for work, and regard for one's impact on one's co-workers. It can be demonstrated in a variety of ways every minute you are in the workplace. Most important is that you do not tolerate the unprofessional behavior of co-workers. This is not to say that you shun the unprofessional worker; instead, you work to demonstrate the benefits of professional behavior.

Professionalism is a benefit both to the employer and the employee. *It is a personal responsibility*. Our industry is what each individual chooses to make of it; choose professionalism and the industry image will follow.

4.3.2 Honesty

Honesty and personal integrity are important traits of the successful professional. Professionals pride themselves on performing a job well and on being punctual and dependable. Each job is completed in a professional way, never by cutting corners or reducing materials. A valued professional maintains work attitudes and ethics that protect property such as tools and materials belonging to employers, customers, and other trades from damage or theft at the shop or job site.

Honesty and success go hand-in-hand. It is not simply a choice between good and bad, but a

---

Ethical Principles for Members of the Construction Trades

Honesty: Be honest and truthful in all dealings. Conduct business according to the highest professional standards. Faithfully fulfill all contracts and commitments. Do not deliberately mislead or deceive others.

Integrity: Demonstrate personal integrity and the courage of your convictions by doing what is right even when there is great pressure to do otherwise. Do not sacrifice your principles for expediency, be hypocritical, or act in an unscrupulous manner.

Loyalty: Be worthy of trust. Demonstrate fidelity and loyalty to companies, employers, fellow craftspeople, and trade institutions and organizations.

Fairness: Be fair and just in all dealings. Do not take undue advantage of another's mistakes or difficulties. Fair people display a commitment to justice, equal treatment of individuals, tolerance for and acceptance of diversity, and open-mindedness.

Respect for others: Be courteous and treat all people with equal respect and dignity regardless of sex, race, or national origin.

Law abiding: Abide by laws, rules, and regulations relating to all personal and business activities.

Commitment to excellence: Pursue excellence in performing your duties, be well-informed and prepared, and constantly endeavor to increase your proficiency by gaining new skills and knowledge.

Leadership: By your own conduct, seek to be a positive role model for others.

Instructor's Notes:

_____

_____

_____

_____

_____

_____

_____

_____

choice between success and failure. Dishonesty will *always* catch up with you. Whether you are stealing materials, tools, or equipment from the job site or simply lying about your work, it will not take long for your employer to find out. Of course, you can always go and find another employer, but this option will ultimately run out on you.

If you plan to be successful and enjoy continuous employment, consistency of earnings, and being sought after as opposed to seeking employment, then start out with the basic understanding of honesty in the workplace and you will reap the benefits.

Honesty means more, however, than just not taking things that do not belong to you. It means giving a fair day's work for a fair day's pay; it means carrying out your side of a bargain; it means that your words convey true meanings and actual happenings. Our thoughts as well as our actions should be honest. Employers place a high value on an employee who is strictly honest.

4.3.3 Loyalty

Employees expect employers to look out for their interests, to provide them with steady employment, and to promote them to better jobs as openings occur. Employers feel that they, too, have a right to expect their employees to be loyal to them—to keep their interests in mind; to speak well of them to others; to keep any minor troubles strictly within the plant or office; and to keep absolutely confidential all matters that pertain to the business. Both employers and employees should keep in mind that loyalty is not something to be demanded; rather, it is something to be earned.

4.3.4 Willingness to Learn

Every office and plant has its own way of doing things. Employers expect their workers to be willing to learn these ways. Also, it is necessary to adapt to change and be willing to learn new methods and procedures as quickly as possible. Sometimes the installation of a new machine or the purchase of new tools makes it necessary for even experienced employees to learn new methods and operations. It is often the case that employees resent having to accept improvements because of the retraining that is involved. However, employers will no doubt think they have a right to expect employees to put forth the necessary effort. Methods must be kept up-to-date in order to meet competition and show a profit. It is this profit that enables the owner to continue in business and that provides jobs for the employees.

4.3.5 Willingness to Take Responsibility

Most employers expect their employees to see what needs to be done, then go ahead and do it. It is very tiresome to have to ask again and again that a certain job be done. It is obvious that having been asked once, an employee should assume the responsibility from then on. Employees should be alert to see boxes that need to be out of the way, stock that should be stacked, or tools that need to be put away. It is true that, in general, responsibility should be delegated and not assumed; once the responsibility has been delegated, however, the employee should continue to perform the duties without further direction. Every employee has the responsibility for working safely.

4.3.6 Willingness to Cooperate

To cooperate means to work together. In our modern business world, cooperation is the key to getting things done. Learn to work as a member of a team with your employer, supervisor, and fellow workers in a common effort to get the work done efficiently, safely, and on time.

4.3.7 Rules and Regulations

People can work together well only if there is some understanding about what work is to be done, when it will be done, and who will do it.

The Customer

When you are on a job site, you should consider yourself to be working for both your employer and your employer's customer. If you are honest and maintain a professional attitude when interacting with customers, everyone will benefit. Your employer will be pleased with your performance, and the customer will be happy with the work that is being done. Try seeing things from a customer's point of view. A good, professional attitude goes a long way toward ensuring repeat business.

Emphasize the importance of good customer relations.

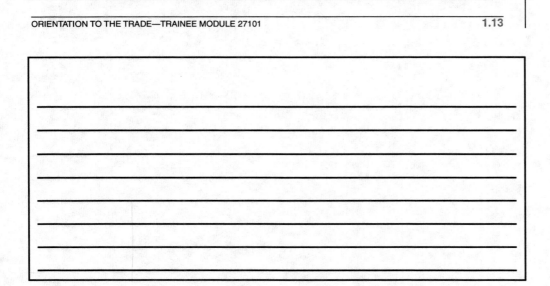

Rules and regulations are a necessity in any work situation and should be so considered by all employees.

4.3.8 Tardiness and Absenteeism

Tardiness means being late for work and absenteeism means being off the job for one reason or another. Consistent tardiness and frequent absences are an indication of poor work habits, unprofessional conduct, and a lack of commitment.

We are all creatures of habit. What we do once we tend to do again unless the results are too unpleasant. The habit of always being late may have begun back in our school days when we found it hard to get up in the morning. This habit can get us into trouble at school, and it can go right on getting us into trouble when we are through school and go to work.

Your work life is governed by the clock. You are required to be at work at a definite time. So is everyone else. Failure to get to work on time results in confusion, lost time, and resentment on the part of those who do come on time. In addition, it may lead to penalties, including dismissal. Although it may be true that a few minutes out of a day are not very important, we must remember that a principle is involved. Our obligation is to be at work at the time indicated. We agree to the terms of work when we accept the job. Perhaps it will help us to see things more clearly if we try to look at the matter from the point of view of the boss. Supervisors cannot keep track of people if they come in any time they please. It is not fair to others to ignore tardiness. Failure to be on time may hold up the work of fellow workers. Better planning of our morning routine will often keep us from being delayed and so prevent a breathless, late arrival. In fact, arriving a little early indicates your interest and enthusiasm for your work, which is appreciated by employers. The habit of being late is another one of those things that stand in the way of promotion.

It is sometimes necessary to take time off from work. No one should be expected to work when sick or when there is serious trouble at home. However, it is possible to get into the habit of letting unimportant and unnecessary matters keep us from the job. This results in lost production and hardship on those who try to carry on the work with less help. Again, there is a principle involved. The person who hires us has a right to expect us to be on the job unless there is some very good reason for staying away. Certainly, we should not let some trivial reason keep us home. We should not stay up nights until we are too tired to go to work the next day. If we are ill, we should use the time at home to do all we can to recover quickly. This, after all, is no more than most of us would expect of a person we had hired to work for us, and on whom we depended to do a certain job.

If it is necessary to stay home, then at least phone the office early in the morning so that the boss can find another worker for the day. Time and again, employees have remained home without sending any word whatever to the employer. This is the worst possible way to handle the matter. It leaves those at work uncertain about what to expect. They have no way of knowing whether you have merely been held up and will be in later, or whether immediate steps should be taken to assign your work to someone else. Courtesy alone demands that you let the boss know if you cannot come to work.

The most frequent causes of absenteeism are illness or death in the family, accidents, personal business, and dissatisfaction with the job. Here we see that some of the causes are legitimate and unavoidable, while others can be controlled. One can usually plan to carry on most personal business affairs after working hours. Frequent absences will reflect unfavorably on a worker when promotions are being considered.

Employers sometimes resort to docking pay, demotion, and even dismissal in an effort to control tardiness and absenteeism. No employer likes to impose restrictions of this kind. However, in fairness to those workers who do come on time and who do not stay away from the job, an employer is sometimes forced to discipline those who will not follow the rules.

Discuss the "Think About It." Emphasize the negative consequences of tardiness and absenteeism.

Late for Work

Showing up on time is a basic requirement for just about every job. Your employer is counting on you to be there at a set time, ready to work. While legitimate emergencies may arise that may cause you to be late for or even miss work, starting a bad habit of consistent tardiness is not something you want to do. What are the possible consequences that you could face as a result of tardiness and absenteeism?

Instructor's Notes:

_____

_____

_____

_____

_____

_____

_____

_____

4.4.0 What You Should Expect from Your Employer

After an applicant has been selected for apprenticeship by the committee, the employer of the apprentice agrees that the apprentice will be employed under conditions that will result in normal advancement. In return, the employer requires the apprentice to make satisfactory progress in on-the-job training and related classroom instruction. The employer agrees that the apprentice will not be employed in a manner that may be considered in violation of the apprenticeship standards. The employer also agrees to pay a prorated share of the cost of operating the apprenticeship program.

4.5.0 What You Should Expect from a Training Program

First and foremost, it is important that the employer you select has a training program. The program should be comprehensive, standardized, and competency-based, not based on the amount of time you spend in a classroom.

When employers take the time and initiative to provide quality training, it is a sign that they are willing to invest in their workforce and improve the abilities of workers. It is important that the training program be national in scope and that transcripts and completion credentials are issued to participants. Construction is unique in that the employers share the workforce. An employee in the trades may work for several contractors throughout their time in the field. Therefore, it is critical that the training program help the worker move from company to company, city to city, or state to state without having to start at the beginning for each move. Ask how many employers in the area use the same program before you enroll. Make sure that you will always have access to transcripts and certificates to ensure your status and level of completion.

Training should be rewarded. The training program should have a well-defined compensation ladder attached to it. Successful completion and mastery of skill sets should be accompanied by increases in hourly wages.

Finally, the curricula should be complete and up-to-date. Any training program has to be committed to maintaining its curricula, developing new delivery mechanisms (CD-ROM, Internet, etc.), and being constantly vigilant for new techniques, materials, tools, and equipment in the workplace.

4.6.0 What You Should Expect from the Apprenticeship Committee

The Apprenticeship Committee is the local administrative body to which the apprentice is assigned and to which the responsibility is delegated for the appropriate training of the individual. Every apprenticeship program, whether state or federal, is covered by standards that have been approved by those agencies. The responsibility of enforcement is delegated to the committee.

The committee is responsible not only for enforcement of standards, but must see to it that proper training is conducted so that a craftsperson graduating from the program is fully qualified in those areas of training designated by the standards.

The committee is the agency that screens and selects individuals for apprenticeship and refers them to participating firms for training.

The committee places apprentices under written agreement for participation in the program.

The committee establishes minimum standards for related instruction and on-the-job training and monitors the apprentice to see that these criteria are adhered to during the training period.

The committee hears all complaints of violations of apprenticeship agreements, whether by employer or apprentice, and takes action within the guidelines of the standards.

The committee notifies the registration agencies of all enrollments, completions, and terminations of apprentices.

5.0.0 ◆ HUMAN RELATIONS

Most people underestimate the importance of working well with others. There is a tendency to pass off human relations as nothing more than common sense. What exactly is involved in human relations? One response would be to say that part of human relations is being friendly, pleasant, courteous, cooperative, adaptable, and sociable.

5.1.0 Making Human Relations Work

As important as the above-noted characteristics are for personal success, they are not enough. Human relations is much more than just getting people to like you. It is also knowing how to handle difficult situations as they arise.

Human relations is knowing how to work with supervisors who are often demanding and sometimes unfair. It is understanding the personality

As time permits, invite an employer to engage in an interactive discussion of employers' obligations to their employees.

Emphasize the importance of human relations and its impact on productivity.

Discuss the "Think About It." Have the trainees share past experiences regarding teamwork and its impact on productivity.

Emphasize the importance of maintaining a positive attitude in the workplace.

Teamwork

Many of us like to follow all sorts of different teams: racing teams, baseball teams, football teams, and soccer teams. Just as in sports, a job site is made up of a team. As a part of that team, you have a responsibility to your teammates. What does teamwork really mean on the job?

traits of others as well as yourself. Human relations is building sound working relationships in situations where others are forced on you.

Human relations is knowing how to restore working relationships that have deteriorated for one reason or another. It is learning how to handle frustrations without hurting others. Human relations is building and maintaining relationships with all kinds of people, whether those people are easy to get along with or not.

5.2.0 Human Relations and Productivity

Effective human relations is directly related to productivity. Productivity is the key to business success. Every employee is expected to produce at a certain level. Employers quickly lose interest in an employee who has a great attitude but is able to produce very little. There are work schedules to be met and jobs that must be completed.

All employees, both new and experienced, are measured by the amount of quality work they can safely turn out. The employer expects every employee to do his or her share of the workload.

However, doing one's share in itself is not enough. If you are to be productive, you must do your share (or more than your share) without antagonizing your fellow workers. You must perform your duties in a manner that encourages others to follow your example. It makes little difference how ambitious you are or how capably you perform. You cannot become the kind of employee you want to be, or the type of worker management wants you to be, without learning how to work with your peers.

Employees must sincerely do everything they can to build strong, professional working relationships with fellow employees, supervisors, and clients.

5.3.0 Attitude

A positive attitude is essential to a successful career. First, being positive means being energetic, highly motivated, attentive, and alert. A positive

attitude is essential to safety on the job. Second, a positive employee contributes to the productivity of others. Both negative and positive attitudes are transmitted to others on the job. A persistent negative attitude can spoil the positive attitudes of others. It is very difficult to maintain a high level of productivity while working next to a person with a negative attitude. Third, people favor a person who is positive. Being positive makes a person's job more interesting and exciting. Fourth, the kind of attitude transmitted to management has a great deal to do with an employee's future success in the company. Supervisors can determine a subordinate's attitude by their approach to the job, reactions to directives, and the way they handle problems.

5.4.0 Maintaining a Positive Attitude

A positive attitude is far more than a smile, which is only one example of an inner positive attitude. As a matter of fact, some people transmit a positive attitude even though they seldom smile. They do this by the way they treat others, the way they look at their responsibilities, and the approach they take when faced with problems.

Here are a few suggestions that will help you to maintain a positive attitude:

- Remember that your attitude follows you wherever you go. If you make a greater effort to be a more positive person in your social and personal lives, it will automatically help you on the job. The reverse is also true. One effort will complement the other.
- Negative comments are seldom welcomed by fellow workers on the job. Neither are they welcome on the social scene. The solution: Talk about positive things and be complimentary. Constant complainers do not build healthy and fulfilling relationships.
- Look for the good things in people on the job, especially your supervisor. Nobody is perfect, but almost everyone has a few worthwhile qualities. If you dwell on people's good features, it will be easier to work with them.

Instructor's Notes:

_____

_____

_____

_____

_____

_____

_____

_____

- Look for the good things where you work. What are the factors that make it a good place to work? Is it the hours, the physical environment, the people, the actual work being done? Or is it the atmosphere? Keep in mind that you cannot be expected to like everything. No work assignment is perfect, but if you concentrate on the good things, the negative factors will seem less important and bothersome.
- Look for the good things in the company. Just as there are no perfect assignments, there are no perfect companies. Nevertheless, almost all organizations have good features. Is the company progressive? What about promotional opportunities? Are there chances for self-improvement? What about the wage and benefit package? Is there a good training program? You cannot expect to have everything you would like, but there should be enough to keep you positive. In fact, if you decide to stick with a company for a long period of time, it is wise to look at the good features and think about them. If you think positively, you will act the same way.
- You may not be able to change the negative attitude of another employee, but you can protect your own attitude from becoming negative.

6.0.0 ◆ EMPLOYER AND EMPLOYEE SAFETY OBLIGATIONS

An obligation is like a promise or a contract. In exchange for the benefits of your employment and your own well-being, you agree to work safely. In other words, you are *obligated* to work safely. You are also obligated to make sure anyone you happen to supervise or work with is working safely. Your employer is also obligated to maintain a safe workplace for all employees. Safety is everyone's responsibility (*Figure 7*).

Some employers will have safety committees. If you work for such an employer, you are then obligated to that committee to maintain a safe working environment. This means two things:

- Follow the safety committee's rules for proper working procedures and practices.
- Report any unsafe equipment and conditions directly to the committee or your supervisor.

Here is a basic rule to follow every working day:

If you see something that is not safe, REPORT IT! Do not ignore it. It will not correct itself. You have an obligation to report it.

Suppose you see a faulty electrical hookup. You know enough to stay away from it, and you do. But then you forget about it. Why should you worry? It is not going to hurt you. Let somebody else deal with it. The next thing that happens is that a co-worker accidentally touches the live wire.

In the long run, even if you do not think an unsafe condition affects you—it does. Do not mess around; report what is not safe. Do not think your employer will be angry because your productivity suffers while the condition is corrected. On the contrary, your employer will be more likely to criticize you for not reporting a problem.

Your employer knows that the short time lost in making conditions safe again is nothing compared with shutting down the whole job because of a major disaster. If that happens, you are out of work anyway. Do not ignore an unsafe condition. In fact, Occupational Safety and Health Administration (OSHA) regulations require you to report hazardous conditions.

101F07.EPS

Figure 7 ◆ Safety is everyone's responsibility.

As time permits, invite a company's safety officer to speak to the trainees about the importance of job site safety.

Emphasize the importance of following OSHA regulations.

Discuss the "Think About It." Emphasize the impact of drug and alcohol use on employee safety and productivity.

This applies to every part of the construction industry. Whether you work for a large contractor or a small subcontractor, you are obligated to report unsafe conditions. The easiest way to do this is to tell your supervisor. If that person ignores the unsafe condition, report it to the next highest supervisor. If it is the owner who is being unsafe, let that person know your concerns. If nothing is done about it, report it to OSHA. If you are worried about your job being on the line, think about it in terms of your life, or someone else's, being on the line.

The U.S. Congress passed the Occupational Safety and Health Act in 1970. This act also created OSHA. It is part of the U.S. Department of Labor. The job of OSHA is to set occupational safety and health standards for all places of employment, enforce these standards, ensure that employers provide and maintain a safe workplace for all employees, and provide research and educational programs to support safe working practices.

OSHA requires each employer to provide a safe and hazard-free working environment. OSHA also requires that employees comply with OSHA rules and regulations that relate to their conduct on the job. To gain compliance, OSHA can perform spot inspections of job sites, impose fines for violations, and even stop any more work from proceeding until the job site is safe.

According to OSHA standards, you are entitled to on-the-job safety training. As a new employee, you must be:

- Shown how to do your job safely
- Provided with the required personal protective equipment
- Warned about specific hazards
- Supervised for safety while performing the work

OSHA was adopted with the stated purpose "to assure as far as possible every working man and woman in the nation safe and healthful working conditions and to preserve our human resources."

The enforcement of this act of Congress is provided by the federal and state safety inspectors who have the legal authority to make employers pay fines for safety violations. The law allows states to have their own safety regulations and agencies to enforce them, but they must first be approved by the U.S. Secretary of Labor. For states that do not develop such regulations and agencies, federal OSHA standards must be obeyed.

These standards are listed in *OSHA Safety and Health Standards for the Construction Industry* (29 CFR, Part 1926), sometimes called *OSHA Standards 1926* (*Figure 8*). Other safety standards that

Drugs and Alcohol

When people use drugs and alcohol, they are putting both themselves and the people around them at serious risk. A construction site can be a dangerous environment, and it is important to be alert at all times. Using drugs and alcohol on the job is an accident waiting to happen. You have an obligation to yourself, your employer, and your fellow employees to work safely. What should you do if you discover someone abusing drugs and/or alcohol at work?

apply to the carpentry trade are published in *OSHA Safety and Health Standards for General Industry* (29 CFR, Parts 1900 to 1910).

The most important general requirements that OSHA places on employers in the construction industry are:

- The employer must perform frequent and regular job site inspections of equipment.
- The employer must instruct all employees to recognize and avoid unsafe conditions, and to know the regulations that pertain to the job so they may control or eliminate any hazards.
- No one may use any tools, equipment, machines, or materials that do not comply with *OSHA Standards 1926*.
- The employer must ensure that only qualified individuals operate tools, equipment, and machines.

101F08.EPS

Figure 8 ◆ OSHA standards govern workplace safety.

CARPENTRY LEVEL ONE—TRAINEE MODULE 27101

Instructor's Notes:

_____

_____

_____

_____

_____

_____

_____

Summary

There are many job and career opportunities for skilled carpenters. An apprenticeship program that combines competency-based, hands-on training with classroom instruction has proven to be the most effective means for a person to learn and advance in the carpentry craft. Developing job skills is only part of the solution; it is just as important to learn good work habits, convey a positive, cooperative attitude to those around you, and practice good safety habits every day.

Review Questions

1. The use of screws as fasteners has been common since _____.
   a. the 1960s
   b. the Stone Age
   c. about 1900
   d. the 1500s

2. The step that follows an apprenticeship is _____.
   a. master carpenter
   b. journeyman
   c. general contractor
   d. estimator

3. A competency-based training program is one that requires the student to _____.
   a. receive at least four years of classroom training
   b. receive on-the-job training for at least two years
   c. demonstrate the ability to perform specific job-related tasks
   d. pass a series of written tests

4. The _____ is most likely to handle the day-to-day operations on the job site.
   a. supervisor
   b. project manager
   c. architect
   d. owner

5. The purpose of the Youth Apprenticeship Program is to _____.
   a. make sure all young people know how to use basic carpentry tools
   b. provide job opportunities for people who quit high school
   c. allow students to start in an apprenticeship program while still in high school
   d. make sure that people under 18 have proper supervision on the job

6. The combined total of on-the-job and classroom training needed for a carpentry apprentice to advance to journeyman is _____ hours.
   a. 2,000
   b. 4,000
   c. 6,000
   d. 8,000

7. Which of the following is true with respect to honesty?
   a. It is okay to borrow tools from the job site as long as you return them before anyone notices.
   b. You are doing your company a favor by using lower-grade materials than those listed in the specifications.
   c. It is okay to take materials or tools from your employer if you feel the company owes you for past efforts.
   d. Being late and not making up the time is the same as stealing from your employer.

8. If one of your co-workers complains about your company, you should _____.
   a. contribute your own complaints to the conversation
   b. agree with the person to avoid conflict
   c. suggest that the person look for another job
   d. find some good things to say about the company

9. If you see an unsafe condition on the job, you should _____.
   a. ignore it because it is not your job
   b. tell a co-worker
   c. call OSHA
   d. report it to a supervisor

10. The purpose of OSHA is to _____.
    a. catch people breaking safety regulations
    b. make rules and regulations governing all aspects of construction projects
    c. ensure that the employer provides and maintains a safe workplace
    d. assign a safety inspector to every project

Have the trainees complete the Review Questions and go over the answers prior to administering the Module Examination.

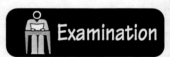

Administer the Module Examination. Be sure to record the results on Craft Training Report Form 200 and submit the results to the Training Program Sponsor.

Nathan Profitt

Project Superintendent
Frank Messer Construction Company
Cincinnati, Ohio

Nathan Profitt began his career with Messer Construction in July of 1978. He started as an unskilled laborer, moved into the position of foreman, and then decided to become a carpenter in 1993. Soon after completing the Commercial Carpentry Program, Nathan was promoted to the position of project engineer. In April of 1999, he was promoted to project superintendent. Messer Construction is one of the largest construction management firms in southwestern Ohio.

How did you choose a career in the carpentry field?
My father retired from Messer Construction as a carpenter foreman, so I believe carpentry has always been in my blood. Even as a laborer, I was doing the things that carpenters typically did. It came naturally for me. I was given good advice by two different men that I should continue my career with Messer as a carpenter because my chances to see my goals realized would become greater. I took that advice.

Tell us about your apprenticeship experience.
Because of my age and my experience on many different jobs, I became an unofficial leader of our class. It was fun helping others get through the course by using my knowledge to encourage others and to lend a hand when someone needed it. I was asked to organize a project for the class to be presented at one of the Construction Expos held at the Cincinnati Convention Center. It was a huge success, and all of the students were proud of the full-scale sectioned replica of a house that we built right there in the Convention Center.

What positions have you held and how did they help you get where you are now?
For fifteen years I worked for Messer as a laborer, ten of those years as a foreman. I worked as a carpenter while I went through the Commercial Carpentry Program. After graduation, I spent seven months as the yard manager. I was then given the opportunity to manage fieldwork as a project engineer, a position I held until April of 1999. At that time, I was promoted to project superintendent, my current position. Each step was aided greatly by the fact that I had completed the Commercial Carpentry Program. Becoming a skilled tradesman has been a solid foundation from which to build my future.

What would you say was the single greatest factor that contributed to your success?
I set a goal. I believed I could get there and beyond to the next goal. I took advantage of the help that was available from my company and from Allied Construction Industries (ACI). ACI is an organization that is available for men and women who want to better themselves. I recognized this early and used ACI's services often.

What does your current job entail?
Currently, I am managing a project with a $2.6 million budget. It is a 48-unit apartment complex in Newport, Kentucky. The four main responsibilities that I focus on are cost, schedule, quality, and safety. Simply put, my job is to complete the construction on time, at or under budget, with the highest level of quality, and most importantly, to see that all who participate in this project do so safely.

What advice do you have for trainees?
Believe in yourself. Recognize what an opportunity you have been given. Realize that this industry needs men and women who want to excel. Work hard and enjoy this business. It will pay big dividends if you do so.

Instructor's Notes:

_____

_____

_____

_____

_____

_____

_____

_____

Trade Terms Introduced in This Module

Building codes: Codes published by state and local governments to establish minimum standards for various types of interior and exterior construction.

Finish carpentry: The portion of the carpentry trade associated with interior and exterior trim, cabinetry, siding, wall finishes, and decorative work.

Rough carpentry: The portion of the carpentry trade associated with framing and other work that will be covered with finish materials.

Takeoff: A list of building materials obtained by analyzing the project drawings (also known as a *material takeoff*).

Samples of NCCER Apprentice Training Recognition

101A02.EPS

Instructor's Notes:

_____

_____

_____

_____

_____

_____

_____

_____

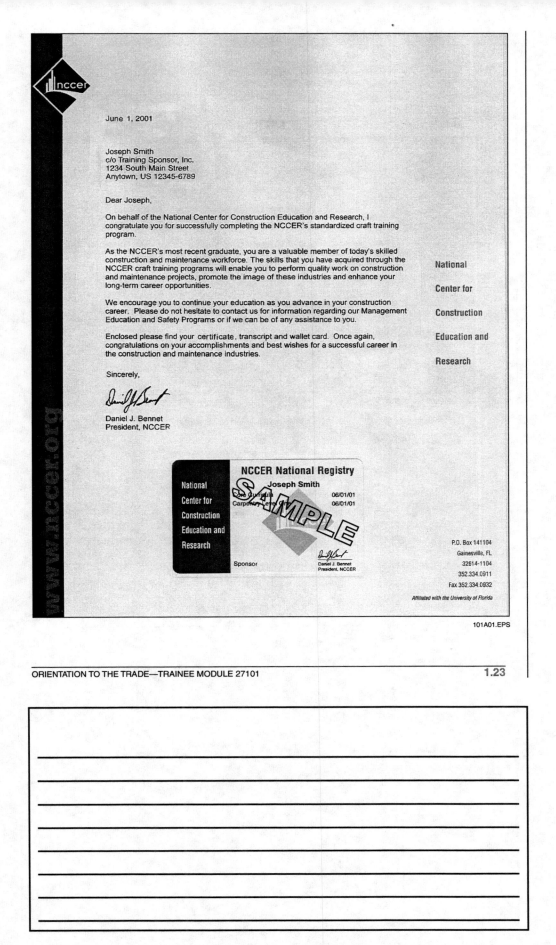

June 1, 2001

Joseph Smith
c/o Training Sponsor, Inc.
1234 South Main Street
Anytown, US 12345-6789

Dear Joseph,

On behalf of the National Center for Construction Education and Research, I congratulate you for successfully completing the NCCER's standardized craft training program.

As the NCCER's most recent graduate, you are a valuable member of today's skilled construction and maintenance workforce. The skills that you have acquired through the NCCER craft training programs will enable you to perform quality work on construction and maintenance projects, promote the image of these industries and enhance your long-term career opportunities.

We encourage you to continue your education as you advance in your construction career. Please do not hesitate to contact us for information regarding our Management Education and Safety Programs or if we can be of any assistance to you.

Enclosed please find your certificate, transcript and wallet card. Once again, congratulations on your accomplishments and best wishes for a successful career in the construction and maintenance industries.

Sincerely,

Daniel J. Bennet
President, NCCER

National

Center for

Construction

Education and

Research

NCCER National Registry
Joseph Smith

Core Curricula 06/01/01
Carpentry Level One 06/01/01

SAMPLE

Sponsor

Daniel J. Bennet
President, NCCER

National
Center for
Construction
Education and
Research

P.O. Box 141104
Gainesville, FL
32614-1104
352.334.0911
Fax 352.334.0932
Affiliated with the University of Florida

101A01.EPS

**NATIONAL CENTER
FOR CONSTRUCTION
EDUCATION AND RESEARCH**
Affiliated with the University of Florida
Post Office Box 141104 Gainesville, Florida 32614-1104
352.334.0911 Fax 352.334.0932 www.nccer.org

06/01/01
Page : 1

Joseph Smith
NCCER
P.O. Box 141104
Gainesville, FL 32614-1104

SSN : 123123123

Course / Description		Instructor/Training Location		Date Completed
00101	Basic Safety	Tami Alsobrooks	Carolinas A.G.C.	01/20/01
00102	Basic Math	Terry Abshire	Becon Construction Co.	01/20/01
00103	Introduction to Hand Tools	Change to 245947728	Associated Builders and Contractors of Alabama	01/20/01
00104	Introduction to Power Tools	Sam Adams	Associated Builders and Contractors of Alabama	01/20/01
00105	Introduction to Blueprints	Sam Adams	Associated Builders and Contractors of Alabama	01/20/01
00106	Basic Rigging	Sam Adams	Associated Builders and Contractors of Alabama	01/20/01
01212	Electric Lighting	Don Jones	Baton Rouge/Pelican Chapter, ABC	01/20/01
01311	Hazardous Locations	Don Jones	Baton Rouge/Pelican Chapter, ABC	01/20/01
01412	Heat Tracing and Freeze Protection	Paul Southerland	Baton Rouge/Pelican Chapter, ABC	01/20/01
08101	Pipefitter Hand Tools	Tami Alsobrooks	ABC Northern New Jersey	01/20/01

SAMPLE

_____
President

101A03.EPS

Instructor's Notes:
_____
_____
_____
_____
_____
_____
_____
_____

Answers to Review Questions

Answer	Section
1. d	2.0.0
2. b	4.0.0
3. c	1.0.0
4. a	4.0.0
5. c	4.2.1
6. d	4.2.2
7. d	4.3.2
8. d	5.4.0
9. d	6.0.0
10. c	6.0.0

The NCCER makes every effort to keep these textbooks up-to-date and free of technical errors. We appreciate your help in this process. If you have an idea for improving this textbook, or if you find an error, a typographical mistake, or an inaccuracy in the NCCER's Craft Training textbooks, please write us, using this form or a photocopy. Be sure to include the exact module number, page number, a detailed description, and the correction, if applicable. Your input will be brought to the attention of the Technical Review Committee. Thank you for your assistance.

Instructors – If you found that additional materials were necessary in order to teach this module effectively, please let us know so that we may include them in the Equipment/Materials list in the Instructor's Guide.

Write: Curriculum Revision and Development Department
National Center for Construction Education and Research
P.O. Box 141104, Gainesville, FL 32614-1104

Fax: 352-334-0932

E-mail: curriculum@nccer.org

Craft _____ Module Name _____

Copyright Date _____ Module Number _____ Page Number(s) _____

Description _____

_____

_____

_____

(Optional) Correction _____

_____

_____

(Optional) Your Name and Address _____

_____

_____

Wood Building Materials, Fasteners, and Adhesives

27102-01

MODULE OVERVIEW

This module introduces the carpentry trainee to wood building materials, fasteners, and adhesives.

PREREQUISITES

Please refer to the Course Map in the Trainee Module. Prior to training with this module, it is suggested that the trainee shall have successfully completed the following modules:

Core Curriculum; Carpentry Level One, Module 27101

LEARNING OBJECTIVES

Upon completion of this module, the trainee will be able to:

1. Explain the terms commonly used in discussing wood and lumber.
2. State the uses of various types of hardwoods and softwoods.
3. Identify various types of imperfections that are found in lumber.
4. Explain how lumber is graded.
5. Interpret grade markings on lumber and plywood.
6. Explain how plywood is manufactured, graded, and used.
7. Identify various types of building boards and identify their uses.
8. Identify the uses of and safety precautions associated with pressure-treated and fire-retardant lumber.
9. Describe the proper method of caring for lumber and wood building materials at the job site.
10. State the uses of various types of engineered lumber.
11. Calculate the quantities of lumber and wood products using industry-standard methods.
12. List the basic nail and staple types and their uses.
13. List the basic types of screws and their uses.
14. Identify the different types of anchors and their uses.
15. Describe the common types of adhesives used in construction work and explain their uses.

PERFORMANCE OBJECTIVES

Under supervision of the instructor, the trainee should be able to:

1. Given a selection of wood building materials, identify a particular material and state its use.
2. Use grade-marking stamps to identify lumber and plywood grades and their uses.
3. Given a selection of lumber, identify various types of naturally-occurring and manufacturing-related lumber defects.
4. Calculate the equivalent board feet for quantities of lumber.
5. Identify the type of fastener required for a specified application.
6. Identify the type of adhesive required for a specific purpose.

NCCER STANDARDIZED CRAFT TRAINING PROGRAM

The National Center for Construction Education and Research (NCCER) provides a standardized national program of accredited craft training. Key features of the program include instructor certification, competency-based training, and performance testing. The program provides trainees, instructors, and companies with a standard form of recognition through a National Craft Training Registry. The program is described in full in the *Guidelines for Accreditation*, published by the NCCER. For more information on standardized craft training, contact the NCCER at P.O. Box 141104, Gainesville, FL 32614-1104, 352-334-0911, visit our Web site at www.NCCER.org, or e-mail info@NCCER.org.

HOW TO USE THIS ANNOTATED INSTRUCTOR'S GUIDE

Each page presents two sections of information. The larger section displays each page exactly as it appears in the Trainee Module. The narrow column ties suggested trainee and instructor actions to each page and provides icons to call your attention to material, safety, audiovisual, or testing requirements. The bottom of each page includes space for your notes.

 If you see the Teaching Tip icon, that means there is a teaching tip associated with this section. Also refer to any suggested teaching tips at the end of the module.

SAFETY CONSIDERATIONS

Ensure that the trainees are equipped with appropriate personal protective equipment.

PREPARATION

Before teaching this module, you should review the Module Outline, Learning and Performance Objectives, and the Materials and Equipment List. Be sure to allow ample time to prepare your own training or lesson plan and gather all required equipment and materials.

MATERIALS AND EQUIPMENT LIST

Materials:
Transparencies
Markers/chalk
Samples of lumber containing:
   Grade stamps
   Natural defects
   Manufacturing defects
Samples of plywood containing grade stamps
Samples of engineered sheet materials
    (OSB, particleboard, etc.)
Samples of engineered lumber (LVL, PSL, glulam, etc.)
Samples of various kinds of:
   Nails
   Screws
   Bolts
   Anchors
   Construction adhesives
Cross section of a tree trunk (optional)
Plywood Specifications and Grade Guide (optional)
Module Examinations*
Performance Profile Sheets*
Job Sheets/Worksheets*

Equipment:
Overhead projector and screen
Whiteboard/chalkboard
Appropriate personal protective equipment
Drill and bits
Hammer
Screwdriver
Calculator

*Packaged with this Annotated Instructor's Guide.

ADDITIONAL RESOURCES

This module is intended to present thorough resources for task training. The following reference works are suggested for both instructors and motivated trainees interested in further study. These are optional materials for continued education rather than for task training.

Building Products Catalog. Atlanta, GA: Georgia-Pacific, Latest Edition.

Carpentry, Leonard Koel. Homewood, IL: American Technical Publishers, 1997.

Carpentry, Gasper J. Lewis. Albany, NY: Delmar Publishers, 2000.

Modern Carpentry, Willis H. Wagner and Howard Bud Smith. Tinley Park, IL: The Goodheart-Willcox Company, Inc., 2000.

TEACHING TIME FOR THIS MODULE

An outline for use in developing your lesson plan is presented below. Note that each Roman numeral in the outline equates to one session of instruction. Each session has a suggested time period of 2½ hours. This includes 10 minutes at the beginning of each session for administrative tasks and one 10-minute break during the session. Approximately 7½ hours are suggested to cover *Wood Building Materials, Fasteners, and Adhesives*. You will need to adjust the time required for hands-on activity and testing based on your class size and resources.

Topic	Planned Time

Session I. Introduction; Lumber Sources and Uses; Lumber Defects; Lumber Grading

A. Introduction _____

B. Lumber Sources and Uses _____

   1. Lumber Cutting _____

   2. General Classifications of Lumber _____

C. Lumber Defects _____

   1. Naturally-Occurring Defects _____

   2. Manufacturing Defects _____

      a. Moisture and Warping _____

      b. Preventing Warping and Splitting _____

D. Lumber Grading _____

   1. Grading Terms _____

      a. Trim _____

   2. Classification of Manufacturing Defects _____

   3. Abbreviations _____

Session II. Plywood; Building Boards; Engineered Wood Products; Pressure-Treated Lumber; Laboratory; Calculating Lumber Quantities

A. Plywood _____

   1. Plywood Sheet Sizes _____

   2. Grading for Softwood Construction Plywood _____

      a. Plywood Glues _____

      b. Plywood Cores _____

      c. Faces _____

      d. Plywood Specification and Grade Guide _____

   3. Plywood Storage _____

   4. Species Used in Plywood _____

B. Building Boards _____

   1. Hardboard _____

   2. Particleboard _____

   3. High-Density Overlay (HDO) and Medium-Density Overlay (MDO) Plywood _____

   4. Oriented Strand Board (OSB) _____

   5. Mineral Fiberboards _____

C. Engineered Wood Products _____

   1. Laminated Veneer Lumber (LVL) _____

   2. Parallel Strand Lumber (PSL) _____

   3. Laminated Strand Lumber (LSL) _____

   4. Wood I-Beams _____

   5. Glue-Laminated Lumber (Glulam) _____

D. Pressure-Treated Lumber _____

E. Laboratory _____

Show the trainees different examples of lumber and building materials. Have them identify each sample (using grade stamps where applicable), state its use, and identify any wood defects. Note the proficiency of each trainee.

F. Calculating Lumber Quantities _____

Session III. Nails; Screws; Anchors; Laboratory; Adhesives; Laboratory; Module Examination and Performance Testing

A. Nails _____

   1. Kinds of Nails _____

   2. Staples _____

     a. Types of Staples _____

B. Screws _____

   1. Wood Screws _____

   2. Sheet Metal Screws _____

   3. Drywall Screws _____

   4. Lag Screws _____

   5. Machine Screws _____

   6. Bolts _____

     a. Stove Bolts _____

     b. Machine Bolts _____

     c. Carriage Bolts _____

C. Anchors _____

   1. Masonry Anchors _____

   2. Hollow-Wall Anchors _____

D. Laboratory _____

Have the trainees identify the correct type of fastener for various applications. Note the proficiency of each trainee.

E. Adhesives _____

   1. Glues _____

   2. Construction Adhesives _____

   3. Mastics _____

   4. Shelf Life _____

F. Laboratory _____

Have the trainees identify the correct type of adhesive for various applications. Note the proficiency of each trainee.

G. Summary _____

   1. Summarize module.

   2. Answer questions.

H. Module Examination _____

   1. Trainees must score 70% or higher to receive recognition from the NCCER.

   2. Record the testing results on Craft Training Report Form 200 and submit the results to the Training Program Sponsor.

I. Performance Testing _____

   1. Trainees must perform each task to the satisfaction of the instructor to receive recognition from the NCCER.

   2. Record the testing results on Craft Training Report Form 200 and submit the results to the Training Program Sponsor.

Module 27102-01

Wood Building Materials, Fasteners, and Adhesives

Instructor's Notes:

_____

_____

_____

_____

_____

_____

_____

_____

Course Map

This course map shows all of the modules in the first level of the Carpentry curriculum. The suggested training order begins at the bottom and proceeds up. Skill levels increase as a trainee advances on the course map. The training order may be adjusted by the local Training Program Sponsor.

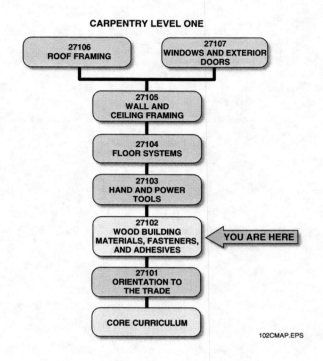

CARPENTRY LEVEL ONE

27106
ROOF FRAMING

27107
WINDOWS AND EXTERIOR DOORS

27105
WALL AND CEILING FRAMING

27104
FLOOR SYSTEMS

27103
HAND AND POWER TOOLS

27102
WOOD BUILDING MATERIALS, FASTENERS, AND ADHESIVES

◁ YOU ARE HERE

27101
ORIENTATION TO THE TRADE

CORE CURRICULUM

102CMAP.EPS

Assign reading of Module 27102.

1.0.0   **INTRODUCTION** 2.1

2.0.0   **LUMBER SOURCES AND USES** 2.3

    2.1.0   Lumber Cutting 2.3

    2.2.0   General Classifications of Lumber 2.4

3.0.0   **LUMBER DEFECTS** 2.7

    3.1.0   Naturally-Occurring Defects .2.7

    3.2.0   Manufacturing Defects 2.9

    3.2.1   Moisture and Warping 2.10

    3.2.2   Preventing Warping and Splitting 2.12

4.0.0   **LUMBER GRADING** 2.13

    4.1.0   Grading Terms 2.14

    4.1.1   Trim 2.16

    4.2.0   Classification of Manufacturing Defects 2.17

    4.3.0   Abbreviations 2.17

5.0.0   **PLYWOOD** 2.17

    5.1.0   Plywood Sheet Sizes 2.17

    5.2.0   Grading for Softwood Construction Plywood 2.18

    5.2.1   Plywood Glues 2.19

    5.2.2   Plywood Cores 2.19

    5.2.3   Faces 2.19

    5.2.4   Plywood Specification and Grade Guide 2.20

    5.3.0   Plywood Storage 2.20

    5.4.0   Species Used in Plywood 2.20

6.0.0   **BUILDING BOARDS** 2.21

    6.1.0   Hardboard 2.21

    6.2.0   Particleboard 2.22

    6.3.0   High-Density Overlay (HDO) and Medium-Density Overlay
           (MDO) Plywood 2.22

    6.4.0   Oriented Strand Board (OSB) .2.23

    6.5.0   Mineral Fiberboards 2.23

7.0.0   **ENGINEERED WOOD PRODUCTS** 2.23

    7.1.0   Laminated Veneer Lumber (LVL) 2.24

    7.2.0   Parallel Strand Lumber (PSL) 2.24

    7.3.0   Laminated Strand Lumber (LSL) 2.24

    7.4.0   Wood I-Beams 2.24

    7.5.0   Glue-Laminated Lumber (Glulam) 2.25

8.0.0   PRESSURE-TREATED LUMBER 2.25

9.0.0   CALCULATING LUMBER QUANTITIES 2.26

10.0.0   NAILS 2.27

        10.1.0   Kinds of Nails 2.27

        10.2.0   Staples 2.31

        10.2.1   *Types of Staples* 2.31

11.0.0   SCREWS 2.31

        11.1.0   Wood Screws 2.31

        11.2.0   Sheet Metal Screws 2.32

        11.3.0   Drywall Screws 2.33

        11.4.0   Lag Screws 2.33

        11.5.0   Machine Screws 2.33

        11.6.0   Bolts 2.34

        11.6.1   *Stove Bolts* 2.34

        11.6.2   *Machine Bolts* 2.34

        11.6.3   *Carriage Bolts* 2.34

12.0.0   ANCHORS 2.35

        12.1.0   Masonry Anchors 2.35

        12.2.0   Hollow-Wall Anchors 2.35

13.0.0   ADHESIVES 2.36

        13.1.0   Glues 2.36

        13.2.0   Construction Adhesives 2.38

        13.3.0   Mastics 2.39

        13.4.0   Shelf Life 2.39

SUMMARY 2.39

REVIEW QUESTIONS 2.40

GLOSSARY 2.42

APPENDIX A 2.43

APPENDIX B 2.45

ANSWERS TO REVIEW QUESTIONS 2.47

ADDITIONAL RESOURCES 2.48

Instructor's Notes:

_____

_____

_____

_____

_____

_____

_____

_____

Figures

Figure 1   Cross section of a tree 2.3
Figure 2   Common lumber cutting methods 2.4
Figure 3   Kinds of knots 2.8
Figure 4   Wane 2.10
Figure 5   Common lumber defects 2.10
Figure 6   Typical lumber grade stamp 2.13
Figure 7   Types of plywood 2.19
Figure 8   OSB panels 2.23
Figure 9   PSL, LVL, and LSL 2.24
Figure 10   Wood I-beams 2.25
Figure 11   Glulam beam construction 2.25
Figure 12   Glulam beam application 2.25
Figure 13   Examples of a board foot 2.26
Figure 14   Nail sizes 2.28
Figure 15   Kinds of nails 2.29
Figure 16   Staples 2.31
Figure 17   Wood screw sizes 2.32
Figure 18   Common screws 2.32
Figure 19   Sheet metal screws 2.33
Figure 20   Drywall screws 2.33
Figure 21   Lag screw and lag shield 2.34
Figure 22   Machine screws 2.34
Figure 23   Bolts 2.34
Figure 24   Masonry anchors 2.35
Figure 25   Examples of masonry anchor applications 2.35
Figure 26   Self-drilling snap-off anchor 2.36
Figure 27   Fastener anchored in epoxy 2.36
Figure 28   Hollow-wall anchors 2.36
Figure 29   Adhesive applicators 2.38

Tables

Table 1   Common Hardwoods and Softwoods 2.5
Table 2   Lumber Appearance Grades 2.14
Table 3   Grades of Dimension Lumber 2.15
Table 4   Nominal and Dressed (Actual) Sizes of Dimension Lumber
        (in Inches) 2.15
Table 5   Guide to APA Performance-Rated Plywood Panels 2.18
Table 6   Classification of Species Used for Plywood Veneers by Groups ...2.21

_____

_____

_____

_____

_____

_____

_____

Wood Building Materials, Fasteners, and Adhesives

Ensure you have everything required to teach the course. Check the Materials and Equipment List at the front of this Instructor's Guide.

Objectives

Upon completion of this module, the trainee will be able to:

1. Explain the terms commonly used in discussing wood and lumber.
2. State the uses of various types of hardwoods and softwoods.
3. Identify various types of imperfections that are found in lumber.
4. Explain how lumber is graded.
5. Interpret grade markings on lumber and plywood.
6. Explain how plywood is manufactured, graded, and used.
7. Identify various types of building boards and identify their uses.
8. Identify the uses of and safety precautions associated with pressure-treated and fire-retardant lumber.
9. Describe the proper method of caring for lumber and wood building materials at the job site.
10. State the uses of various types of engineered lumber.
11. Calculate the quantities of lumber and wood products using industry-standard methods.
12. List the basic nail and staple types and their uses.
13. List the basic types of screws and their uses.
14. Identify the different types of anchors and their uses.
15. Describe the common types of adhesives used in construction work and explain their uses.

Prerequisites

Successful completion of the following Task Modules is recommended before beginning study of this Task Module: Core Curriculum; Carpentry Level One, Module 27101.

Required Trainee Materials

1. Trainee Task Module
2. Appropriate personal protective equipment

Note
One of the main purposes of this module is to introduce the apprentice carpenter to the language of carpentry, especially the terms associated with the types and grades of wood building materials and the many types of fasteners. Many such terms are introduced in this module; they are defined in detail as they occur.

1.0.0 ◆ INTRODUCTION

Many kinds of lumber and wood building products are used in construction. Different wood products suit different purposes, so it is very important for a carpenter to know all the types and grades of wood building materials, along with their uses and limitations. When selecting lumber, the carpenter must know what type best fits the application. A carpenter must also know what type and size of fastener or what type of adhesive to use in any situation. Therefore, the carpenter must be familiar with the many types of nails, screws, anchors, and adhesives used in construction. This module provides an introduction to that subject.

When selecting lumber, it is also important to be able to recognize defects in the material and to know how to properly handle and store wood products to prevent damage from occurring. When ordering lumber, the carpenter must know how to specify amounts. Sometimes, material is

Show Transparency 1, Course Objectives.

Show Transparency 2, Performance Profile Tasks.

Emphasize that carpenters must be familiar with a number of different building materials, fasteners, and adhesives.

Teaching Tip

Obtain photographs of wood and metal framing, and pass them around for the trainees to examine.

sold by the linear foot; other times, it is sold by the board foot. Still other material is sold in sheets. The carpenter must know when and how to use the various methods of ordering.

Although wood framing was the norm for many years, the use of metal framing has become increasingly popular in both commercial and residential construction. The modules in this level focus on wood frame construction. However, metal framing is discussed in each of the applicable modules of this level and will be covered in more detail in subsequent levels.

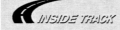

Building a New Home

If wood is used, you can expect to see a lot of it on the job site. It takes approximately 14,000 board feet of lumber to build an average size house.

102P0201.EPS

Instructor's Notes:

_____

_____

_____

_____

_____

_____

_____

_____

2.0.0 ◆ LUMBER SOURCES AND USES

2.1.0 Lumber Cutting

Throughout the ages, people have used wood for a variety of purposes: as fuel, as building materials, in weapons, and in transportation. In this module, it is important to distinguish lumber from wood. Lumber refers to the boards, timbers, etc., produced from sawmills, whereas wood refers to the material itself, which comes from many species of trees.

Wood has several advantages:

- It is easily worked.
- It has durability and beauty.
- It has great ability to absorb shocks from sudden loads.
- It is free from rust and corrosion, comparatively light in weight, and adaptable to a countless variety of purposes.

Wood is useful as a building material because of the manner in which a tree forms its fibers: growing by the addition of new material to the outer layer just under the bark and preserving its old fibers as it adds new ones.

Viewing a cross section of a tree, as shown in *Figure 1,* one can see that the trunk consists of a series of concentric rings covered by a layer of bark. Each annular ring represents one year of tree growth. This growth takes place just under the bark in the *cambium layer.* The cells that are formed there are long; their tubular fibers are composed mainly of cellulose. They are bound together by a substance known as *lignin,* which connects them into bundles. In wood, lignin is the material that acts as a binder between the cells, adding strength and stiffness to the cell walls. The bundles of fibers run the length of the tree and carry food from the roots to the branches and leaves.

Running at right angles, the fibers from the outer layer inward are another group of cells known as *medullary rays.* These rays carry food from the inner bark to the cambium layer and act as storage tanks for food. The rays are more pronounced in some species than others (such as oak) but are present in all trees. As one layer of wood succeeds another, the cells in the inner layer die (cease to function as food storage) and become useful only to give stiffness to the tree. This older wood, known as *heartwood,* is usually darker in color, drier, and harder than the living layer (*sapwood*).

Heartwood and sapwood of equivalent character have equal strength, so it is not necessary to specify heartwood when strength alone is the governing factor. However, heartwood is more durable than sapwood. If wood is to be exposed to decay-producing conditions without the benefit of a preservative, a minimum percentage of heartwood will be specified.

Sapwood takes preservative treatment more readily than heartwood, and it is equally durable when treated.

In many kinds of trees, each annual ring is divided into two layers. The inner layer consists of cells with relatively large cavities. This heavier, harder, stronger material is called the *summerwood.* The amount of summerwood is one measure of the density of the wood; the higher the density, the greater the strength.

Springwood is the portion of the annual growth ring formed during the early part of the yearly growth period. It is lighter in color, less dense, and not as strong as summerwood.

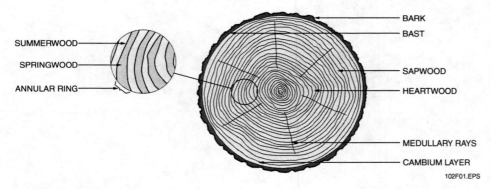

Figure 1 ◆ Cross section of a tree.

Obtain a cut portion of an actual tree trunk and allow the trainees to examine it. Identify the different portions of the tree's cross section.

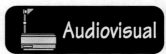

Show Transparency 3 (Figure 1).

Explain the impact of different cutting methods on the lumber produced.

Show Transparency 4 (Figure 2).

Review the characteristics and uses of various types of wood.

Summerwood forms during the latter part of the yearly growth period. It is darker, denser, and stronger than springwood.

Most trees are sawed so that the growth rings form an angle of less than 45 degrees with the surface of the boards produced. Such lumber is called *flat-grained* in softwoods and *plain-sawed* in hardwoods (*Figure 2*). This is the least expensive method, but the lumber produced by this method is more likely to shrink or warp.

There is another method in which the wood is cut with the growth rings at an angle greater than 45 degrees. Lumber cut by this method is known as *edge-grained* or *vertical-grained* in softwoods and *quarter-sawed* in hardwoods. The lumber it produces is usually more durable and less likely to warp or shrink. Quarter-sawed lumber is often used for hardwood floors.

In order to obtain the most lumber from the log, most logs are cut using a combination of the two methods.

2.2.0 General Classifications of Lumber

Lumber falls into one of two general classifications: hardwood or softwood. Hardwoods come from deciduous (leaf-bearing) trees such as oak and maple. Softwoods come from coniferous (cone-bearing) trees such as pine and fir. (The term *softwood* does not mean it is not strong or durable. On the contrary, some softwoods are harder than certain hardwoods.) Descriptions and uses of various hardwoods and softwoods are in *Table 1*. Study this table to become familiar with the woods and their uses.

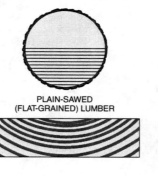

PLAIN-SAWED
(FLAT-GRAINED) LUMBER

QUARTER-SAWED
(EDGE-GRAINED) LUMBER

102F02.EPS

Figure 2 ◈ Common lumber cutting methods.

2.4

Instructor's Notes:

_____

_____

_____

_____

_____

_____

_____

_____

Table 1 (part 1 of 2)   Common Hardwoods and Softwoods

Tree	Use	Class (H—Hard, S—Soft)
Alder	Used in cabinets and furniture. Alder is a less expensive substitute for cherry. It is also known as *red alder* and *fruitwood.*	H
Ash	Coarse-grained, hard, and heavy. The color is reddish brown; the heartwood and sapwood are almost white. Used for furniture, cabinets, trim, tool handles, and plywood.	H
Basswood	One of the softest hardwoods. It has an open, coarse grain. The color is creamy white. Used for veneer core and moldings.	H
Beech	Strong, heavy, hard. Takes a good polish, works easily. The color is white to brown. Also used for tool handles, furniture, and plywood.	H
Birch	Very strong, heavy, and fine-grained. It takes an excellent finish. Can be beautifully hand polished. The sapwood is a soft yellow; the heartwood is brown, sometimes slightly colored with red. Due to the climate, birch has a wide variety of color changes. Used for furniture, cabinets, flooring, and plywood.	H
Western red cedar	The thin sapwood layer is a pale yellow. The heartwood varies from a deep dark brown to a very light pink. The wood is light and soft, but lacks strength. It is straight-grained and has high resistance to changes in moisture. After seasoning, it retains its shape and size exceptionally well. Used for shingles, shakes, siding, planking, building logs, and yard lumber.	S
Tennessee cedar	This product is very light, soft, and close-grained. The sapwood is white in color. The heartwood is reddish brown. This cedar has a very strong odor that acts as an insect repellent. Used for trunk linings and closets.	S
Cherry	Cherry is a close-grained and light wood. It takes a good finish and works easily. The sapwood is an off-white. The heartwood is reddish brown. This product can be used for cabinets, interior trim, and furniture.	H
Chestnut	This beautiful, open-grained wood is worked very easily. The sapwood is light brown in color. The heartwood is a dark tan. Can be used for cabinets or plywood and paneling.	H
Cypress	Cypress is a very durable, light product, and is very easily worked. The heartwood is a brilliant yellow; the sapwood is light brown in color. It is used for interior trim and can also be used for cabinet work and exterior trim.	S
Elm	This strong, majestic tree is very hard and the lumber is very heavy. The color of this tree is light to medium brown, with slight variations of red and gray. Used primarily for tool handles and heavy construction purposes.	H
Douglas fir	Fir is very strong and durable. The sapwood is slightly off-white and the heartwood varies in color from a light red to a deep yellow. Fir is used for all types of construction and in the manufacturing of plywood.	S
Western hemlock	This lumber is straight-grained and fine-textured, and it works easily. It has very good qualities for gluing. It is almost totally free from pitch and is often interchanged with fir. Hemlock is used for general construction.	S
Gum	This soft-textured, close-grained tree is very durable. The color of the heartwood varies from red to brown. The sapwood is yellowish white. It closely resembles walnut in color. This wood works easily, but has a tendency to warp. It is sometimes substituted for walnut in the manufacture of furniture.	H
Larch	This product is strong, fairly hard, and resembles Douglas fir more than any other softwood. This lumber is also suitable for general construction purposes.	H
Mahogany	This lumber is an open-pored, strong, durable product. Mahogany can be worked easily. The color is basically a reddish brown. Poor-grade mahogany has small gray spots or flecks. Much mahogany is from overseas. Can be used for paneling, furniture, plywood, and interior trim.	H
Sugar maple	This lumber is very difficult to work with, but it is very strong and can take an excellent finish. The product is very close-grained and light brown to dark brown in color. Furniture and flooring are some of its uses.	H

Table 1 (part 2 of 2) Common Hardwoods and Softwoods

Tree	Use	Class (H—Hard, S—Soft)
Silver maple	This lumber is not very strong or durable. It is soft and light, and it works easily. This lumber is frequently used for turning and interior trim.	H
White oak	White oak is very hard and heavy. It has a close grain with open pores. The product is very difficult to work with, but it takes an excellent finish. The heartwood is tan in color, with a light sapwood. Can be used for flooring, plywood, interior trim, and furniture.	H
Red oak	Red oak has a coarser grain than white oak and is also a little softer. The variation in color is from a light tan to a medium brown. The uses for red oak are furniture, plywood, bearing posts, interior trim, and paneling. It could also be used for cabinets.	H
White pine	This lumber is very easily worked. It is soft, light, and very durable. The heartwood is a very slightly yellowish white. Can be used for construction (code permitting) and is also used in furniture and millwork.	S
Sugar pine	This is a light and very soft lumber. It is uniform in texture and works easily. The sapwood is yellow. The heartwood is light red to medium brown. Sugar pine is mostly quarter-sawed and has specks of reddish brown, which makes it easy to recognize. Can be used for inexpensive furniture and interior millwork.	S
Ponderosa pine	This pine is fairly soft and uniform in texture. Besides being strong, it works smoothly without splintering. The sapwood is a light pale yellow with a darker heartwood. Ponderosa pine is ideal for woodworking in the shop and is common in door and window frames and moldings.	S
Lodgepole pine	This pine lumber seasons very easily and is of moderate durability. This product is soft and straight-grained with a fine, uniform texture. The sapwood is almost white. This product is used largely in making propping timbers for miners.	S
Southern yellow pine	This lumber is also called *southern pine* or *yellow pine*. It is a very difficult product to work with. It is strong, hard, and tough. The heartwood is orange-red in color, with the sapwood being lighter in color. This product is used for heavy structural purposes such as floor planks, beams, and timbers.	S
Poplar	This product is uniform in texture, light, and soft. The color varies from white to yellow. This lumber is easy to work with. Can be used for inexpensive furniture, crates, plywood, and moldings.	H
Redwood	The redwood is the largest tree in North America. Redwood is easily worked, light, and coarse-grained. The color is dull red. Although very beautiful, redwood lacks strength. It is used primarily for interior and exterior trim, planking for patio decks, and shingles.	S
White spruce	This lumber is soft, works very easily, and is very lightweight. There is very little contrast between the heartwood and sapwood. The color is a pale yellow. It is not very durable, but among the softwoods it is one of medium strength. In certain situations, it is vulnerable to decay. White spruce has value as pulp wood and is used in light construction.	S
Engelmann spruce	This type of spruce takes a good finish. The lumber is very light and straight-grained. This species is usually larger than white spruce and has a larger percentage of clear lumber with little or no defects. It is used for oars, paddles, sounding boards for musical instruments, and construction work.	S
Sitka spruce	These trees have a straight grain and grow very tall. This species is a very tough, strong wood. The color is a very creamy white with a pinkish or light red tinge. This species resists shattering and splintering. It is used primarily in constructing masts, spars, scaffolding, and general construction (code permitting).	S
Walnut	This product is durable, very hard, strong, and it can easily be finished to a beautiful luster. The sapwood is a light brown to a medium brown. The heartwood is a reddish brown to a dark brown. This product is used in the manufacture of fine cabinets, furniture, flooring, plywood, and interior trim.	H

Instructor's Notes:

_____

_____

_____

_____

_____

_____

_____

_____

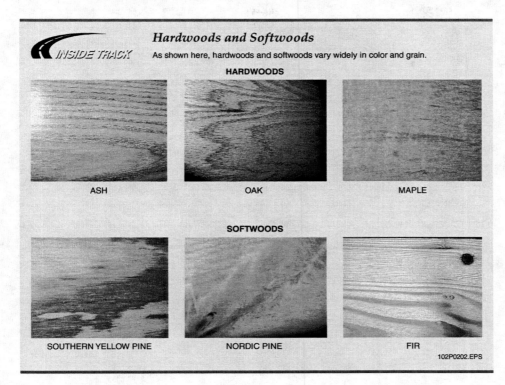

HARDWOODS

ASH   OAK   MAPLE

SOFTWOODS

SOUTHERN YELLOW PINE   NORDIC PINE   FIR

102P0202.EPS

3.0.0 ◆ LUMBER DEFECTS

Before we discuss lumber grading, which is to a large extent based on the amount and types of defects in the lumber, it is necessary to understand the various kinds of defects. Lumber defects can occur naturally (e.g., knots). Defects can also occur during the cutting process. Further defects can occur while the lumber is drying, or because of improper storage or handling. Defects in lumber can affect the lumber's appearance, strength, and usability. In most common uses, a certain amount of defects are permitted; unless it is severely damaged, some use can be found for just about any piece of lumber. That is why lumber is graded and why there are many grades.

3.1.0 Naturally-Occurring Defects

Common defects that occur during the tree's growth process are described below. Some of these defects are actually considered desirable for some uses. One example is burled walnut used in high-quality wood finishes such as those in luxury cars. Heavily-knotted pine, while not accept-

Classroom

Point out that lumber grading is based largely on the amount and types of defects in lumber.

Discuss the types of natural defects found in lumber.

able for stress-bearing uses, is commonly used for decorative paneling.

- *Burl*—A burl in wood and wood veneer is a localized severe distortion of the grain, generally rounded in outline and swirling in appearance, from ½" to several inches in diameter and usually resulting from the overgrowth of dead branch stubs.
- *Compression wood*—Abnormal wood that forms on the underside of leaning or crooked coniferous trees. It is characterized, aside from its distinguishing color, by being hard and brittle and by its relatively lifeless appearance. It is not permitted, in readily identifiable and damaging form, in stress grades nor where specifically limited. It is commonly used by artists for wood carving because of its different grains and architectural aesthetics.
- *Decay*—A disintegration of the wood substance due to the action of wood-destroying fungi. It is also known as *dote, rot,* and *unsound wood.* Types of decay include:
  - Heart center decay: a localized decay developing along the pith in some species, readily identifiable and easily detected by visual inspection.
  - White specks: small white pits or spots in wood caused by a fungus.
  - Honeycomb: similar to white speck, but the pockets are larger.
  - Peck: channeled or pitted areas or pockets, as sometimes found in cedar and cypress.
- *Holes*—Holes may extend partially or entirely through a piece and may be from many causes. Holes that extend only partially through the piece may also be designated as surface pits.

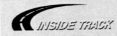

Amber

The popular yellowish gemstone known as *amber* is actually fossilized pitch.

- *Knots*—A portion of a branch or limb that has become incorporated in a piece of lumber. In lumber, knots are classified by the form, size, quality, and occurrence (*Figure 3*). A red knot is one that results from a live branch growth in the tree and is intergrown with the surrounding wood. A black knot is one that results from a dead branch, which the wood growth of the tree has surrounded. Architects will specify how many and what kinds of knots are allowed.
- *Pitch*—An accumulation of sticky resinous material. There are three degrees of pitch:
  - Light pitch is the light but evident presence of pitch.
  - Medium pitch is a somewhat more evident presence of pitch than light pitch.
  - Heavy pitch is a very evident accumulation of pitch shown by its color and consistency.
- *Pitch streak*—A well-defined accumulation of pitch in the wood cells in a more or less regular streak. It should not be confused with dark grain. A pitch seam is a split that contains pitch.

Show Transparency 5 (Figure 3).

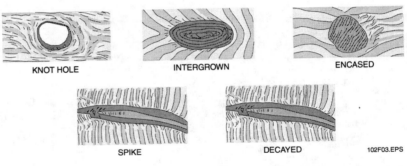

Figure 3 ◆ Kinds of knots.

Instructor's Notes:

_____

_____

_____

_____

_____

_____

_____

- *Pocket*—A well-defined opening between the rings of annular growth, which develops during the growth period of a tree. It usually contains pitch or bark.
- *Stained wood*—Stained heartwood or firm red heart is a marked variation from the natural color. It may range from pink to brown. It should not be confused with natural red heart. Natural color is usually uniformly distributed through certain annular rings, whereas stains are usually in irregular patches. In grades where it is permitted, it has no more effect on the intended use of the piece than other characteristics permitted in the grade. Stained sapwood similarly has no effect on the intended use of the pieces in which it is permitted, but it affects appearance in varying degrees:
  - Light-stained sapwood is so slightly discolored that it does not materially affect natural finishes.
  - Medium-stained sapwood has a pronounced difference in coloring, which sometimes affects its usefulness for natural finishes, but not for paint finishes.
  - Heavy-stained sapwood has so pronounced a difference in color that the grain may be obscured, but the lumber containing it is acceptable for painted finishes.

Discoloration through exposure to the elements is admitted in all grades of framing and sheathing lumber, provided that exposure has not occurred over a long period.

3.2.0 Manufacturing Defects

Lumber can be damaged while it is being cut from the log at the sawmill. It can also suffer damage from warping and splitting while the cut lumber is drying. The common manufacturing defects and their effects on lumber grading are explained in the following.

- *Chipped grain*—A barely noticeable irregularity in the surface of a piece caused when particles of wood are chipped or broken below the line of cut. It is too small to be classified as torn grain and, as usually found, not considered unless it exceeds 25 percent of the surface involved.
- *Torn grain*—An irregularity in the surface of a piece where wood has been torn or broken out by surfacing. Torn grain is described as follows:
  - Very light torn grain (not over ¼₄" deep)
  - Light torn grain (not over ½₂" deep)
  - Medium torn grain (not over ¹⁄₁₆" deep)
  - Heavy torn grain (not over ⅛" deep)
  - Very heavy torn grain (over ⅛" deep)

- *Raised grain*—The unevenness between springwood and summerwood on the surface of dressed lumber. Slight raised grain is an unevenness somewhat less than ¼₄". Very light raised grain is not over ½₂". Medium raised grain is not over ¹⁄₁₆". Heavy raised grain is not over ⅛".
- *Loosened grain*—A grain separation or loosening between springwood and summerwood without displacement. Loosened grain is described as follows:
  - Very light loosened grain (not over ¼₄" of separation)
  - Light loosened grain (not over ½₂" of separation)
  - Medium loosened grain (not over ¹⁄₁₆" of separation)
  - Heavy loosened grain (not over ⅛" of separation)
  - Very heavy loosened grain (over ⅛" of separation)
- *Skips*—Areas on a piece that failed to surface clean. Skips are described as follows with equivalent areas being permissible:
  - Very light skip is not over ¼₄" deep.
  - Light skip is not over ½₂" deep.
  - Medium skip is not over ¹⁄₁₆" deep.
  - Heavy skip is not over ⅛" deep.
- *Hit and miss*—A series of skips not over ¹⁄₁₆" deep with surfaced areas in between. Hit or miss means completely surfaced, partly surfaced, or entirely rough.

Wood Defects

When lumber is properly stored at a job site, there is a good chance that most of it will be usable. You may encounter situations in which the lumber may be too damaged to use, at least for its intended purpose. Experienced carpenters can find other uses for damaged and defective lumber. Wood with cosmetic defects, like knots, may be desirable for paneling purposes. Some warped lumber can be used in joists and rafters if the proper precautions are taken to account for the distortion. It is a good idea, however, to consult with an experienced carpenter or supervisor when you encounter these types of situations.

Discuss the types of manufacturing defects found in lumber.

- *Mismatch*—An uneven fit in worked lumber when adjoining pieces do not meet tightly at all points of contact or when the surfaces of adjoining pieces are not in the same plane.
- *Machine burn*—A darkening of the wood caused by overheating by machine knives or rolls when pieces are stopped in the machine.
- *Machine bite*—A depressed cut of the machine knives at the end of a piece.
- *Machine gouge*—A groove cut by the machine below the desired line.
- *Machine offset*—An abrupt dressing variation in the edge surface, which usually occurs near the end of the piece without reducing the width or changing the plane of the wide surface.
- *Chip marks*—Shallow depressions or indentations on or in the surface of dressed lumber caused by shavings or chips getting embedded in the surface during dressing.
- *Knife marks*—The imprints or markings of the machine knives on the surface of dressed lumber. Slight knife marks are readily visible, but no unevenness to the touch is evident.
- *Wavy dressing*—Involves more uneven dressing than knife marks. *Very slight* wavy dressing is barely perceptible to the touch. *Slight* wavy dressing is perceptible to the touch.
- *Wane*—A condition in which there is bark or an absence of wood at the edge of the lumber (*Figure 4*).

Figure 4 ◆ Wane.

102F04.EPS

INSIDE TRACK

Wanes

Wanes occur when lumber is cut too close to the outer surface of the tree.

3.2.1 Moisture and Warping

Trees contain a large amount of moisture. For example, a newly cut 10' length of 2 × 10 lumber can contain more than four gallons of water. Because of its high moisture content, freshly cut (*green*) lumber cannot be used until it is dried. If green lumber is used, it can cause cracked ceilings or floors, squeaky floors, sticking doors, and other problems. Some of the drying is done by heating lumber in kilns at the sawmill; however, the lumber continues to dry for a long time after that. As it dries, the cut lumber will shrink. This can cause warping and splitting because the lumber dries unevenly. A carpenter must be able to tell when lumber contains too much moisture. Moisture meters can be used on site for that purpose. A simple way to check for moisture is to hit the lumber with a hammer. If moisture comes to the surface in the dent left by the hammer, the lumber is too wet and should not be used. *Figure 5* shows the various kinds of damage that can occur during the drying of lumber. Most of these conditions are preventable if the lumber is properly dried and stored. That subject will be dealt with in the next section.

The terms *check*, *split*, and *shake* are used to define separations at the end of a length of lumber. A split, as shown in *Figure 5*, is a crack that extends

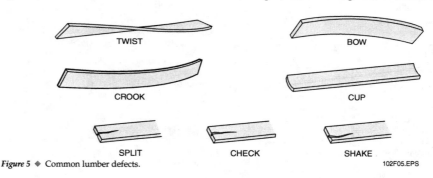

Figure 5 ◆ Common lumber defects.

102F05.EPS

Instructor's Notes:

_____

_____

_____

_____

_____

_____

_____

all the way through the wood. A check is a crack along the growth rings. A shake is a hollow separation between the growth rings. These terms are sometimes used interchangeably, so you have to be careful how you interpret them. Someone who tells you the lumber contains checks may be describing lumber that is completely split open at the end.

Drying Lumber

There are two methods of drying lumber. Kiln drying is the most common method, but air drying is also used. In air drying, the lumber is stacked outdoors and allowed to dry for up to three months, as compared with only a few days for kiln drying. Even though the process is much longer, air-dried lumber is still likely to have a higher moisture content than kiln-dried lumber.

Lumber Defects

Some lumber defects such as knots occur naturally, but most defects occur either during the manufacturing and drying process or as a result of improper storage and handling. The following pictures show various types of lumber defects.

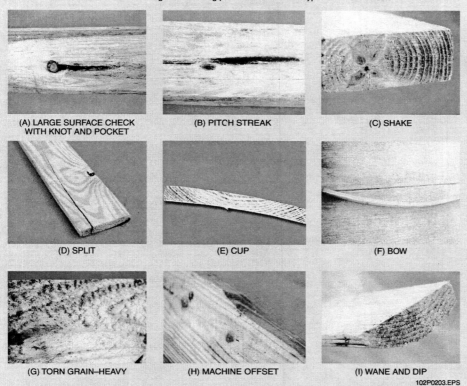

(A) LARGE SURFACE CHECK WITH KNOT AND POCKET

(B) PITCH STREAK

(C) SHAKE

(D) SPLIT

(E) CUP

(F) BOW

(G) TORN GRAIN–HEAVY

(H) MACHINE OFFSET

(I) WANE AND DIP

102P0203.EPS

3.2.2 Preventing Warping and Splitting

Building materials may be delivered to a job site weeks before they are used. Unless the lumber, plywood, and other materials are properly stacked and protected, they can be seriously damaged. Although a contractor allows for a certain amount of waste and spoilage in the estimate, such an allowance does not account for damage due to improper storage. It is up to the work crew to make sure the lumber is properly taken care of.

When a complete framing package for a residence is delivered to the job site, the framing package looks great. All the material is neatly packed. Some material is strapped and some neatly tied. However, when the load is dumped, it will shift. What was a neat load suddenly becomes a pile of lumber spread out in different directions. Some lumber is split or broken. Other material may be partially buried in sand, mud, or snow. If the framing package is left the way it is, it can cause you more problems than you could imagine. For example:

- The lumber is exposed to the elements. Sun, wind, rain, and snow can create serious problems.
- What would have been good lumber may now be crooked, warped, or cupped.
- Some lumber will crack when strain is placed against it; there is a good possibility of splits.
- It will cost the contractor money to replace the materials.
- The time involved in getting additional material delivered to the job site could affect the schedule.
- There is a safety hazard associated with using damaged materials.
- Additional time required in working with defective materials is costly.

A simple way to avoid these problems is to have knowledgeable workers at the site when the materials are delivered. It takes a few hours to stack the lumber and protect it from the elements, saving your employer the cost of replacing the damaged materials. Also, delivery people are more likely to be careful if an observer is present.

There are several factors to consider when unloading and stacking building materials at the job site. These include:

- The order in which the material will be used
- The weight of the material
- The weather conditions

You should get instructions from your supervisor before deciding how and where to stack the materials. Here are some general guidelines.

Be sure to consider the order in which the material will be used. For example, do not put the siding on top of the sheathing.

All material should be stacked on a level surface or as close to level as possible. If stacking material indoors, the amount of material you stack on the subfloor should not exceed its weight-bearing capacity. If stacking outdoors, try not to block the foundation or slab. The next step is to find some scrap lumber and place it flat on the ground. Two or three 2 × 4s approximately 4' apart from each other will work nicely. Stack the widest lumber on top of this scrap lumber. Be sure the good lumber does not come in contact with the ground. Ensure that there is enough support under the lumber so that it will not sag.

Separate the lumber, keeping the pressure-treated material from the regular material.

Keep your materials banded until it is time to use them. Try to avoid using lumber that has been exposed to heavy rain.

Plywood can be stacked vertically or horizontally. Since some plywood is only ½" thick, it should be laid horizontally on top of scrap pieces as described above. Fiberboard should be stacked in a flat position; do not let the fiberboard touch the ground.

Protect the lumber from the elements by covering the materials with a waterproof material (e.g., roofing paper or polyethylene). Cover the lumber to protect it from the elements, but not to prevent air from circulating around the stack. Preventing air circulation could cause twisting or warping, depending on the moisture content of the air.

Special care is needed for interior finish materials. If you are using knock-down (KD) jambs (either wood or metal) for the doors, these usually come tied together. Always ensure that the jambs are stacked flat and off the floor or concrete. Some moldings will come wrapped or tied. Take care to lay them flat and off the floor or concrete. Wall paneling should be treated the same way. Extreme caution should be taken not to damage the edges or mar the face of the panel.

Stacking interior finish materials on top of scrap lumber serves two purposes:

- It prevents moisture in the concrete floor from coming in contact with the bottom piece.
- Although the finish material is under cover, there is always a possibility of a pipe bursting and water flowing on the floor.

Emphasize the importance of properly handling and storing all building materials.

Explain that terms shown in bold (blue) are defined in the Glossary at the back of this module.

Instructor's Notes:

4.0.0 ◆ LUMBER GRADING

Lumber is graded by inspectors who examine the lumber after it is planed. Grading is based on the number, size, and types of defects in the lumber, which determine its load-carrying capacity. The more defects, the weaker the wood. The defects considered are those described in the preceding section. Grading methods and standards are established by the American Lumber Standards Committee. Basically, the grade describes the type, size, and number of defects allowed in the worst board of that grade.

Inspection criteria are established by the U.S. Department of Commerce and published as product standards. Inspections are performed by regional agencies and associations, such as the Southern Pine Inspection Bureau, Western Wood Products Association, California Redwood Association, etc. Hardwood grades are regulated by the National Hardwood Lumber Association. Different associations may have slightly different standards for the numbers and types of defects permitted within a particular grade. Inspectors from accredited inspection agencies examine the lumber, determine the grade in accordance with the association's specifications, and apply the

applicable grade stamp. Keep in mind that since grading is done right after the board is planed, it cannot account for warping that might occur if the board is improperly stored.

Carpenters must be familiar with the various lumber grades. The architect who designs the building will specify the lumber grades to be used on the job. It is absolutely essential that the carpenter or contractor who buys the lumber use the specified grade. In order to do that, you must be able to read the grade stamp. *Figure 6* shows and explains a typical grade stamp.

Lumber is graded based on its strength, stiffness, and appearance. The highest grades have very few defects; the lowest grades may have knots, splits, and other problems. The primary agencies for lumber grading are the Western Wood Products Association and the Southern Pine Council. Although the end results of the grading processes are the same, their grade designations are somewhat different. It is important that you learn to interpret the grade stamps for the grading association in your area.

Softwood lumber is divided into three major grades: boards, dimension, and framing. Within the board classification, there are several appearance grades, as shown in *Table 2*.

Discuss lumber grading. Show examples of lumber containing grading stamps. Explain the meaning of the stamps.

Show Transparency 7 (Figure 6).

(A) The trademark indicates agency quality supervision.

(B) Mill Identification – firm name, brand, or assigned mill number.

(C) Grade Designation – grade name, number, or abbreviation.

(D) Species Identification – indicates species individually or in combination.

(E) Condition of seasoning at time of surfacing:

S-Dry – 19% maximum moisture content

MC15 – 15% maximum moisture content

S-GRN – over 19% moisture content (unseasoned)

102F06.EPS

Figure 6 ◆ Typical lumber grade stamp.

 **Grading Lumber**

An understanding of lumber grading is very important to proper building construction. For example, it allows you to communicate effectively about the details of the construction process with others on the job site. More importantly, the building drawings specify grades of lumber that should be used for various parts of the structure. You will have to interpret those drawings in order to construct the building properly.

Table 2 Lumber Appearance Grades

Appearance	Grades	Boards
Selects	B and better C select D select	Selects are used for walls, ceilings, trim, and other areas where appearance is important but the use of hardwood is not desirable because of cost or other considerations.
Finish	Superior Prime E	
Paneling	Clear (any select or finish grade) No. 2 common (selected for knotty paneling) No. 3 common (selected for knotty paneling)	
Siding (bevel, bungalow)	Superior Prime	
Boards, sheathing	No. 1 common No. 2 common No. 3 common No. 4 common	Common boards are used for shelving, decking, window trim, and other applications where appearance is less critical, or where on-site selection and cutting is practical when appearance is a consideration.
Alternate board grades	Select merchantable Construction Standard Utility	

Summarize the basic classifications used in grading lumber. Refer the trainees to *Appendix A* in the Trainee Module.

4.1.0 Grading Terms

Among the different agencies, criteria for grading may vary somewhat. Generally, all grading agencies use five basic size classifications as follows:

- *Boards*—Consists of members up to 1½" thick and 2" and wider.
- *Light framing (L.F.)*—Consists of members 2" to 4" thick and 2" to 6" wide.
- *Joists and planks (J&P)*—Consists of members 2" to 4" thick and 6" or wider.
- *Beams and stringers (B&S)*—Consists of members 5" and thicker by 8" and wider.
- *Posts and timbers (P&T)*—Consists of members 5" × 5" and greater, approximately square.

For each species and size classification, the grading agencies establish several stress-quality grades, for example: Select structural, No. 1, No. 2, No. 3, etc., and assign allowable stresses for each (see *Table 3*).

Some other lumber terms used in grading include:

- *Nominal size*—The size by which it is known and sold in the market (e.g., 2 × 4), as opposed to the dressed size.

- *Dressed (actual) size*—The dimensions of lumber after surfacing with a planing machine. The dressed size is usually ½" to ¾" less than the nominal or rough size. A 2 × 4 stud, for example, usually measures about 1½" by 3½". The American Lumber Standard lists standard dressed sizes. *Table 4* compares rough cut (nominal) lumber sizes to dressed sizes for plainsawn and quarter-sawn lumber. Additional data on softwood and hardwood lumber dimensions are provided in *Appendix A*.

- *Dressed lumber*—Lumber that is surfaced by a planing machine on one side (S1S), two sides (S2S), one edge (S1E), two edges (S2E), or any combination of sides and edges (S1S1E, S1S2E, or S4S). Dressed lumber may also be referred to as *planed* or *surfaced*.

- *Dimension lumber*—Lumber is supplied in nominal 2", 3", or 4" thicknesses with standard widths. Light framing, studs, joists, and planks are classified as dimension lumber.

- *Matched lumber*—Lumber that is edge- or end-dressed and shaped to make a close tongue-and-groove (T & G) joint at the edges or ends when laid edge to edge or end to end.

Instructor's Notes:

_____

_____

_____

_____

_____

_____

_____

Table 3 Grades of Dimension Lumber

Dimension		
Light framing 2" to 4" thick 2" to 4" wide	Construction Standard Utility Economy	Light framing lumber is used where great strength is *not* required, such as for studs, plates, and sills.
Studs 2" to 4" thick	Stud Economy stud	This optional all-purpose grade is limited to 10 feet in length and is suitable for all stud uses, including loadbearing walls.
Structural light framing 2" to 4" thick 2" to 4" wide	Select structural No. 1 No. 2 No. 3 Economy	Structural lumber is used for light framing and forming where greater bending strength is required. Typical uses include trusses, concrete pier wall forms, etc.
Structural joist and planks 2" to 4" thick 6" and wider	Select structural No. 1 No. 2 No. 3 Economy	These grades are designed especially for lumber 6 inches and wider and are suitable for use as joists and rafters and for general framing.
Timbers		
Timbers	Select structural No. 1 No. 2 (No. 1 mining) No. 3 (No. 2 mining)	Timbers are used for stringers, beams, posts, and other support members.

Table 4 Nominal and Dressed (Actual) Sizes of Dimension Lumber (in Inches)

Nominal	Dressed
2 × 2	1½ × 1½
2 × 4	1½ × 3½
2 × 6	1½ × 5½
2 × 8	1½ × 7¼
2 × 10	1½ × 9¼
2 × 12	1½ × 11¼

Calculating Dressed Softwood Lumber Sizes

As you can see in *Table 4*, nominal lumber sizes don't represent the actual, or dressed, dimensions of the board or lumber. When calculating dressed sizes of softwood boards, there is a general rule that can be applied. Boards with a nominal width of 1" or less usually have dressed dimensions that are ¼" smaller. Boards with a nominal width of 2" to 6" usually have dressed dimensions that are ½" smaller. For boards with widths greater than 6", you can subtract ¾" from the nominal dimensions to get the dressed sizes. Remember, this is a general rule and may not be accurate in every case. The only way to be certain is to perform an actual measurement.

- *Patterned lumber*—Lumber that is shaped to a pattern in addition to being dressed, matched, or shiplapped, or any combination of these workings.

- *Rough lumber*—Lumber as it comes from the sawmill prior to any dressing operation.

- *Stress-grade lumber*—Lumber grades having assigned working stress and elasticity values in accordance with basic accepted principles of strength grading. Stress is the force exerted on a unit area of lumber, usually expressed in terms of *pounds per square inch.* Allowable stress is the maximum stress established by the applicable building codes. Allowable stress is always less than the ultimate stress (the amount of stress that can be withstood before the material fails) for obvious structural and safety reasons.

- *Surfaced lumber*—Same as dressed lumber.

- *Framing lumber*—Lumber used for the structural members of a building, such as studs and joists.

- *Finish lumber*—Lumber suitable for millwork or for the completion of the interior of a building, chosen because of its appearance or ability to accept a high-quality finish.

- *Select lumber*—In softwoods, a general term for lumber of good appearance and finishing qualities.

4.1.1 Trim

Trimming is the act of crosscutting a piece of lumber to a given length.

- Double-end-trimmed (DET) lumber is trimmed reasonably square by a saw on both ends.
- Precision-end-trimmed (PET) lumber is trimmed square and smooth on both ends to uniform lengths with a manufacturing tolerance of $\frac{1}{16}$" over or under (+/−) in length in 20 percent of the pieces.
- Square-end-trimmed lumber is trimmed square, permitting only a slight manufacturing tolerance of $\frac{1}{64}$" for each nominal 2" of thickness or width.

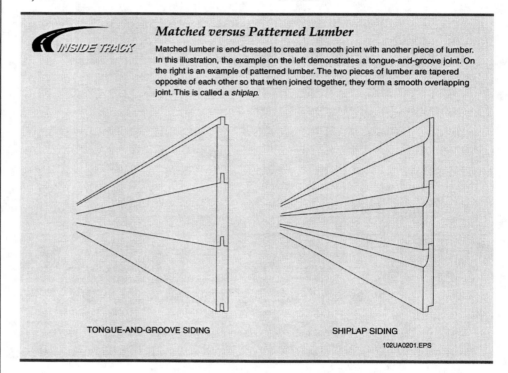

Matched versus Patterned Lumber

INSIDE TRACK

Matched lumber is end-dressed to create a smooth joint with another piece of lumber. In this illustration, the example on the left demonstrates a tongue-and-groove joint. On the right is an example of patterned lumber. The two pieces of lumber are tapered opposite of each other so that when joined together, they form a smooth overlapping joint. This is called a *shiplap*.

TONGUE-AND-GROOVE SIDING

SHIPLAP SIDING

102UA0201.EPS

Instructor's Notes:

_____

_____

_____

_____

_____

_____

_____

_____

4.2.0 Classification of Manufacturing Defects

The various types of defects that can occur while lumber is being cut, dressed, and dried were discussed in an earlier section. The following describes how the defects are considered in grading lumber:

- *Standard A*—Very light torn grain; occasional light chip marks; very slight knife marks.
- *Standard B*—Very light torn grain; very light raised grain; very light loosened grain; slight chip marks; average of one slight chip mark per lineal foot but not more than two in any lineal foot; very slight knife marks; slight mismatch.
- *Standard C*—Medium torn grain; light raised grain; light loosened grain; very light machine bite; very light machine gouge; very light machine offset; light chip marks if well scattered; occasional medium chip marks; very slight knife marks; slight mismatch.
- *Standard D*—Heavy torn grain; medium raised grain; very heavy loosened grain; light machine bite; light machine gouge; light machine offset; medium chip marks; slight knife marks; very light mismatch.
- *Standard E*—Very heavy torn grain; raised grain; very heavy loosened grain; medium machine bite; machine gouge; medium machine offset; chip marks; knife marks; light wavy dressing; light mismatch.
- *Standard F*—Very heavy torn grain; raised grain; very heavy loosened grain; heavy machine bite; machine gouge; heavy machine offset; chip marks; knife marks; medium wavy dressing; medium mismatch.
- *Standard G*—Loosened grain; raised grain; torn grain; machine bite; machine burn; machine gouge; machine offsets; chip marks; medium wavy dressing; mismatch.

4.3.0 Abbreviations

Most contract documents and construction drawings use abbreviations. *Appendix B* lists the common abbreviations used in the Southern Pine Industry. Many of them also hold for the other wood species.

5.0.0 ◆ PLYWOOD

Plywood is made of layers (plies) of wood veneers. A layer may be 1/16" to 5/8" thick. The center layer is known as the *core*. As layers are added on either side of the core, they are placed alternately at right angles, which increases the strength of the sheet. Layers with the grain at right angles to the core are called *crossbands*. The outer exposed layer is known as the *face veneer*.

Veneers are made by the rotary cutting of trees. First, logs are cut to a specific length, then softened with a hot water or steam bath. The bark is then removed and the log locked into its center on a lathe. As the lathe turns the log, a long knife slices off a thin veneer. A steel roller at the rear of the knife assists in keeping the veneer intact and maintains a uniform thickness of approximately 1/16" to 5/16".

This continuous veneer is trimmed into smaller sheets that are fed into a hot oven or dryer to reduce the moisture content. The moisture content is reduced to a range from 3 percent to 8 percent. At this point, the veneers are separated into core and crossband materials. The veneer that will be used for the face now goes through a patching machine to remove and correct defects in the veneer. The patching machine will also match face grains. After all this, the materials go through one final step called *splicing* prior to becoming a sheet of plywood.

In the final manufacturing process, hot glue is applied by machine to the core, crossbands, faces, and plies. The rough plywood sheet will now go into a hot press. This hot press will apply a great amount of pressure to the pieces to squeeze out the excess glue and will compress the rough plywood to its approximate final thickness. This process takes from two to twenty minutes. Once the process is complete, the sheets are cut and sanded to their final thickness.

5.1.0 Plywood Sheet Sizes

The average or standard size of plywood is 4'-0" × 8'-0". This is standard in the construction field. A few companies produce plywood from 6' to 8' widths and up to 16' in length. Sheathing-grade plywood is nominally sized by the manufacturer to allow for expansion; that is, 4'-0" × 8'-0" is really 47¾" × 95¾".

The thickness of plywood will vary from 3/16" to 1¼". The common sizes are ¼", ½", and ⅝" for finish paneling and ⅜", ½", ⅝", and ¾" for some structural purposes. There are three types of edges on plywood:

- Butt joints (two standard pieces joined)
- A shiplap cut or edge
- Interlocking tongue-and-groove

Opposite edges or all four edges may be cut to match.

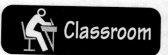

Classroom

Show examples of lumber with various manufacturing defects. Discuss the resulting grade of the lumber.

Emphasize the importance of being able to interpret common abbreviations. Refer the trainees to *Appendix B* in the Trainee Module.

Show examples of plywood and explain how it is manufactured.

Performance Profile Test

Have each trainee complete to your satisfaction Performance Profile Tasks 2 and 3 on grade marking stamps and lumber defects. Fill out Performance Profile Sheets for each trainee.

Homework

Assign reading of Sections 5.0.0–9.0.0 for the next class session.

Show examples of plywood with grading stamps. Explain the gradings.

5.2.0 Grading for Softwood Construction Plywood

After the plywood has been manufactured, the finished product is graded. The grading proce- dure is similar to the grading of lumber. The trade stamp that appears on each sheet lists the grade of the plywood (*Table 5*).

Plywood is rated for interior or exterior use. Exterior-rated plywood is used for sheathing, sid-

Table 5 Guide to APA Performance-Rated Plywood Panels *American Plywood Association*

```
_____APA_____
PANEL GRADE ──────► RATED SHEATHING
SPAN RATING ──────► 32/16   15/32 INCH ◄────── THICKNESS
                        SIZED FOR SPACING
EXPOSURE
DURABILITY ───────► EXPOSURE 1
CLASSIFICATION
                  _____000_____ ◄────── MILL NUMBER
              ► PS1-83        NER-108 ◄
  CODE RECOGNITION OF           APA'S PERFORMANCE RATED
  APA AS A QUALITY              PANEL STANDARD
  ASSURANCE AGENCY
```

GRADE DESIGNATION	DESCRIPTION AND COMMON USES	TYPICAL TRADEMARKS
APA RATED SHEATHING EXP 1 or 2	Specially designed for subflooring and wall and roof sheathing, but can also be used for a broad range of other construction and industrial applications. Can be manufactured as conventional veneered plywood, as a composite, or as a nonveneered panel. For special engineered applications, including high load requirements and certain industrial uses, veneered panels conforming to PS 1 may be required. Specify Exposure 1 when long construction delays are anticipated. Common thicknesses: 5/16, 3/8, 7/16, 15/32, 1/2, 19/32, 5/8, 23/32, 3/4.	____APA____ RATED SHEATHING 32/16 15/32 INCH SIZED FOR SPACING EXPOSURE 1 ___000___ NER-108
APA STRUCTURAL 1 RATED SHEATHING EXP 1	Unsanded all-veneer plywood grades for use where strength properties are of maximum importance; structural diaphragms, box beams, gusset plates, stressed-skin panels, containers, pallet bins. Made only with exterior glue (Exposure 1). Common thicknesses: 5/16, 3/8, 15/32, 1/2, 19/32, 5/8, 23/32, 3/4.	____APA____ RATED SHEATHING STRUCTURAL 1 48/24 23/32 INCH SIZED FOR SPACING EXTERIOR ___000___ PS 1-83 CC NER-108
APA RATED STURD-I-FLOOR EXP 1 OR 2	For combination subfloor-underlayment. Provides smooth surface for application of carpet and possesses high concentrated and impact load resistance. Can be manufactured as conventional veneered plywood, as a composite, or as a nonveneered panel. Available square edge or tongue-and-groove. Specify Exposure 1 when long construction delays are anticipated. Common thicknesses: 19/32, 5/8, 23/32, 3/4.	____APA____ RATED STURD-I-FLOOR 20 oc 19/32 INCH SIZED FOR SPACING EXPOSURE 1 ___000___ NER-108
APA RATED STURD-I-FLOOR 48 OC (2-4-1) EXP 1	For combination subfloor-underlayment on 32- and 48-inch spans and for heavy timber roof construction. Provides smooth surface for application of resilient floor coverings and possesses high concentrated and impact load resistance. Manufactured only as conventional veneered plywood and only with exterior glue (Exposure 1). Available square edge or tongue-and-groove. Thickness: 1⅛.	____APA____ RATED STURD-I-FLOOR 48 oc 1-1/8 inch SIZED FOR SPACING 2-4-1 EXPOSURE 1 TAG ___000___ UNDERLAYMENT PS1-83 NER-108

102T05.TIF

Instructor's Notes:

_____

_____

_____

_____

_____

_____

_____

ing, and other applications where there may be exposure to moisture or wet weather conditions. Exterior plywood panels are made of high-grade veneers bonded together with a waterproof glue that is as strong as the wood itself.

Interior plywood uses lower grades of veneer for the back and inner plies. Although the plies can be bonded with a water-resistant glue, water-proof glue is normally used. The lower-grade veneers reduce the bonding strength, however, which means that interior-rated panels are not suitable for exterior use.

5.2.1 Plywood Glues

Although most plywood today is rated for interior use, most of the glue used on plywood is exterior glue. The grade stamp will indicate whether the plywood is interior or exterior and if the glue is waterproof or not.

There are plywood sheets designated as structural grade or panels. These panels are manufactured for heavy-duty purposes. This type of plywood is seldom used in residential construction. It may be either interior or exterior plywood.

5.2.2 Plywood Cores

The plywood manufactured for industrial use is often called *soft plywood*. This product has a veneer core. This core is manufactured by the same procedure used to make other plies. It may

be of single ply or two plies having the grain running in the same direction and glued together to form a single core or layer (for example, ½" plywood may consist of veneer, but only three plies). Therefore, the number of plies is typically an odd number. The face veneers on sheets of plywood always run in the same direction.

The plywood used in the manufacture of cabinets, doors, furniture, and finished components is known as *hardwood plywood*. This type of plywood may have any of four different types of core:

- A lumber core
- A particleboard core
- A veneer core (common in construction plywood)
- A fiberboard core

The type of core will not be indicated on the grade stamp, but one can easily tell by looking at the edge of the plywood (*Figure 7*).

5.2.3 Faces

On the top of most plywood grade stamps are two letters separated by a hyphen. Only the letters A, B, C, D, and N are used. These letters indicate the quality of the veneer. The quality or grade of the face of the front panel is indicated by the first letter. The quality of grade of the face of the back panel is indicated by the second letter. The best quality that is free from defects carries an N grade.

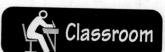

Selecting Plywood

Plywood that is expressly manufactured for either interior or exterior use may be used for other purposes in certain situations. Some local codes may require the use of pressure-treated plywood for exterior construction, in bathrooms, or in other high-moisture areas of a house. Pressure-treated plywood can withstand moisture better than interior plywood. Always check the local code(s) before beginning any construction project.

LUMBER CORE   PARTICLEBOARD CORE   VENEER BOARD   FIBERBOARD CORE

102F07.TIF

Figure 7 ◆ Types of plywood.

WOOD BUILDING MATERIALS, FASTENERS, AND ADHESIVES—TRAINEE MODULE 27102   2.19

This grade should be selected to receive a finish that is natural and exposed to view.

N-grade veneers are typically used in high-quality applications where a natural finish is desired. N-grade veneers are not normally found in construction plywood.

A plywood face that is smooth, has no open defects, but could have some heat repairs is an A grade. This type of veneer accepts paint readily. The next grade is B. This veneer offers a surface that is solid with splits no larger than ⅛". Some defects may be larger than ⅛" and may be fixed or repaired with smooth plugs. In most cases, tight knots are allowed. Sanding may create minor flaws that are allowed. This product may be treated like the A grade, but it will not be as smooth. All A and B faces are sanded.

In the C grade, the veneer may have splits to a maximum of ½" and up to 1½" knot holes; however, they must not affect the required strength of the panel. On C grade panels, you may find some sanding defects; these are permitted. The next grade is C Plugged. This veneer is an improved C. The veneer has tighter limits on splits and knot holes. The C Plugged grade has a fully sanded surface. C faces are sanded only if noted.

The poorest grade is the D grade. It has less strength and a poor appearance because of the defects. The D grade is typically used for the cores and backs of interior plywood.

Plywood that is designated for sheathing, subflooring, concrete forms, and sheets used for special structural purposes is called *engineered plywood*. The veneers are either unsanded or lightly sanded. The plywood sheets carry slightly different grade markings. The face veneers are either C-C or C-D.

5.2.4 Plywood Specification and Grade Guide

The U.S. Department of Commerce *Plywood Specification and Grade Guide* is the standard for plywood grading. Always be sure to check the specifications for grades. Different associations can certify plywood. Some mills certify their own plywood, and this certification may not be accepted by the grading associations or agencies.

Be sure to find out who is certifying plywood as a certain grade.

Plywood made from hardwoods such as birch and oak is used in making furniture and cabinets. Hardwood plywood uses a different grading system than the one described previously.

5.3.0 Plywood Storage

Remember, when stacking plywood only two methods may be used. Sheets may be stacked flat on a solid surface and off the floor or ground, or the sheets may be stacked in a full vertical position between two posts. Keep the plywood off the floor or ground. Storing plywood at an angle will cause the material to warp or twist.

5.4.0 Species Used in Plywood

In the manufacture of plywood, a total of 70 species of wood are used. The majority of plywood made today is produced from native softwoods, but there are some hardwoods used. Part of this material is grown in the United States and part of the material is imported from the Far East.

The species are put into five different groups. The strongest types are in Group 1. In some cases, the group will appear on the grade stamp. When

What's wrong with this picture?

102P0204.EPS

INSIDE TRACK

Plywood Safety

Plywood is awkward to carry. Remember, carry only one sheet of plywood at a time, and do not hold it over your head. In strong winds, the plywood could act like a wooden sail and injure you and others on the job site.

Instructor's Notes:

_____

_____

_____

_____

_____

_____

_____

the selection of species is important, the grade stamp can verify the species. Some stamps may have a class such as Class I or Class II. This indicates the strength of the plywood and the stresses the plywood can endure beyond simple bending. By this we mean tension, compression, shear, and the ability of the plywood to hold nails.

Refer to *Table 6* for a classification of species used for plywood veneers by groups.

6.0.0 ◆ BUILDING BOARDS

The ingenuity and technology that helped develop the plywood industry also assisted in the development of other materials in sheet form. The manufacturing processes for these materials is similar to the process for plywood. The main ingredients for these products, known as *building boards*, are vegetable or mineral fibers. After mixing these ingredients with binder, the mixture becomes very soft.

At this point, the mixture passes through a press, which uses heat and pressure to determine the thickness and density of the finished board.

Sawdust, wood chips, and wood scraps are the major waste materials at a sawmill. These scrap materials are softened with heat and moisture, mixed with a binder and other ingredients, and then run through presses that produce the density and thickness desired by the manufacturer.

The finished wood products that come off the presses are classified as *hardboard, particleboard, oriented strand board (OSB), overlay plywood,* or *fiberboard.*

6.1.0 Hardboard

Hardboard is a manufactured building material, sometimes called *tempered board* or *pegboard.* Hardboards are extremely dense. The common thicknesses for hardboards are ³⁄₁₆", ¼", and ⁵⁄₁₆". The standard sheet size for hardboards is 4'-0" × 8'-0".

Show examples of hardboard. Describe the three grades of hardboard.

Table 6 Classification of Species Used for Plywood Veneers by Groups

Group 1	Group 2	Group 3	Group 4	Group 5
Apitong	Cedar, Port Orford	Alder, red	Aspen,	Basswood
Beech, American	Cypress	Birch, paper	Bigtooth	Poplar, balsam
Birch,	Douglas fir 2	Cedar, Alaska	Quaking	
Sweet	Fir,	Fir, subalpine	Cativo	
Yellow	Balsam	Hemlock, eastern	Cedar,	
Douglas fir 1	California red	Maple, bigleaf	Incense	
Kapur	Grand	Pine,	Western red	
Keruing	Noble	Jack	Cottonwood,	
Larch, western	Pacific silver	Lodgepole	Eastern	
Maple, sugar	White	Ponderosa	Black	
Pine,	Hemlock, western	Spruce	(Western poplar)	
Caribbean	Lauan,	Redwood	Pine,	
Ocote	Almon	Spruce,	Eastern white	
Pine, southern	Bagtikan	Engelmann	Sugar	
Loblolly	Mayapis	White		
Longleaf	Red lauan			
Shortleaf	Tangile			
Slash	White lauan			
Tanoak	Maple, black			
	Mengkulang			
	Meranti, red			
	Mersawa			
	Pine,			
	Pond			
	Red			
	Virginia			
	Western white			
	Spruce,			
	Black			
	Red			
	Sitka			
	Sweet gum			
	Tamarack			
	Yellow poplar			

_____
_____
_____
_____
_____
_____
_____

However, they can be made in widths up to 6' and lengths up to 16' for specialized uses.

These boards are resistant to water, but are susceptible to the edges breaking if they are not properly supported. Holes must be pre-drilled for nailing; direct nailing into the material would cause it to fracture.

Three grades of hardboard are manufactured. The first is known as *standard*. This hardboard is light brown in color, smooth, and can easily accept paint. Standard hardboard is suitable only for interior use such as cabinets.

The second grade of hardboard is called *tempered hardboard (Masonite)*. This hardboard is the same as standard grade except that in the manufacturing process, the standard board is coated with oils and resins and heated or baked to a dark brown color. In going through this process, the tempered hardboard becomes denser, stronger, and more brittle.

Tempered hardboard is suitable for either interior or exterior uses such as siding, wall paneling, and other decorating purposes. One type of tempered hardboard is called *perforated hardboard* or *pegboard*. It has holes punched in it 1" apart. Special hooks are made to fit into the holes to support different types of items (shelves, brackets, and hangers for workshops and kitchens, for example).

The third type of hardboard is called *service grade*. This hardboard is not as dense, strong, or heavy as standard grade. The surfaces are not as smooth as standard grade, and it costs less than other hardboards. It can be used for basically everything for which standard or tempered hardboard is used. Service-grade hardboard is manufactured for items such as cabinets, parts of furniture, and perforated hardboard.

6.2.0 Particleboard

The main composition of this type of material is small particles or flakes of wood. Some manufacturing plants may use southern pine materials. Others may use stock selected for its species, fiber characteristics, moisture content, and color. There are two types of particleboard: Type I and Type II.

- Type I is basically a mat-formed particleboard generally made with vera-formaldehyde resin, suitable for interior applications. Refer to the material safety data sheet (MSDS) for handling. This type comes in three classes: A, B, and C. The classes are density grades. Class A is a high-density board of 50 lbs. per cu. ft. or over. Class B is a medium-density board of 37 to 50 lbs. per cu. ft. Class C is a low-density board of 37 lbs. per cu. ft. and lower.

- Type II is a mat-formed particleboard made with durable, moisture- and heat-resistant binders. This makes the particleboard suitable for interior and certain exterior applications when so labeled. This type comes in two classes: Class A is a high-density board of 50 lbs. per cu. ft. or over; Class B is a medium-density board of 37 to 50 lbs. per cu. ft.

In addition to the two types and the classes in each type, particleboard is further broken down into still another class of either Class 1 or Class 2. These are strength classifications based on properties of the panels.

Each manufacturer of particleboard has specification sheets that contain the technical information previously discussed. The information sheets are usually available from your local lumber dealer.

Particleboard is pressed under heat into panels. The sheets range in size from ¼" to 1½" in thickness and from 3' to 8' in width. There are also thicknesses of 3" and lengths ranging up to 24' for special purposes. Particleboard has no grain, is smoother than plywood, is more resilient, and is less likely to warp. Particleboard has many uses such as shelving and cabinets. There are cases when ¼" particleboard is used as a plywood core.

Some types of particleboard can be used for underlayment if the local building codes permit. If particleboard is used as underlayment, it is suggested that it be laid with the long dimension across the joists and the edges staggered. Particleboard can be nailed, although some types will crumble or crack when nailed close to the edges.

6.3.0 High-Density Overlay (HDO) and Medium-Density Overlay (MDO) Plywood

High-density overlay (HDO) plywood panels have a hard, resin-impregnated fiber overlay heat-bonded to both surfaces. HDO panels are abrasion- and moisture-resistant and can be used for concrete forms, cabinets, countertops, and similar high-wear applications. HDO also resists damage from chemicals and solvents. HDO is available in five common thicknesses: ⅜", ½", ⅝", ¾", and 1".

Medium-density overlay (MDO) panels are coated on one or both surfaces with a smooth, opaque overlay. MDO accepts paint well and is very suitable for use as structural siding, exterior decorative panels, and soffits. MDO panels are available in eight common thicknesses ranging from ¹¹⁄₃₂" to ²³⁄₃₂".

Classroom

Show examples of particleboard. Describe the two types of particleboard.

Show examples of HDO and MDO plywood and discuss their applications.

Instructor's Notes:

Both HDO and MDO panels are manufactured with waterproof adhesive and are thus suitable for exterior use. If MDO panels are to be used outdoors, however, the panels should be edge-sealed with one or two coats of a good-quality exterior housepaint primer. This is easier to do when the panels are stacked.

6.4.0 Oriented Strand Board (OSB)

Oriented strand board (OSB) is a manufactured structural panel used for wall and roof sheathing and single-layer floor construction. *Figure 8* shows two kinds of OSB panels. OSB consists of compressed wood strands arranged in five or more cross-banded layers and bonded with phenolic resin. Some of the qualities of OSB are dimensional stability, stiffness, fastener holding capacity, and no core voids. Some OSB that is used for roof sheathing, for example, is manufactured with a reflective coating on the underside. This gives this particular type of board a higher fire rating. Before cutting OSB, be sure to check the applicable MSDS for safety hazards. The MSDS is the most reliable source for safety information.

6.5.0 Mineral Fiberboards

The building boards just covered are classified as vegetable fiberboards. Mineral fiberboards fall into the same category as vegetable fiberboards. The main difference is that they will not support combustion and will not burn. Glass and gypsum rock are the most common minerals used in the manufacture of these fiberboards. Fibers of glass

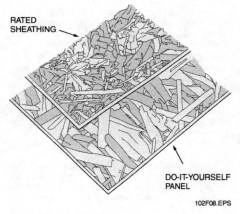

RATED SHEATHING

DO-IT-YOURSELF PANEL

102F08.EPS

Figure 8 ◆ OSB panels.

or gypsum powder are mixed with a binder and pressed or sandwiched between two layers of asphalt-impregnated paper, producing a rigid insulation board.

There are some types of chemical foams mixed with glass fibers that will also make a good, rigid insulation. However, this mineral insulation will crush and should not be used when it must support a heavy load.

> **CAUTION**
>
> Whenever you work with older materials that may be made with asbestos, contact your supervisor for the company's policies on safe handling of the material. State and federal regulations require specific procedures to follow prior to removing, cutting, or disturbing any suspect materials. Also, some materials emit a harmful dust when cut. Check the MSDS before cutting.

7.0.0 ◆ ENGINEERED WOOD PRODUCTS

In the past, the primary source of structural beams, timbers, joists, and other weight-bearing lumber was old-growth trees. These trees, which need many decades to mature, are tall and thick and can produce a large amount of high-quality, tight-grained lumber. Extensive logging of these trees to meet demand resulted in higher prices and conflict with forest conservation interests.

The development of wood laminating techniques by lumber producers has permitted the use of younger-growth trees in the production of structural building materials. These materials are given the general classification of *engineered lumber products*.

Engineered lumber products fall into five general classifications: laminated veneer lumber (LVL), parallel strand lumber (PSL), laminated strand lumber (LSL), wood I-beams, and glue-laminated lumber (glulam). These materials provide several benefits:

- They can be made from younger, more abundant trees.
- They can increase the yield from a tree by 30 percent to 50 percent.
- They are stronger than the same size of structural lumber. Therefore, the same size piece can bear more weight. Or, looked at another way, a smaller-dimensioned piece of engineered lumber can bear equal weight.
- Greater strength allows the engineered lumber to span a greater distance.

Show examples of OSB and discuss its applications.

Show examples of mineral fiberboards and discuss their applications.

Emphasize the safety precautions for working with materials that might contain asbestos.

Identify the different classifications of engineered wood products. Emphasize the benefits of using these products.

Show examples of LVL, PSL, and LSL and discuss their applications. Discuss the use of wood I-beams.

WOOD BUILDING MATERIALS, FASTENERS, AND ADHESIVES—TRAINEE MODULE 27102   2.23

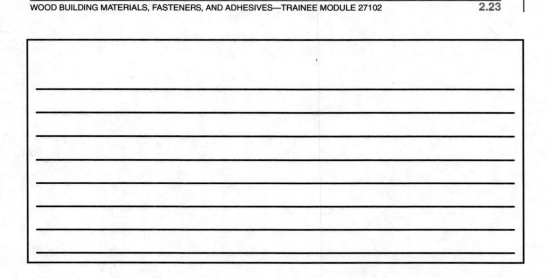

- A length of engineered wood is lighter than the same length of solid lumber. It is therefore easier to handle.
- They are dimensionally accurate and do not warp, crown, or twist.

Figure 9 shows three of the main types of engineered lumber.

PSL   102F09A.EPS

LVL   102F09B.EPS

LSL   102F09C.EPS

Figure 9 ◆ PSL, LVL, and LSL.

7.1.0 Laminated Veneer Lumber (LVL)

Like plywood, LVL is made from laminated wood veneer (*Figure 9*). Douglas fir and southern pine are the primary sources. Thin ($\frac{1}{10}$" to $\frac{3}{16}$") sheets are peeled from the tree in widths of 27" or 54". The veneers are *laid-up* in a staggered pattern with the veneers overlapping to increase strength. Unlike plywood, the grain of each layer runs in the same direction as the other layers. The veneers are bonded with an exterior-grade adhesive, then pressed together and heated under pressure.

LVL is used for floor and roof beams and for headers over windows and doors. It is also used in scaffolding and concrete forms. No special cutting tools or fasteners are required.

7.2.0 Parallel Strand Lumber (PSL)

PSL is made from long strands of Douglas fir and southern pine. The strands are about $\frac{1}{8}$" or $\frac{1}{10}$" thick and are bonded together with adhesive in a special heating process.

PSL is used for beams, posts, and columns. It is manufactured in thicknesses up to 7". Columns can be up to 7" wide; beams range up to 18" in width.

7.3.0 Laminated Strand Lumber (LSL)

LSL can be made from small logs of almost any kind of wood. Aspen, red maple, and poplar that cannot be used for standard lumber are commonly used. In the manufacturing process, the logs are cut into short strands, which are bonded together and pressed into long blocks (billets) up to $5\frac{1}{2}$" thick, 8' wide, and 40' long.

LSL is used for millwork such as doors and windows and any other product that requires high-grade lumber. However, LSL will not support as much of a load as a comparable size of PSL because PSL is made from stronger wood.

7.4.0 Wood I-Beams

Wood I-beams (*Figure 10*) consist of a web with flanges bonded to the top and bottom. This arrangement, which mimics the steel I-beam, provides exceptional strength. The web can be made of OSB or plywood. The flanges are grooved to fit over the web.

Wood I-beams are used as floor joists, rafters, and headers. Because of their strength, wood I-beams can be used in greater spans than a comparable length of dimension lumber. Lengths of up to 60' are available.

Instructor's Notes:

_____

_____

_____

_____

_____

_____

_____

_____

Figure 10 ◆ Wood I-beams.

102F11.EPS

Figure 11 ◆ Glulam beam construction.

102F12.EPS

Figure 12 ◆ Glulam beam application.

7.5.0 Glue-Laminated Lumber (Glulam)

Glulam is made from lengths of solid, kiln-dried lumber that have been glued together. It is popular in architectural applications where exposed beams are used (*Figure 11*). Because of its exceptional strength and flexibility, glulam can be used in areas subject to high winds or earthquakes. Glulam is available in three appearance grades: industrial, architectural, and premium. Industrial grade is used in open buildings such as warehouses and garages where appearance is not a priority or where beams are not exposed. Architectural grade is used where beams are exposed and appearance is important. Premium, the highest grade, is used where the highest quality appearance is needed (*Figure 12*).

Glulam beams are available in widths from 2½" to 8¾". Depths range from 5⅛" to 28½". They are available in very long lengths. They are used for many purposes, including ridge beams; basement beams; headers of all types; stair treads, supports, and stringers; and cantilever and vaulted ceiling applications. Because glulam beams are laminated, they can be formed into arches and any number of curved configurations. Glulam beams are especially popular for use in churches.

8.0.0 ◆ PRESSURE-TREATED LUMBER

Pressure-treated lumber is softwood lumber protected by chemical preservatives forced deep into the wood through a vacuum-pressure process. Pressure-treated lumber has been used for many years in on-ground and below-ground applications such as landscape timbers, sill plates, and foundations. In some parts of the country, it is also extensively used in the building of decks, porches, docks, and other outdoor structures subject to decay from exposure to the elements. It is popular for these uses in areas where structures are exposed to insect or fungus attack. A major advantage of pressure-treated lumber is its relatively

Provide an example of glulam and describe its qualities and applications.

Show examples of pressure-treated lumber and discuss its applications.

Uses of Engineered Wood Products

Engineered wood products are used in a wide array of applications that were once exclusively served by cut lumber. For example, PSL is used for columns, ridge beams, girders, and headers. LVL is also used for form headers and beams. Wood I-beams are used to frame roofs as well as floors. An especially noteworthy application is the use of LSL studs, top plates, and soleplates in place of lumber to frame walls.

Fire-Retardant Building Materials

Lumber and sheet materials are sometimes treated with fire-retardant chemicals. The lumber can either be coated with the chemical in a non-pressure process or impregnated with the chemical in a pressure-treating process. Fire-retardant chemicals react to extreme heat, releasing vapors that form a protective coating around the outside of the wood. This coating, known as *char*, delays ignition and inhibits the release of smoke and toxic fumes.

low price in comparison with redwood and cedar. When natural woods such as these are used, only the more expensive heartwood will resist decay and insects.

WARNING!

Because the chemicals used in pressure-treated lumber represent some hazard to people and the environment, special precautions apply:

- When cutting pressure-treated lumber, always wear eye protection and a dust mask.
- Wash any skin that is exposed while cutting or handling the lumber.
- Wash clothing that is exposed to sawdust separately from other clothing.
- Do not burn pressure-treated lumber, as the ash poses a health hazard. Check regulations for proper disposal.
- Be sure to read and follow the manufacturer's safety instructions.

Pressure-treated lumber is available in three grades. The three grades are designated by their pounds per cubic foot of preservative retention. Above ground (.25 lb./cu. ft.) grade is to be used only 18" or more above ground. Ground contact grade (.40 lb./cu. ft.) is used when there is contact with water or soil on or below ground. The third grade (.60 lb./cu. ft.) is used when structural reliability is required, such as in wooden foundations and power poles.

9.0.0 ◆ CALCULATING LUMBER QUANTITIES

Large quantities of lumber are normally ordered by the *board foot*. A board foot is equivalent to a piece of lumber that is 1" thick, 12" wide, and 1' long. Each of the boards in *Figure 13* represents one board foot. When calculating board feet, always make sure you use the nominal lumber dimensions. The following formula is used to calculate board feet:

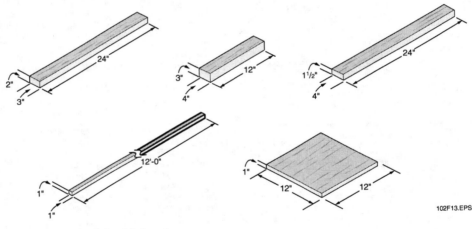

Figure 13 ◆ Examples of a board foot.

Instructor's Notes:

_____

_____

_____

_____

_____

_____

_____

_____

Measure Twice, Cut Once

An apprentice carpenter was helping his boss remodel a basement. The boss asked the apprentice to measure and cut the studs for the walls. The apprentice, looking to impress the boss, attacked the job with gusto. Instead of trying out the first stud, he cut all of them at once, based on a quick measurement. When all the studs were cut, they started to place them, only to find out that each stud was 3¼" short. The apprentice had forgotten to add the measuring tape case to the tape reading.

Assign reading of Sections 10.0.0–13.4.0 for the next class session.

Discuss the applications of various types of nails. Pass around examples for the trainees to examine.

Show Transparency 10 (Figure 14).

Show Transparency 11 (Figure 15).

Board feet = number of pieces × thickness in inches × width in inches × length in feet ÷ 12

For example, 20 pieces of 2 × 6 lumber that are each 8' long equals 160 board feet.

$$\frac{20 \times 2 \times 6 \times 8}{12} = \frac{1,920}{12} = 160 \text{ board feet}$$

To calculate the amount of sheet materials such as plywood, particleboard, drywall, and sheathing you need, calculate the area to be covered (length × width = area). The area will always be the square of the values used (square feet, square inches, etc.). For example, to figure the amount of plywood decking needed for a 20' by 41' roof:

$$20 \times 41 = 820 \text{ square feet}$$

A standard 4' by 8' sheet contains 32 square feet. Therefore, if you divide 820 by 32, it yields 25⅝ sheets, which is then rounded to 26 sheets.

Trim and moldings are priced by the lineal foot and are ordered by the dimension of the piece (for example, 150' of 1" quarter round molding). *Appendix A* contains lumber conversion tables.

That completes the coverage of wood building materials. In the remaining sections of this module, you will learn about the fasteners and adhesives used in the construction industry.

10.0.0 ◆ NAILS

Nailing is the most common method used for attaching two pieces of lumber. Some nails are made specifically to be driven with a hammer; others are made for use with pneumatic or cordless nailers. You will often hear nails referred to as *8-penny, 10-penny*, etc. This method of designating the size of a nail dates back many years. In written form, the penny designation appears as a lowercase *d* (*Figure 14*); an 8-penny nail would read 8d, and so forth. Nails above 16d are referred to as *spikes*. Spikes range from 20d (4") to 60d (6"). The

thickness (gauge) of the nail shank increases as the length increases.

10.1.0 Kinds of Nails

Several kinds of nails are discussed below (*Figure 15*).

- *Common nails*—These are the most frequently used of all nails. They have a flat head, smooth, round shank (shaft), and diamond point. They are available from 2d to 60d in length. They are used when appearance is not of prime importance, such as in rough framing or building concrete forms.
- *Box nails*—These are very similar to the common nail in appearance, except they have a thinner head and shank and less tendency to split the wood. They often have a resin coating to resist rust and create more holding power. They are available from 2d to 40d.
- *Finish nails*—Finish nails have a small barrel-shaped head with a slight indentation in the top to receive a nail set tool so the nail can be driven slightly below the surface of the wood. The hole above the head is then filled with a special putty before paint or finish is applied to conceal the nail. Finish nails are used for installing millwork and trim where the final appearance is important. Their small heads reduce their holding power considerably. If holding power is important, finish nails can be obtained with galvanized or japanned coatings. Finish nails are available in 2d to 20d lengths.
- *Casing nails*—Casing nails look a lot like finish nails; however, they have a conical-shaped head and the shank is larger. Also, they come in lengths from 2d to 40d as opposed to the 20d length of a regular finish nail. The casing nail has more holding power because of the larger shank and the shape of the head.

WOOD BUILDING MATERIALS, FASTENERS, AND ADHESIVES—TRAINEE MODULE 27102   2.27

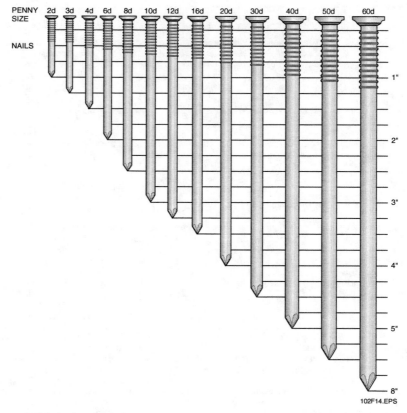

Figure 14 ◆ Nail sizes.

- *Duplex or scaffold (doublehead) nail*—The duplex nail is a specially-designed nail for building scaffolds, concrete forms, and other temporary work that will later be disassembled. The advantage of this nail is that it can be driven to the first head, which gives it sufficient holding power. It can be easily removed at a later time with a claw hammer or wrecking bar without damaging the wood.
- *T-nails*—This is a specially-designed nail used in pneumatic nailing guns. It is coated with resin and comes in strips for insertion into the gun. T-nails come in lengths of 1¼", 1½", 1¾", 2", 2⅜", and 2½". These nails are also available with galvanized coatings. T-nailing reduces labor time as it only takes a pull of the trigger to drive them home.

- *Drywall (ratchet) nail*—This is another specially-designed nail with an annular or ring shank. Its principal use is to fasten gypsum or drywall to wood studs. The ring shanks increase holding power and prevent the nail from backing out of the stud. Lengths are from 1" to 2½".
- *Masonry nails*—Masonry nails are available in lengths of ½" to 4". The shanks on the longer ones have grooves with a sinker head. They are case-hardened and used to fasten metal or wood to masonry or concrete.
- *Cut nails*—Cut nails could also be grouped in with masonry nails, because in today's construction they are very seldom used for other than masonry purposes. They are rectangular in shape and have a blunt point, with little tendency to bend. They are obtainable up to about 5" long.

Instructor's Notes:

A Penny's Worth

Back in the fifteenth century, an Englishman could buy 100 three-inch nails for 10 pennies, so the three-inch nail became known as a *10-penny nail*. Other nails were known as *8-penny nails*, *12-penny nails*, and so on, depending on the cost per hundred. Although a penny isn't worth much these days, the designation is still used to identify the nails. The "d" that is used in the written form (8d, 10d, etc.) is said to represent the *denarius*, a small Roman coin used in Britain that was equal to a penny.

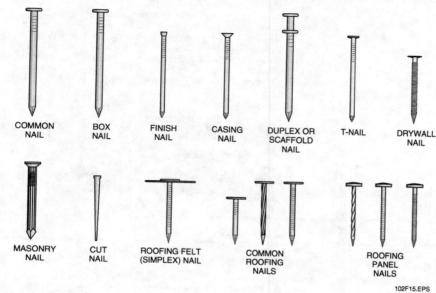

102F15.EPS

Figure 15 ◆ Kinds of nails.

WARNING!

Wear a face shield when driving cut nails, as they are subject to breakage, which could do grave damage to an eye.

- *Roofing nails*—Most roofing nails are galvanized to resist rusting. There are several types of roofing nails:
  - Roofing fasteners (often called *Simplex fasteners*) are commonly used to install roofing felt. They consist of a square metal or plastic plate with a screw driven through the center of the plate.

  - Common roofing nails come in lengths of ¾" to 2". All have a thin head but different shanks. They are primarily used to apply asphalt or fiberglass shingles. They are also used to fasten insulation sheathing board to studs. The longer ones are used for re-roofing jobs, while the shorter ones are used on new roofs.
  - Roofing panel nails have a neoprene washer. They range in length from 1¾" to 2½", with either a helical (spiral) or plain shank. The helical shank has greater holding power. Roofing panel nails are used on fiberglass, steel, or aluminum roofing panels. The nylon washer serves as a weatherproofing seal.

Warn the trainees to wear a face shield when driving cut nails.

Selecting the Right Nail

As a rule of thumb, select a nail that is three times longer than the thickness of the material you are fastening. However, the nail length should not exceed the total thickness of the two pieces being fastened together. For example, if you are nailing two 2 × 4 joists where they overlap on a girder, three times the thickness of one joist would be 4½". Since the total thickness of the two joists is only 3", however, it would not make sense to use a 4½" nail.

Common nails are used primarily for rough framing. Box nails, which are thinner and have slightly smaller heads than common nails, are used when working near the edge of the lumber (toenailing studs, for example) because they are less likely to split the wood. Box nails are also used for fastening exterior insulation board and siding.

There are three types of nails used for trim. The finish nail is used primarily for fastening interior trim. The casing nail, which is thicker and has a slightly larger head than the finish nail, is used to fasten heavy trimwork and exterior finish material. The wire brad is thinner than finish and casing nails and is used for very light trimwork such as small moldings. These trim nails all have a small head with a slight indentation, which makes it easy to drive the nail head below the surface of the wood using a nail set (shown here).

102P0205.EPS

Panel Nails

Panel nails are available in a variety of colors to match the paneling. These nails have annular rings to provide more holding capacity. Nails used to fasten paneling should be long enough to penetrate ¾" into the framing.

Listed below are some additional types of nails you will occasionally encounter:

- *Hardboard siding nails*—These nails are designed for installing aluminum, vinyl, and steel siding.
- *Insulated siding nails*—These nails have a zinc coating and come in various colors to match the color of the siding being installed.
- *Panel nails*—These also come in various colors. They come in round- and ring-shank design. The head is small and blends in with the paneling for good appearance.

- *Wire brads*—Their appearance is similar to that of regular finish nails. However, they are available with smaller shanks than finish nails, thus reducing splitting tendencies. They are used for fine finish work and come in lengths from ³⁄₁₆" to 3".
- *Escutcheon pins*—These specially-designed nails generally come as part of a hardware kit. They have an oval head and vary in length from ³⁄₁₆" to 2". They are used for such things as metal trim on store cabinets or nailing up house numbers. They add a nice appearance to the finish trim and generally match the color of the hardware being installed.

Instructor's Notes:

10.2.0 Staples

This module will deal only with the various types of staples used in building construction.

There are hundreds of different types of staples manufactured to do specific jobs. Refer to the different manufacturers' catalogs for this information. The factors that should be considered are:

- The type of point
- The crown (or top) width
- The length
- What type of metal they are made of
- The various coatings that are available

10.2.1 Types of Staples

The type of staple is determined by the type of point it has (*Figure 16*).

- *Chisel*—Recommended for grainy woods; keeps legs parallel to the entire leg length.
- *Crosscut chisel*—Legs penetrate straight and parallel through cross-grain wood. Good for general tacking and nailing purposes.
- *Outside chisel*—Good for clinching inward after penetrating the material being fastened.
- *Inside chisel*—Good for clinching outward after penetrating the material being fastened.
- *Spear*—Good for penetrating dense materials; will deflect easily when hitting an obstruction.
- *Divergent*—After start of penetration, legs diverge to allow use of longer legs in thin material; very good for wallboard insulation.
- *Outside chisel divergent*—Legs diverge, then cross, locking the staple; provides excellent penetration ability.

The most common crown widths vary from ⅜" to 1". The wider crowns are used mostly for mill-work, crating, and application of asphalt shingles. The most common wire gauge (thickness) is 14 to 16 and the length is normally ½" to 2".

Staples are glued together in strips for insertion into a stapling tool. Some stapling tools are manually-operated. Others are powered by pneumatics or electricity.

11.0.0 ◆ SCREWS

The three types of screws that we will focus on in this module are wood, sheet metal, and machine.

Screws have a neat appearance and can be decorative. They have much more holding power than nails and can easily be removed without damaging the materials involved. However, the cost factor of the screw and the time involved in installation must be considered.

11.1.0 Wood Screws

Wood screws are commonly made in 20 different stock thicknesses and lengths. They range in length from ¼" to 5". The screw gauge (diameter) ranges from 0 to 24. The higher the number, the larger the diameter. (See the wood screw chart in *Figure 17.*)

The two most common types of screw heads are slotted head and Phillips head. Screw heads may have any of several different shapes (*Figure 18*).

When using wood screws, it is generally necessary to drill pilot holes, particularly in hardwoods, to receive the screw. This does four things:

- Ensures pulling the materials tightly together
- Makes the screw easier to drive
- Prevents splitting the wood
- Prevents damage to the screw

CHISEL STAPLE   CROSSCUT CHISEL STAPLE   OUTSIDE CHISEL STAPLE   INSIDE CHISEL STAPLE

DIVERGENT STAPLE   OUTSIDE CHISEL DIVERGENT STAPLE   SPEAR STAPLE

Figure 16 ◆ Staples.

102F16.EPS

Discuss the applications of various types of staples. Pass around examples for the trainees to examine.

Show Transparency 12 (Figure 16).

Discuss the applications of various types of wood screws. Pass around examples for the trainees to examine.

Show Transparencies 13 and 14 (Figures 17 and 18).

Emphasize the importance of drilling pilot holes.

For decorative purposes, screws can be obtained in many different finishes (e.g., antique copper or brass, gold- or silver-plated, chromium, lacquered, etc.).

11.2.0 Sheet Metal Screws

Sheet metal screws are thread-cutting or thread-forming screws used to fasten light-gauge sheet metal (*Figure 19*). The threads of sheet metal screws are deeper than those of wood screws. This allows the two pieces of metal being fastened to be drawn tightly together. Because their threads are deeper and hold better, sheet metal screws are also recommended for use with softer building

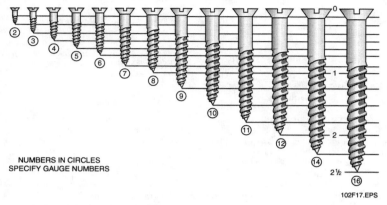

NUMBERS IN CIRCLES
SPECIFY GAUGE NUMBERS

102F17.EPS

Figure 17 ◆ Wood screw sizes.

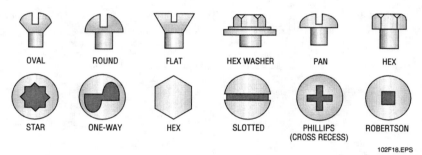

OVAL   ROUND   FLAT   HEX WASHER   PAN   HEX

STAR   ONE-WAY   HEX   SLOTTED   PHILLIPS (CROSS RECESS)   ROBERTSON

102F18.EPS

Figure 18 ◆ Common screws.

Drilling a Pilot Hole

When drilling a pilot hole, use a drill bit that is the same size as the shank of the screw. Hold the drill bit lengthwise over the screw. You should be able to see the wings, or threads, of the screw. If you can't, the drill bit is too large. Coat the screw with soap, paraffin, or beeswax to make it easier to drive.

2.32

CARPENTRY LEVEL ONE—TRAINEE MODULE 27102

Instructor's Notes:

materials like particleboard and hardboard. Sheet metal screws come in about the same range of sizes as wood screws.

Self-drilling sheet metal screws have a cutting edge on their point that eliminates the need for pre-drilling. These screws are driven with a power driver and are commonly used to fasten metal framing.

11.3.0 Drywall Screws

Drywall screws are used for wood and metal as well as gypsum drywall (*Figure 20*). These screws have a Phillips head and are designed to be driven with a powered screwdriver. When used correctly for drywall, the bugle head makes a depression in the surface without breaking the paper.

Type W screws have a sharp point and a thread designed for easy penetration and holding power. They are effective for use with wood. Type S screws are self-drilling for use with metal. Type G screws have a deeper thread designed for use with drywall panels. Drywall screws should never be used in damp or outdoor applications because they will rust.

11.4.0 Lag Screws

Lag screws, also called *lag bolts*, have a square or hexagon-shaped head and are designed to be driven with a wrench rather than a screwdriver (*Figure 21*). A lag screw is basically a heavy-duty wood screw. It is often used to fasten heavy material to wood. A pilot hole is recommended to prevent the wood from splitting.

The lag screw can be combined with a lag shield for use in concrete. The lag shield is an insert made of lead that is placed into a pre-drilled hole in the concrete. When the lag screw is driven into the insert, the insert expands, anchoring the lag screw.

11.5.0 Machine Screws

Carpenters use machine screws to fasten butt hinges to metal jambs or door closers to their brackets, and to install lock sets. It is often necessary to drill and tap holes in metal to receive machine screws as fasteners for wood or various kinds of trim. They have four basic head designs, as shown in *Figure 22*.

Discuss the applications of various types of drywall screws. Pass around examples for the trainees to examine.

Show Transparency 16 (Figure 20).

Point out that a lag screw is basically a heavy-duty wood screw.

Discuss machine screws and their applications. Pass around examples for the trainees to examine.

Show Transparency 17 (Figure 22).

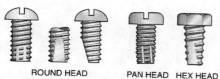

ROUND HEAD   PAN HEAD   HEX HEAD

THREAD-CUTTING SCREWS

STANDARD THREAD-FORMING SCREW   SELF-DRILLING SCREW

THREAD-FORMING SCREWS

102F19.EPS

Figure 19 ◆ Sheet metal screws.

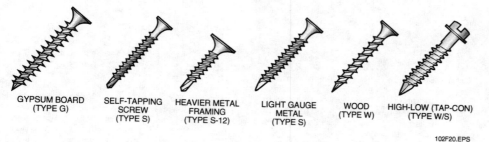

GYPSUM BOARD (TYPE G)   SELF-TAPPING SCREW (TYPE S)   HEAVIER METAL FRAMING (TYPE S-12)   LIGHT GAUGE METAL (TYPE S)   WOOD (TYPE W)   HIGH-LOW (TAP-CON) (TYPE W/S)

102F20.EPS

Figure 20 ◆ Drywall screws.

Figure 21 ◆ Lag screw and lag shield.

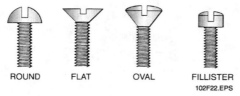

ROUND   FLAT   OVAL   FILLISTER

102F22.EPS

Figure 22 ◆ Machine screws.

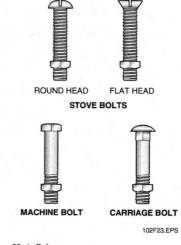

ROUND HEAD   FLAT HEAD

STOVE BOLTS

MACHINE BOLT   CARRIAGE BOLT

102F23.EPS

Figure 23 ◆ Bolts.

Machine screws can be obtained with a straight slot or Phillips head. They come in diameters ranging from ⅟₁₆" to ⅜" and lengths of ⅛" to 3". The most common machine screws are made of steel or brass, but they can be obtained in other metals as well. They may be obtained with coarse or fine thread.

11.6.0 Bolts

Bolts are often used by the carpenter in attaching one unit or member to another. There are many different designs and types for special jobs, but we will discuss only the most common in this module (*Figure 23*).

11.6.1 Stove Bolts

Stove bolts come in either roundhead or flathead design and lengths from ⅜" to 6". The shanks are threaded all the way to the head on those up to 2" long. If they are longer than 2", they are threaded to a maximum of 2". They can be obtained in several different materials such as steel, brass, and copper. They are used in lighter types of construction assemblies.

11.6.2 Machine Bolts

Machine bolts come with nuts. The nut is normally the same thickness as the diameter of the bolt. Specially-designed nuts are obtainable for specific purposes such as self-locking or cap nuts for safety or appearance. The bolts normally have square or hexagon heads with nuts to match. They come in lengths from ¾" to 30". The most common diameters range from ¼" to 2" and are obtainable in either regular or fine thread. For a decorative appearance, they can be ordered with special finishes such as antique or chrome.

11.6.3 Carriage Bolts

The carriage bolt is similar to the machine bolt except for the design of the head. The head is oval and the shank is square. It is designed that way so that the bolt can be driven or drawn into the wood, and the nut tightened without the bolt turning. Also, other members can be fastened over the heads with little or no interference. They come in lengths of ¾" to 20" with either standard or fine threads. Normal diameters range from ¼" to ¾". They can also be obtained in different materials and finishes, such as brass or galvanized.

2.34

Instructor's Notes:

_____

_____

_____

_____

_____

_____

_____

_____

12.0.0 ◆ ANCHORS

Anchors are special types of fasteners used to secure parts such as beams to concrete or to secure material to hollow walls and ceilings. In this section, we will cover some of the common types of anchors you will use on the job.

12.1.0 Masonry Anchors

Figure 24 shows several types of masonry anchors. Most of these anchors are designed to be inserted into a pre-drilled hole in the masonry, then expanded. Their main purpose is to allow something to be fastened securely to the masonry surface. A good example is securing the wooden sill plate to the building foundation. *Figure 25* shows the use of an anchor bolt for this purpose. In this

case, the anchor bolt is set into the concrete while the concrete is still wet. The concrete screw, also shown in *Figure 25*, is another anchoring device commonly used with masonry. Use a matching drill bit to drill a pilot hole before installing the concrete screw.

With the exception of the anchor bolt and nail anchor, all the anchors shown in *Figure 24* are placed or driven into a pre-drilled hole and expanded by turning a nut or screw. The nail anchor expands when the nail is struck with a hammer.

Figure 26 shows the use of a self-drilling snap-off anchor. This anchor has a cutting sleeve that is first used as a drill bit, then becomes the expander. A hole is drilled in the concrete using the cutting sleeve as a drill bit. Then the sleeve and expander plug are driven into the hole, causing the anchor to expand. The anchor is then sheared off flush with the surface and a threaded bolt can be screwed into the anchor to secure the workpiece.

Some anchors are secured with epoxy adhesive (discussed in the next section). As shown in *Figure 27*, the epoxy is mixed, then poured into the hole. The bolt is set into the epoxy and leveled. After the epoxy hardens, the workpiece can be secured to the anchor with a nut.

12.2.0 Hollow-Wall Anchors

Screws and nails cannot obtain a secure grip on soft building material such as gypsum wallboard. They are likely to pull out when weight is applied. Several types of hollow-wall anchors are available for these applications. Some are shown in *Figure 28*. These anchors are designed to expand within the hole or behind the wall as a screw is tightened.

The wings on a toggle bolt are spring-loaded. They are collapsed by hand when the bolt is inserted into the hole. Once the toggle assembly is

WEDGE   SLEEVE   STUD   DOUBLE EXPANSION

NAIL ANCHOR   DROP-IN ANCHOR   ANCHOR BOLT

102F24.EPS

Figure 24 ◆ Masonry anchors.

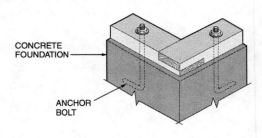

CONCRETE FOUNDATION

ANCHOR BOLT

CONCRETE SCREW

102F25.EPS

Figure 25 ◆ Examples of masonry anchor applications.

WOOD BUILDING MATERIALS, FASTENERS, AND ADHESIVES—TRAINEE MODULE 27102   2.35

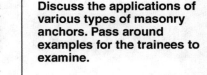

through the hole, the wings spring open. When the screw is tightened, the toggle assembly is drawn up against the back of the wall. If the screw is removed completely, the toggle assembly will fall off.

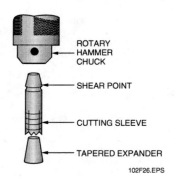

Figure 26 ◆ Self-drilling snap-off anchor.

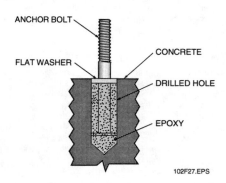

Figure 27 ◆ Fastener anchored in epoxy.

The plastic toggle works on the same principle as the toggle bolt. The toggle is compressed for insertion into the hole, then expands as the screw is tightened.

The screw anchor or *molly* screw also expands as the screw is tightened. The advantage of this fastener is that the screw can be removed once the fastener is installed because there is a collar on the outside of the fastener. The screw anchor selected must match the thickness of the wall or the anchor will not be effective. The type of anchor bolt shown requires a pre-drilled hole. Other types are pointed so they can be driven into the wall.

The conical screw or *auger* is driven into the wall. A screw of a smaller diameter is then driven into the head of the auger to secure the fixture.

Fiber and plastic inserts are placed into pre-drilled holes. The legs of the fastener expand outward as a screw is driven into the opening.

Molly screws, toggles, and augers are typically used in various drywall applications.

13.0.0 ◆ ADHESIVES

The term *adhesives* describes a variety of products that are used to attach one surface to another. Construction adhesives are used to attach paneling and drywall to framing, install subflooring, and attach ceiling tiles. Special types of adhesives are used to install ceramic tiles, floor coverings, and carpeting. There are also many types of glues used in a household or workshop environment. Different glues have different characteristics. Some are quick-drying; others are slow. Some are waterproof; some are not. Some must be dried under pressure, etc. Therefore, it is important for the carpenter to know what adhesive to use in a particular situation. At some point in your career, knowing what adhesive to use in a given situation will become second nature. In the meantime, ask a professional such as a building supply specialist. Always read the instructions on the label before

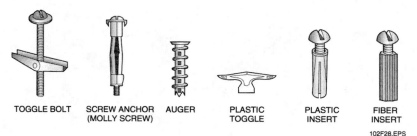

TOGGLE BOLT   SCREW ANCHOR   AUGER   PLASTIC   PLASTIC   FIBER
                (MOLLY SCREW)   TOGGLE   INSERT   INSERT

102F28.EPS

Figure 28 ◆ Hollow-wall anchors.

2.36

Instructor's Notes:
_____
_____
_____
_____
_____
_____
_____
_____

Adhesive Safety

Here are some general safety precautions to follow when using adhesives:

- Be sure to provide good ventilation.
- Keep all flames and lighted tobacco products out of the area.
- Make sure switches, electric tools, or other sparking devices are not used in the area.
- Close off any area where people that are smoking could wander through.
- Do not breathe vapors for any length of time.
- Immediately remove any adhesive that comes in contact with your skin.
- Wear protective goggles and clothing when required.
- Follow the manufacturer's MSDS instructions and check the label on the container for specific safety precautions.
- When in doubt, ask questions.

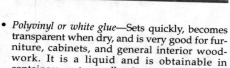

Emphasize the safety procedures required when working with adhesives.

using any adhesive. Also keep in mind that new products are being introduced all the time, so it is important to keep up with what is going on in the industry.

Finally, keep in mind that most adhesives are made from chemicals, some of which are toxic and/or flammable. Read and follow the safety instructions on the label and check the MSDS. The MSDS for any toxic or flammable substance must be available at the job site. In general, it is a good idea to work with adhesives in a well-ventilated area away from open flames and to wear a face mask and eye protection.

13.1.0 Glues

Glues are used in laminated construction such as arches, curved members, and building beams, and in the construction of cabinets and millwork. Carpenters may carry a plastic bottle of liquid glue in their toolboxes for making some interior trim joints or cabinet work on the job.

Glues come in two forms: liquid or dry. The dry forms are mixed into a liquid on the job. Some glues come in two or more liquid forms (for example, epoxy glue). When mixed together in directed proportions, they become very strong adhesives. In all cases where mixing is involved, the manufacturer's MSDS instructions must be followed, both for the quality desired and safety.

Some of the most common types of glue the carpenter might use are:

- *Animal or hide glue*—Made from animal hides, hooves, and bones. It has a long setting time, which makes it suitable when extra assembly time is needed. It can be used on furniture and other wood products, but is not waterproof.

- *Polyvinyl or white glue*—Sets quickly, becomes transparent when dry, and is very good for furniture, cabinets, and general interior woodwork. It is a liquid and is obtainable in containers up to a gallon in size, or by special order in larger quantities.

- *Casein glue*—Comes in a powder form and is mixed with water. It can be used in temperatures down to 35°F. It is water resistant. It can be used on oily woods. Under normal circumstances, it will last for years. However, the powder tends to deteriorate with age.

- *Urea formaldehyde or plastic resin glue*—Comes in powder form and is mixed with water. It is slow setting and has good resistance to heat and water. It is not recommended for oily woods. It is often used for laminated timbers and general woodworking where moisture is present.

- *Resorcinol resin or waterproof glue*—An expensive glue that comes in two liquid chemical parts. One part acts as a catalyst on the other part when mixed. It is excellent for use in exterior woodwork or laminated timbers that are exposed to extremely cold temperatures. Eye goggles and a respirator are needed when mixing.

- *Contact cement*—A type of rubber cement. It is a liquid and comes ready-mixed in containers from a pint to a 55-gallon drum. It is not a high-strength glue. It is used extensively in applying plastic laminates to cabinets and countertops. It is applied to both surfaces and allowed to dry to the touch before contact is made between the two materials. Upon contact, the bonding is instantaneous and movement or sliding into a different position cannot be done.

Discuss the applications of various types of glues.

Warn the trainees that many glues and cements are flammable, explosive, or otherwise hazardous and must be handled accordingly.

Discuss various types of construction adhesives and their applications.

Show Transparency 24 (Figure 29).

 WARNING!
Observe all MSDS precautions for this material. Extreme caution must be taken when using contact cement, as it is highly flammable. It should be used in a well-ventilated area, as the fumes can overcome a worker.

13.2.0 Construction Adhesives

Construction adhesives are available for many types of heavy-duty construction uses. They are made for specific applications such as installing paneling, structural decking, and drywall. In addition to providing strong bonding between materials, construction adhesives offer such advantages as sound deadening and the bonding of dissimilar materials such as wood, gypsum, glass, metal, and concrete.

Construction adhesives are usually applied from a cartridge with a manual ratchet-type gun or pneumatic gun (*Figure 29*). Cartridges range in capacity from about 10 ounces to 30 ounces.

There are several types of adhesives that can be used to fasten sheet materials to framing members. They include the following:

- *Construction adhesive*—Used to apply wood or gypsum to framing. It has a solvent base. The tool needed for the application of construction adhesive is a caulking gun or cartridge.

- *Neoprene adhesive*—Can be used to apply wood to wood and gypsum to wood or metal. Neoprene adhesive will also bond to concrete. Like construction adhesive, neoprene adhesive is also applied with a caulking gun. Neoprene adhesive has a solvent base.
- *Contact cement*—Application is by brush or roller. Contact cement will bond to metal. It can also bond wallboard to framing and wallboard to plywood. It is also used to bond plastic laminates. Contact cement should be nearly dry before joining the two materials together.
- *Drywall adhesive*—Although suitable for lamination, it is used primarily for applying wallboard to framing. A caulking gun is used for application. Like contact cement, it is available as a water-based product; however, solvent-based adhesive is more widely used.
- *Instant-bond glue*—Known by trade terms such as Krazy Glue® and Super Glue®. They work well on non-porous materials such as glass, ceramics, metal, and many plastics. Some types of instant-bond glue are designed for porous materials such as wood and paper. When using this glue, do not touch your fingers to each other or to anything else. Acetone or nail polish remover will dissolve the glue, but be sure to keep it away from your eyes.

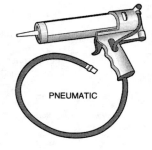

PNEUMATIC

HAND-OPERATED

HAND-OPERATED GUN FOR TWO-PART PRODUCTS

102F29.EPS

Figure 29 ◆ Adhesive applicators.

Construction Adhesive

When using construction adhesive, the material that is being joined should be applied while the adhesive is still wet. If you get this adhesive on your skin, you can use suntan lotion or paint thinner to help remove it.

2.38

CARPENTRY LEVEL ONE—TRAINEE MODULE 27102

Instructor's Notes:

_____
_____
_____
_____
_____
_____
_____
_____

- *Epoxies*—Made by mixing a resin and a hardening agent at the time of use. Epoxies are good for bonding dissimilar materials. Fast-acting epoxies set almost instantly. They usually come in a special syringe-type applicator that keeps the two components separate. They blend when the plungers are depressed. Slower-setting epoxies are mixed in a container at the job site.

WARNING!

Many adhesives and cements are flammable and/or explosive. Make sure they are stored in a well-ventilated area and always read all warning labels on the adhesive packaging. Take the proper precautions to avoid an explosion or fire.

13.3.0 Mastics

Mastics are generally used to apply floor coverings, roofing materials, ceramic tiles, or wall paneling. Most mastics have a synthetic rubber or latex base. Some are thick pastes and can be used where moisture is not present. Others are waterproof and can be used in bathrooms, kitchens, and laundries, or in direct contact with concrete. Some require hot application with special equipment to keep them fluid while being applied. Others are applied with brushes, trowels, or rollers. Some come in tubes and are applied with hand- or air-operated caulking guns.

Keep in mind that good coverage with no voids is needed for satisfactory bonding. Also, if applying floor tiles or paneling, caution must be taken to prevent mastic from squeezing up between the joints or cracks and thus staining or marring the finish surface. To avoid this, place each piece without sliding after contact is made. When applying the mastic, make sure it is evenly spread. For satisfactory bonding, smooth, clean, dry surfaces are

a must. Flaking plaster or old paint must be pretreated with special chemicals or removed. Leveling cements or compounds may be obtained and applied to overcome unevenness in walls or concrete floors. If moisture is likely to be encountered (e.g., concrete on basement walls or floors), a waterproofing material must be applied before the mastic. Only waterproof mastic should be used in these circumstances. Under adverse conditions, check the manufacturer's recommendations for secure installation. Many manufacturers recommend the type of adhesive to be used with their product.

13.4.0 Shelf Life

All adhesives have a limited life span that ranges from 12 to 24 months. If they are not used within that time, they will begin to lose their bonding ability. Most manufacturers claim a shelf life of at least two years. Keep in mind, however, that three to six months or more may have elapsed from the time it was manufactured until it was purchased. When you buy adhesive, write the date on the container and do not use it beyond the shelf life listed on the label.

Summary

It is important to know how to select the right material for any given situation. There are many types and grades of lumber and sheet materials. Each has specific uses. More importantly, each has certain limitations that the user must know in order to avoid harmful or costly errors.

Likewise, there is a correct size and type of fastener for every fastening situation and an adhesive that is just right for occasions when an adhesive is needed. This module provided an introduction to the many types of fasteners and adhesives a carpenter uses on the job. As your training progresses, you will learn more about how and when to use them.

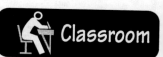

Discuss the applications of various types of mastics.

Have the trainees identify the correct type of adhesive for various applications.

Have each trainee complete to your satisfaction Performance Profile Task 6 on adhesive identification and use. Fill out Performance Profile Sheets for each trainee.

Have the trainees complete the Review Questions and go over the answers prior to administering the Module Examination. Answer any questions the trainees may have.

Review Questions

1. Edge-grained lumber is another name for _____ lumber.
   a. quarter-sawed
   b. flat-grained
   c. plain-sawed
   d. sapwood

2. Which of these is a softwood?
   a. Cherry
   b. Oak
   c. Cedar
   d. Maple

3. Which of the following is a defect that occurs naturally in wood?
   a. Compression wood
   b. Shake
   c. Cup
   d. Check

4. The term *dote,* when used in reference to lumber, means _____ .
   a. a type of knot
   b. bark at the edge of the cut lumber
   c. a type of decay
   d. a pitch streak

5. In the grading stamp shown below, the term *S-GRN* means that the lumber _____ .
   a. is made from short-grain wood
   b. contains more than 19% moisture
   c. should be painted with green primer before use
   d. is southern grown

12 **STAND**

ABC **S-GRN** ⏥ D FIR

102UA0202.EPS

6. The *nominal* size of a piece of lumber is the _____ .
   a. actual size
   b. size before it has fully dried
   c. dressed size
   d. size by which it is known and sold

7. If lumber is graded as double-end-trimmed, its ends are trimmed _____ .
   a. square within a tolerance of ⅛" for every 2"
   b. twice at each end
   c. reasonably square on both ends
   d. square and smooth on both ends within a tolerance of ⅛" for the entire length

8. The product known as *soft plywood* has a _____ core.
   a. lumber
   b. fiber
   c. veneer
   d. particleboard

9. *Standard* hardboard is suitable only for _____ .
   a. cabinet making
   b. siding
   c. exterior sheathing
   d. soundproofing

10. When you want to obtain safety information about an adhesive, the best source of reliable information is _____ .
    a. your supervisor
    b. a co-worker
    c. the applicable MSDS
    d. a building supply dealer

11. Which of these engineered wood products is commonly used to make doors and other mill-work?
    a. Glulam
    b. LVL
    c. PSL
    d. LSL

12. A glulam beam consists of _____ .
    a. laminated veneers bonded together under pressure
    b. solid, kiln-dried lumber glued together
    c. veneers bonded to a core with glue
    d. long strands of veneer bonded together with adhesive

13. Which of these statements about pressure-treated lumber is *not* true?
    a. It is often used to make landscape beams.
    b. Scrap material should be burned.
    c. Eye protection and a dust mask must be worn when cutting it.
    d. It is less expensive than redwood.

14. How many board feet are there in (30) 2" × 10" × 15' joists?
    a. 750
    b. 480
    c. 1,250
    d. 640

Instructor's Notes:

_____

_____

_____

_____

_____

_____

_____

15. How many sheets of plywood would you need in order to provide subflooring for a 25' by 30' room?
    a. 28
    b. 25
    c. 30
    d. 24

16. An 8-penny nail is approximately _____ long.
    a. 1"
    b. 2½"
    c. 4"
    d. 6"

17. The type of staple point that clinches outward after it penetrates the material is the _____ .
    a. crosscut chisel
    b. outside chisel
    c. spear
    d. inside chisel

18. The type of anchor that is threaded into a lead shield using a wrench is known as a _____ .
    a. molly screw
    b. toggle bolt
    c. snap-off anchor
    d. lag screw

19. The type of screw that does not end in a sharp point is a _____ screw.
    a. wood
    b. machine
    c. lag
    d. drywall

20. The adhesive that requires two separate components that are mixed together at the time of use is _____ .
    a. epoxy
    b. polyvinyl glue
    c. contact cement
    d. mastic

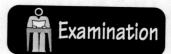

Examination

Administer the Module Examination. Record the results on Craft Training Report Form 200 and submit the results to the Training Program Sponsor.

Performance Profile Test

Ensure that all Performance Profile Tests have been completed and Performance Profile Sheets for each trainee are filled out. Be sure to record the results of the Testing on Craft Training Report Form 200 and submit the results to the Training Program Sponsor.

Trade Terms Introduced in This Module

Butt joint: The joint formed when one square-cut edge of a piece of material is placed against another material.

Cantilever: A beam, truss, or slab (floor) that extends past the last point of support.

Catalyst: A substance that causes a reaction in another substance.

Galvanized: Protected from rusting by a zinc coating.

Gypsum: A chalky material that is a main ingredient in plaster and drywall.

Header: A piece used to span an opening (door, window, etc.). It is typically designed to carry a load.

Jambs: The sides and tops of window and door frames.

Japanned: Coated with a glossy finish.

Joist: Generally, equally-spaced framing members that support floors and ceilings.

Material safety data sheets (MSDS): Information that details any toxic, chemical, or potentially harmful substances that are contained in a product.

Millwork: Manufactured wood products such as doors, windows, and moldings.

Nail set: A punch-like metal tool used to recess finishing nails.

Pith: The soft core at the center of a tree.

Rafter: A sloping structural member of a roof frame to which sheathing is secured.

Resins: Protective natural or synthetic coatings.

Sheathing: Sheet material or boards to which roofing material or siding is secured.

Shiplap: A method of cutting siding in which each board is tapered and grooved so that the upper piece fits tightly over the lower piece.

Sill plate: A horizontal timber that supports the framework of a building on the bottom of a wall or box joist. It is also called a *sole plate.*

Stringer: A structural timber or board used as a support for other members such as stair treads.

Tongue-and-groove: A joint made by fitting a tongue on the edge of a board into a matching groove on the adjoining board.

Vaulted ceiling: A high, open ceiling that generally follows the roof pitch.

Instructor's Notes:

_____

_____

_____

_____

_____

_____

_____

APPENDIX A

Lumber Conversion Tables

Nominal and Actual Softwood Lumber Sizes

Nominal (Rough)	Actual (S4S)
1" × 4"	¾" × 3½"
1" × 6"	¾" × 5½"
1" × 8"	¾" × 7¼"
1" × 10"	¾" × 9¼"
1" × 12"	¾" × 11¼"
2" × 4"	1½" × 3½"
2" × 6"	1½" × 5½"
2" × 8"	1½" × 7¼"
2" × 10"	1½" × 9¼"
2" × 12"	1½" × 11¼"
4" × 4"	3½" × 3½"
4" × 6"	3½" × 5½"

Nominal and Actual Hardwood Lumber Sizes

Nominal (Rough)	Actual (Plain)	Actual (S4S)
½"	5/16"	5/16"
⅝"	7/16"	7/16"
¾"	9/16"	9/16"
1"	25/32"	25/32"
1¼"	1 1/16"	1 1/32"
1½"	1 5/16"	1 9/32"
2"	1¾"	1 11/16"
2½"	2¼"	2¼"
3"	2¾"	2¾"
4"	3¾"	3¾"

Rapid Calculation of Board Measure

Width	Thickness	Board Feet
3"	1" or less	¼ of the length
4"	1" or less	⅓ of the length
6"	1" or less	½ of the length
9"	1" or less	¾ of the length
12"	1" or less	Same as the length
15"	1" or less	1¼ of the length

Board Feet

Nominal Size (Inches)	Actual Length (in Feet)								
	8	10	12	14	16	18	20	22	24
1 × 2		1⅔	2	2⅓	2⅔	3	3⅓	3⅔	4
1 × 3		2½	3	3½	4	4½	5	5½	6
1 × 4	2¾	3⅓	4	4⅔	5⅓	6	6⅔	7⅓	8
1 × 5		4⅙	5	5⅚	6⅔	7½	8⅓	9⅙	10
1 × 6	4	5	6	7	8	9	10	11	12
1 × 7		5⅚	7	8⅙	9⅓	10½	11⅔	12⅚	14
1 × 8	5⅓	6⅔	8	9⅓	10⅔	12	13⅓	14⅔	16
1 × 10	6⅔	8⅓	10	11⅔	13⅓	15	16⅔	18⅓	20
1 × 12	8	10	12	14	16	18	20	22	24
1¼ × 4		4⅙	5	5⅚	6⅔	7½	8⅓	9⅙	10
1¼ × 6		6¼	7½	8¾	10	11¼	12½	13¾	15
1¼ × 8		8⅓	10	11⅔	13⅓	15	16⅔	18⅓	20
1¼ × 10		10⁵⁄₁₂	12½	14⁷⁄₁₂	16⅔	18¾	20⅚	22¹¹⁄₁₂	25
1¼ × 12		12½	15	17½	20	22½	25	27½	30
1½ × 4	4	5	6	7	8	9	10	11	12
1½ × 6	6	7½	9	10½	12	13½	15	16½	18
1½ × 8	8	10	12	14	16	18	20	22	24
1½ × 10	10	12½	15	17½	20	22½	25	27½	30
1½ × 12	12	15	18	21	24	27	30	33	36
2 × 4	5⅓	6⅔	8	9⅓	10⅔	12	13⅓	14⅔	16
2 × 6	8	10	12	14	16	18	20	22	24
2 × 8	10⅔	13⅓	16	18⅔	21⅓	24	26⅔	29⅓	32
2 × 10	13⅓	16⅔	20	23⅓	26⅔	30	33⅓	36⅔	40
2 × 12	16	20	24	28	32	36	40	44	48
3 × 6	12	15	18	21	24	27	30	33	36
3 × 8	16	20	24	28	32	36	40	44	48
3 × 10	20	25	30	35	40	45	50	55	60
3 × 12	24	30	36	42	48	54	60	66	72
4 × 4	10⅔	13⅓	16	18⅔	21⅓	24	26⅔	29⅓	32
4 × 6	16	20	24	28	32	36	40	44	48
4 × 8	21⅓	26⅔	32	37⅓	42⅔	48	53⅓	58⅔	64
4 × 10	26⅔	33⅓	40	46⅔	53⅓	60	66⅔	73⅓	80
4 × 12	32	40	48	56	64	72	80	88	96

Instructor's Notes:

_____

_____

_____

_____

_____

_____

_____

_____

Common Abbreviations Used
in the Lumber Industry

AD	Air Dried
ADF	After Deducting Freight
ALS	American Lumber Standard
AST	Anti-Stain Treated
AVG	Average
B & B	B and Better
B & S	Beams and Stringers
Bd.	Board
BF or Bd. Ft.	Board Feet
Bdl.	Bundle or Bundled
B/L	Bill of Lading
Bev.	Bevel or Beveled
CB1S	Center Bead on One Side
CB2S	Center Bead on Two Sides
CC	Cubical Content
Cft. or Cu. Ft.	Cubic Feet or Foot
CG2E	Center Groove on Two Edges
CIF	Cost, Insurance, and Freight
CIFE	Cost, Insurance, Freight, and Exchange
Clg.	Ceiling
Com.	Common
Csg.	Casing
Ctr.	Center
CV1S	Center V on One Side
CV2S	Center V on Two Sides
D & M	Dressed and Matched
DB Clg.	Double-Beaded Ceiling
DB Part	Double-Beaded Partition
DET	Double-End-Trimmed
Dim.	Dimension
Dkg.	Decking
D/sdg	Drop Siding
EB1S	Edge Bead on One Side
EB2S	Edge Bead on Two Sides
EE	Eased Edges
EG	Edge Vertical or Rift Grain
EM	End Matched
EV1S	Edge V on One Side
EV2S	Edge V on Two Sides

E & CB1S	Edge and Center Bead on One Side
E & CB2S	Edge and Center Bead on Two Sides
E & CV1S	Edge and Center V on One Side
E & CV2S	Edge and Center V on Two Sides
FA	Facial Area
Fac.	Factory
FBM	Feet Board Measure
FG	Flat (slash) Grain
Flg.	Flooring
FOB	Free on Board
FOHC	Free of Heart Centers
FOK	Free of Knots
Frt.	Freight
Ft.	Foot or Feet
G	Girth
GM	Grade Marked
G/R	Grooved Roofing
HB	Hollow Back
Hrt.	Heart
H & M	Hit and Miss
in.	Inch or Inches
Ind.	Industrial
J & P	Joints and Planks
KD	Kiln Dried
Lbr.	Lumber
LCL	Less than Carload
Lft.	Lineal Foot or Feet
Lin.	Lineal
LL	Longleaf
Lng.	Lining
M	Thousand
MBM	Thousand (feet) Board Measure
M.C.	Moisture Content
Merch.	Merchantable
MG	Medium Grain
MLDG.	Molding
Mft.	Thousand Feet

No.	Number	SSND	Sap Stained No Defects (stained)
N1E or N2E	Nosed on One or Two Edges	STR	Structural
Ord.	Order	SYP	Southern Yellow Pine
Par.	Paragraph	S & E	Side and Edge (surfaced on)
Part.	Partition	S1E	Surfaced on One Edge
Pat.	Pattern	S2E	Surfaced on Two Edges
Pcs.	Pieces	S1S	Surfaced on One Side
PE	Plain End	S2S	Surfaced on Two Sides
P & T	Post and Timbers	S1S1E	Surfaced on One Side and One Edge
P1S and P2S	See SIS and S25		
Reg.	Regular	S2S2E	Surfaced on Two Sides and Two Edges
Rfg.	Roofing		
Rgh.	Rough	S2S1E	Surfaced on Two Sides and One Edge
R/L	Random Length		
RES	Resawn	S2S & CM	Surfaced on Two Sides and Center Matched
Sdg.	Siding		
S E	Square Edge	S2S - SL	Surfaced on Two Sides and Shiplapped
Sel.	Select		
S E & S	Square Edge and Sound	S2S & SM	Surfaced on Two Sides and Standard Matched
SL	Shiplap		
SM	Surface Measure	S4S	Surfaced on Four Sides
Specs.	Specifications	T & G	Tongued and Grooved
SR	Stress Rated	Wdr.	Wider
Std.	Standard	Wt.	Weight
Std. Lgths.	Standard Lengths	YP	Yellow Pine
STD M	Standard Matched		

Instructor's Notes:

_____

_____

_____

_____

_____

_____

_____

_____

Answers to Review Questions

Answer	Section
1. a	2.1.0
2. c	2.2.0, Table 1
3. a	3.1.0
4. c	3.1.0
5. b	4.0.0, Figure 6
6. d	4.1.0
7. c	4.1.1
8. c	5.2.2
9. a	6.1.0
10. c	13.0.0
11. d	7.3.0
12. b	7.5.0
13. b	8.0.0
14. a	9.0.0
15. d	9.0.0
16. b	10.0.0, Figure 14
17. d	10.2.1
18. d	11.4.0
19. b	11.5.0, Figure 22
20. a	13.2.0

Additional Resources

This module is intended to present thorough resources for task training. The following reference works are suggested for further study. These are optional materials for continued education, rather than for task training.

Building Products Catalog, Latest Edition. Atlanta, GA: Georgia-Pacific.

Carpentry, Latest Edition. Homewood, IL: American Technical Publishers.

Carpentry, Latest Edition. Albany, NY: Delmar Publishers.

Modern Carpentry, Latest Edition. Tinley Park, IL: Goodheart-Willcox Company, Inc.

Instructor's Notes:

_____

_____

_____

_____

_____

_____

_____

_____

The NCCER makes every effort to keep these textbooks up-to-date and free of technical errors. We appreciate your help in this process. If you have an idea for improving this textbook, or if you find an error, a typographical mistake, or an inaccuracy in the NCCER's Craft Training textbooks, please write us, using this form or a photocopy. Be sure to include the exact module number, page number, a detailed description, and the correction, if applicable. Your input will be brought to the attention of the Technical Review Committee. Thank you for your assistance.

Instructors – If you found that additional materials were necessary in order to teach this module effectively, please let us know so that we may include them in the Equipment/Materials list in the Instructor's Guide.

Write: Curriculum Revision and Development Department
National Center for Construction Education and Research
P.O. Box 141104, Gainesville, FL 32614-1104

Fax: 352-334-0932

E-mail: curriculum@nccer.org

Craft _____ Module Name _____

Copyright Date _____ Module Number _____ Page Number(s) _____

Description _____

_____

_____

_____

(Optional) Correction _____

_____

_____

(Optional) Your Name and Address _____

_____

_____

_____

Hand and Power Tools

27103-01

MODULE OVERVIEW

This module expands upon the hand and power tool information provided in the Core Curriculum and introduces the carpentry trainee to additional tools used in the carpentry trade.

PREREQUISITES

Please refer to the Course Map in the Trainee Module. Prior to training with this module, it is suggested that the trainee shall have successfully completed the following modules:

Core Curriculum; Carpentry Level One, Modules 27101 and 27102

LEARNING OBJECTIVES

Upon completion of this module, the trainee will be able to:

1. Identify the hand tools commonly used by carpenters and describe their uses.
2. Use hand tools in a safe and appropriate manner.
3. State the general safety rules for operating all power tools, regardless of type.
4. State the general rules for properly maintaining all power tools, regardless of type.
5. Identify the portable power tools commonly used by carpenters and describe their uses.
6. Use portable power tools in a safe and appropriate manner.
7. Identify the stationary power tools commonly used by carpenters and describe their uses.
8. Use stationary power tools in a safe and appropriate manner.

PERFORMANCE OBJECTIVES

Under the supervision of the instructor, the trainee should be able to:

1. Identify the hand and power tools used by carpenters and state their uses.
2. From the following list of hand tools, select at least five and demonstrate or describe their safe and proper use:
   - Screwdrivers
   - Pliers
   - Chisels
   - Levels
   - Squares
   - Planes
   - Clamps
   - Saws
3. Demonstrate or describe the safe and proper use of the following portable power tools:
   - Circular saw
   - Jig saw
   - Power plane
   - Pneumatic fastener
   - Powder-actuated tool
4. From the following list of stationary power tools, select at least five and demonstrate or describe their safe and proper use:
   - Table saw
   - Radial arm saw
   - Table band saw
   - Power miter/compound miter saw
   - Frame and trim saw
   - Combination belt-disc sander
   - Drill press
   - Jointer
   - Planer
   - Shaper

NCCER STANDARDIZED CRAFT TRAINING PROGRAM

The National Center for Construction Education and Research (NCCER) provides a standardized national program of accredited craft training. Key features of the program include instructor certification, competency-based training, and performance testing. The program provides trainees, instructors, and companies with a standard form of recognition through a National Craft Training Registry. The program is described in full in the *Guidelines for Accreditation,* published by the NCCER. For more information on standardized craft training, contact the NCCER at P.O. Box 141104, Gainesville, FL 32614-1104, 352-334-0911, visit our Web site at www.NCCER.org, or e-mail info@NCCER.org.

HOW TO USE THIS ANNOTATED INSTRUCTOR'S GUIDE

Each page presents two sections of information. The larger section displays each page exactly as it appears in the Trainee Module. The narrow column ties suggested trainee and instructor actions to each page and provides icons to call your attention to material, safety, audiovisual, or testing requirements. The bottom of each page includes space for your notes.

   If you see the Teaching Tip icon, that means there is a teaching tip associated with this section. Also refer to any suggested teaching tips at the end of the module.

SAFETY CONSIDERATIONS

Ensure that the trainees are equipped with appropriate personal protective equipment.

PREPARATION

Before teaching this module, you should review the Module Outline, the Learning and Performance Objectives, and the Materials and Equipment List. Be sure to allow ample time to prepare your own training or lesson plan and gather all required equipment and materials.

MATERIALS AND EQUIPMENT LIST

Materials:

Transparencies
Markers/chalk
Soapstone
Yard-long lengths of 1" reinforcing rod
1 × 4 stock about 18" to 24" long
2 × 4s 18" to 24" long
2 × 4s 4' long
6" × 12" pieces of ¾" plywood
Pieces of crown molding 4' long
Angle iron, steel rod, or pipe for cutting
Steel stock for grinding and shaping
Assorted sanding belts and discs for rough and fine sanding
Stock to be sanded
Wood stock for routing
Laminate samples
Stock to be drilled and/or beveled
Stock to be bored

Blocks of scrap wood
Stock to be planed
Stock to be shaped
Fasteners (nails and staples) designed for the pneumatic fastener being used
Sheet metal stock
Lubricant for powder-actuated tool being used
Supply of fasteners for use with powder-actuated tool being used
Supply of boosters for use with powder-actuated tool being used
Stock to be fastened using the powder-actuated tool
Copies of Worksheet 1*
Copies of Job Sheets 1 through 13*
Module Examinations*
Performance Profile Sheets*

*Packaged with this Annotated Instructor's Guide.

Equipment:

Overhead projector and screen
Whiteboard/chalkboard
Appropriate personal protective equipment
Folding rule or steel tape
Assorted screwdrivers
Tongue-and-groove pliers (channel-lock pliers)
Clamping pliers (vise grips)
Flooring and brick chisels
Levels:
    Line
    Water
    Builder's
    Transit
    Laser
Squares:
    Try
    Sliding T-bevel
    Speed square
    Miter
    Framing
    Adjustable T-square
Planes:
    Block
    Jack
    Smoothing
    Jointer
Clamps:
    Web
    Hand-screw
    Bar
    Spring
    Locking C
    Pipe
Saws:
    Hacksaw and replacement blades
    Backsaw
    Dovetail
    Compass
    Coping
Chalkline
Wood chisel

Oil stone and oil
Clamping device
Portable circular saw
Circular saw protractor
Table saw
Radial arm saw
Miter/compound miter saw
Frame and trim saw
Reciprocating saw
Portable jig saw
Table band saw
Portable band saw
Miter gauge
Ripping fence for table saw, table band saw,
    portable circular saw, and portable jig saw
Push stick
Sawhorses or other solid support
Belt-disc sander
Portable belt and disc sanders
Drill press and chuck key
Drills (various sizes) and forester or multispur bit
Drill vise or clamps
Jointer with fence
Thickness planer and blades
Shaper with fence, cutters, necessary wrenches,
    and miter head
Push block and hold-down devices
Portable power plane and blades
Portable drills and drill bits
Hammer drill and drill bits
Screwguns and screws
Power metal shears
Bench grinder
Router and router bits
Laminate trimmer and bits
Pneumatic fastener and manufacturer's
    instruction manual
Electric air compressor with air hose
Powder-actuated tool and manufacturer's
    instruction manual

ADDITIONAL RESOURCES

This module is intended to present thorough resources for task training. The following reference works are suggested for both instructors and motivated trainees interested in further study. These are optional materials for continued education rather than for task training.

Carpentry. Leonard Koel. Homewood, IL: American Technical Publishers, 1997.

Carpentry. Gasper J. Lewis. Albany, NY: Delmar Publishers, 2000.

Modern Carpentry. Willis H. Wagner and Howard Bud Smith. Tinley Park, IL: The Goodheart-Willcox Company, Inc., 2000.

Popular Science Complete Book of Power Tools. R. J. DeCristoforo. New York, NY: Black Dog and Leventhal, 1998.

TEACHING TIME FOR THIS MODULE

An outline for use in developing your lesson plan is presented below. Note that each Roman numeral in the outline equates to one session of instruction. Each session has a suggested time period of 2½ hours. This includes 10 minutes at the beginning of each session for administrative tasks and one 10-minute break during the session. Approximately 20 hours are suggested to cover *Hand and Power Tools*. You will need to adjust the time required for hands-on activity and testing based on your class size and resources.

Topic	Planned Time

Session I. Introduction; Hand Tools

A. Introduction _____

B. Hand Tools _____

   1. Screwdrivers _____

   2. Pliers _____

   3. Chisels _____

   4. Levels _____

   5. Squares _____

   6. Laboratory _____

   Hand out Job Sheet 27103-1. Under your supervision, have the trainees perform the tasks on the Job Sheet. Note the proficiency of each trainee.

   7. Planes _____

   8. Laboratory _____

   Hand out Job Sheet 27103-2. Under your supervision, have the trainees perform the tasks on the Job Sheet. Note the proficiency of each trainee.

   9. Clamps _____

  10. Saws _____

  11. Laboratory _____

   Hand out Job Sheet 27103-3. Under your supervision, have the trainees perform the tasks on the Job Sheet. Note the proficiency of each trainee.

Session II. Guidelines for Using All Power Tools; Power Saws

A. Guidelines for Using All Power Tools _____

   1. Safety Rules Pertaining to All Power Tools _____

   2. Guidelines Pertaining to the Care of All Power Tools _____

B. Power Saws _____

   1. Circular Saws _____

   2. Laboratory _____

   Hand out Worksheet 27103-1 and Job Sheets 27103-4 and -5. Under your supervision, have the trainees complete the circular saw safety test prior to performing the related tasks on the Job Sheets. Note the proficiency of each trainee.

   3. Table Saws _____

   4. Laboratory _____

   Hand out Worksheet 27103-1 and Job Sheets 27103-4 and -5. Under your supervision, have the trainees complete the table saw safety test before performing the related tasks on the Job Sheets. Note the proficiency of each trainee.

Session III. Power Saws (Continued)

A. Power Saws _____

   1. Radial Arm Saws _____

2. Laboratory

Hand out Worksheet 27103-1 and Job Sheets 27103-4 and -5. Under your supervision, have the trainees complete the radial arm saw safety test before performing the related tasks on the Job Sheets. Note the proficiency of each trainee.

3. Power Miter Saws/Compound Miter Saws
4. Laboratory

Hand out Worksheet 27103-1 and Job Sheet 27103-6. Under your supervision, have the trainees complete the compound miter saw safety test before performing the related tasks on the Job Sheet. Note the proficiency of each trainee.

5. Frame and Trim Saws
6. Laboratory

Hand out Worksheet 27103-1 and Job Sheet 27103-6. Under your supervision, have the trainees complete the frame and trim saw safety test before performing the related tasks on the Job Sheet. Note the proficiency of each trainee.

7. Circular Saw Blades

Session IV. Power Saws; Jointers

A. Power Saws

1. Reciprocating Saws
2. Laboratory

Hand out Worksheet 27103-1. Under your supervision, have the trainees complete the reciprocating saw safety test before using the reciprocating saw. Note the proficiency of each trainee.

3. Jig Saws
4. Laboratory

Hand out Worksheet 27103-1 and Job Sheets 27103-4 and -5. Under your supervision, have the trainees complete the jig saw safety test before performing the related tasks on the Job Sheets. Note the proficiency of each trainee.

5. Band Saws
6. Laboratory

Hand out Worksheet 27103-1 and Job Sheets 27103-4 and -5. Under your supervision, have the trainees complete the portable and table band saw safety test before performing the related tasks on the Job Sheets. Note the proficiency of each trainee.

B. Jointers

1. Laboratory

Hand out Worksheet 27103-1 and Job Sheet 27103-7. Under your supervision, have the trainees complete the jointer safety test before performing the related tasks on the Job Sheet. Note the proficiency of each trainee.

Session V. Thickness Planer; Shapers; Sanders and Grinders

A. Thickness Planer

1. Laboratory

Hand out Worksheet 27103-1 and Job Sheet 27103-8. Under your supervision, have the trainees complete the thickness planer safety test before performing the related tasks on the Job Sheet. Note the proficiency of each trainee.

B. Shapers

    1. Laboratory

    Hand out Worksheet 27103-1 and Job Sheet 27103-9. Under your supervision, have the trainees complete the shaper safety test prior to performing the related tasks on the Job Sheet. Note the proficiency of each trainee.

C. Sanders and Grinders

    1. Belt-Disc Sander

    2. Laboratory

    Hand out Worksheet 27103-1 and Job Sheet 27103-10. Under your supervision, have the trainees complete the belt-disc sander safety test before performing the related tasks on the Job Sheet. Note the proficiency of each trainee.

    3. Portable Sander/Grinders

    4. Laboratory

    Hand out Worksheet 27103-1. Under your supervision, have the trainees complete the palm sander, half-sheet sander, belt sander, and rotary grinder-sander safety tests prior to using these tools. Note the proficiency of each trainee.

Session VI. Drill Press; Bench Grinders; Routers/Laminate Trimmers

A. Drill Press

    1. Laboratory

    Hand out Worksheet 27103-1 and Job Sheet 27103-11. Under your supervision, have the trainees complete the drill press safety test prior to performing the related tasks on the Job Sheet. Note the proficiency of each trainee.

B. Bench Grinders

    1. Laboratory

    Hand out Worksheet 27103-1. Under your supervision, have the trainees complete the bench grinder safety test before operating the bench grinder. Note the proficiency of each trainee.

C. Routers/Laminate Trimmers

    1. Laboratory

    Hand out Worksheet 27103-1. Under your supervision, have the trainees complete the router and laminate trimmer safety test before operating these tools. Note the proficiency of each trainee.

Session VII. Portable Power Planes; Portable Drills and Screwguns; Power Metal Shears; Pneumatic/Cordless Nailers and Staplers

A. Portable Power Planes

    1. Laboratory

    Hand out Worksheet 27103-1. Under your supervision, have the trainees complete the power plane safety test before operating these tools. Note the proficiency of each trainee.

B. Portable Drills and Screwguns

    1. Laboratory

    Hand out Worksheet 27103-1. Under your supervision, have the trainees complete the portable drill, hammer drill, and screwgun safety tests before operating these tools. Note the proficiency of each trainee.

C. Power Metal Shears

   1. Laboratory

   Hand out Worksheet 27103-1. Under your supervision, have the trainees complete the power metal shears safety test before using the shears. Note the proficiency of each trainee.

D. Pneumatic/Cordless Nailers and Staplers

   1. Laboratory

   Hand out Worksheet 27103-1 and Job Sheet 27103-12. Under your supervision, have the trainees complete the pneumatic fasteners safety test before performing the related tasks on the Job Sheet. Note the proficiency of each trainee.

Session VIII. Powder-Actuated Fastening Tools; Module Examination and Performance Testing

A. Powder-Actuated Fastening Tools

   1. Laboratory

   Hand out Worksheet 27103-1 and Job Sheet 27103-13. Under your supervision, have the trainees complete the powder-actuated fasteners safety test before performing the related tasks on the Job Sheet. Note the proficiency of each trainee.

B. Summary

   1. Summarize module.

   2. Answer questions.

C. Module Examination

   1. Trainees must score 70% or higher to receive recognition from the NCCER.

   2. Record the testing results on Craft Training Report Form 200 and submit the results to the Training Program Sponsor.

D. Performance Testing

   1. Trainees must perform each task to the satisfaction of the instructor to receive recognition from the NCCER.

   2. Record the testing results on Craft Training Report Form 200 and submit the results to the Training Program Sponsor.

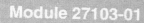

Module 27103-01

Hand and Power Tools

_____
_____
_____
_____
_____
_____
_____
_____

Course Map

This course map shows all of the modules in the first level of the Carpentry curriculum. The suggested training order begins at the bottom and proceeds up. Skill levels increase as a trainee advances on the course map. The training order may be adjusted by the local Training Program Sponsor.

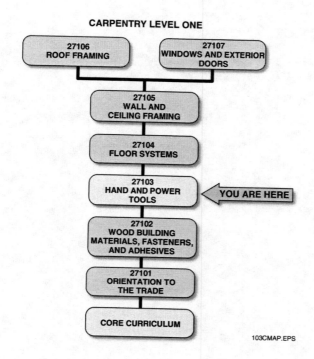

CARPENTRY LEVEL ONE

Homework

Assign reading of Module 27103.

103CMAP.EPS

1.0.0   INTRODUCTION 3.1
2.0.0   HAND TOOLS 3.1
    2.1.0   Screwdrivers 3.1
    2.2.0   Pliers 3.1
    2.3.0   Chisels 3.2
    2.4.0   Levels 3.2
    2.5.0   Squares 3.4
    2.6.0   Planes 3.6
    2.7.0   Clamps 3.7
    2.8.0   Saws 3.8
3.0.0   POWER TOOLS 3.9
    3.1.0   Guidelines for Using All Power Tools 3.9
    3.1.1   Safety Rules Pertaining to All Power Tools 3.9
    3.1.2   Guidelines Pertaining to the Care of All Power Tools 3.10
    3.2.0   Power Saws 3.10
    3.2.1   Circular Saws 3.10
    3.2.2   Table Saws 3.11
    3.2.3   Radial Arm Saws 3.13
    3.2.4   Power Miter Saws/Compound Miter Saws 3.14
    3.2.5   Frame and Trim Saws 3.15
    3.2.6   Circular Saw Blades 3.15
    3.2.7   Reciprocating Saws 3.16
    3.2.8   Jigsaws 3.17
    3.2.9   Band Saws 3.18
    3.3.0   Jointers 3.19
    3.4.0   Thickness Planer 3.21
    3.5.0   Shapers 3.21
    3.6.0   Sanders and Grinders 3.22
    3.6.1   Belt-Disc Sanders 3.22
    3.6.2   Portable Sanders/Grinders 3.23
    3.7.0   Drill Press 3.25
    3.8.0   Bench Grinders 3.27
    3.9.0   Routers/Laminate Trimmers 3.27
    3.10.0 Portable Power Planes 3.28
    3.11.0 Portable Drills and Screwguns .3.29
    3.12.0 Power Metal Shears 3.30
    3.13.0 Pneumatic/Cordless Nailers and Staplers 3.31
    3.14.0 Powder-Actuated Fastening Tools 3.33

SUMMARY 3.34
REVIEW QUESTIONS 3.35
PROFILE IN SUCCESS 3.37
GLOSSARY 3.38
ANSWERS TO REVIEW QUESTIONS 3.39
ADDITIONAL RESOURCES 3.40

Figures

Figure 1   Special screwdrivers 3.2
Figure 2   Groove-joint pliers 3.2
Figure 3   Special chisels 3.3
Figure 4   Levels 3.3
Figure 5   Squares 3.5
Figure 6   Planes 3.6
Figure 7   Clamps 3.7
Figure 8   Special saws 3.8
Figure 9   Mortise and tenon joint 3.9
Figure 10   Circular saw 3.10
Figure 11   Table saw 3.12
Figure 12   Radial arm saw 3.13
Figure 13   Compound miter saw 3.14
Figure 14   Frame and trim saw 3.15
Figure 15   Circular saw blades 3.16
Figure 16   Reciprocating saw 3.16
Figure 17   Jigsaw 3.18
Figure 18   Table band saw 3.18
Figure 19   Portable band saw 3.19
Figure 20   Jointer-planer 3.20
Figure 21   Thickness planer 3.21
Figure 22   Shaper 3.22
Figure 23   Belt-disc sander 3.23
Figure 24   Palm and half-sheet sanders 3.24
Figure 25   Belt sander 3.24
Figure 26   Rotary grinder-sander 3.25
Figure 27   Drill press 3.25
Figure 28   Bench grinder 3.27
Figure 29   Power router 3.27
Figure 30   Laminate trimmer 3.28
Figure 31   Power plane 3.29
Figure 32   Portable drills 3.29
Figure 33   Power-driven screwdriver (drywall gun) 3.30

Instructor's Notes:

_____

_____

_____

_____

_____

_____

_____

_____

Figure 34   Hammer drill 3.30
Figure 35   Power shears 3.31
Figure 36   Pneumatic nailers and stapler 3.32
Figure 37   Impulse cordless nail gun 3.32
Figure 38   Powder-actuated fastening tool 3.33
Figure 39   Major parts of a powder-actuated fastening tool 3.33

Table

Table 1   Powder Charge Color-Coding System 3.34

Hand and Power Tools

 **Materials**

Ensure you have everything required to teach the course. Check the Materials and Equipment List at the front of this Instructor's Guide.

Objectives

Upon completion of this module, the trainee will be able to:

1. Identify the hand tools commonly used by carpenters and describe their uses.
2. Use hand tools in a safe and appropriate manner.
3. State the general safety rules for operating all power tools, regardless of type.
4. State the general rules for properly maintaining all power tools, regardless of type.
5. Identify the portable power tools commonly used by carpenters and describe their uses.
6. Use portable power tools in a safe and appropriate manner.
7. Identify the stationary power tools commonly used by carpenters and describe their uses.
8. Use stationary power tools in a safe and appropriate manner.

Prerequisites

Successful completion of the following Task Modules is recommended before beginning study of this Task Module: Core Curriculum; Carpentry Level One, Modules 27101 and 27102.

Required Trainee Materials

1. Trainee Task Module
2. Appropriate personal protective equipment

1.0.0 ◆ INTRODUCTION

In the Core Curriculum, you were introduced to many of the hand and power tools that are commonly used by all trades. This module builds on that information and introduces many new tools, especially the power tools widely used by carpenters on a construction site. The focus of this module is on familiarizing you with each of these tools. Safety rules for the use of the power tools are also covered. While under the supervision of your instructor and/or supervisor, you will learn how to properly operate and use each of the tools described in this module.

2.0.0 ◆ HAND TOOLS

In this module, you will be introduced to some additional hand tools and their uses. Keep in mind that this material is introductory; you will receive specific training in the use of these tools as your training progresses.

2.1.0 Screwdrivers

The carpenter will generally use either the common screwdriver or the Phillips-head (cross-recessed head) screwdriver. Other types of screwdrivers, such as those shown in *Figure 1*, are designed for special types of screws such as those found in furniture and cabinetry.

2.2.0 Pliers

In addition to the basic pliers covered in the Core Curriculum, there are two other types you will need on the job:

- *Groove-joint pliers*—Sometimes called *tongue-and-groove pliers*, these pliers are used to hold or turn large, round parts. The jaw opening can be adjusted by moving the pivot to a different channel (see *Figure 2*).
- *Clamping pliers*—Also known as *vise grips*, these pliers will lock on the work when the handles are squeezed hard enough. A lever inside the handles is used to release the jaws.

 **Audiovisual**

Show Transparency 1, Course Objectives.

Show Transparency 2, Performance Profile Tasks.

 **Classroom**

Provide a brief overview of hand tools.

 **Audiovisual**

Show Transparency 3 (Figure 1).

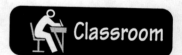 **Classroom**

Show various types of screwdrivers and pliers and discuss their applications. Pass around examples for the trainees to examine.

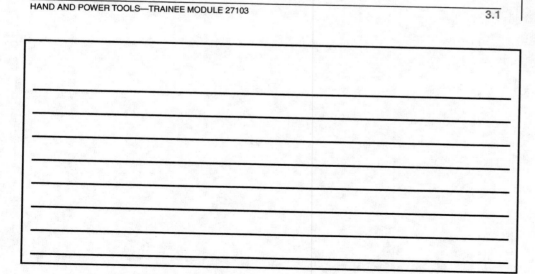

CLUTCH-HEAD
SCREWDRIVER

TORX-HEAD
SCREWDRIVER

SQUARE-HEAD
SCREWDRIVER

103F01.EPS

Figure 1 ◆ Special screwdrivers.

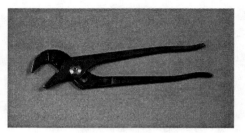

103F02.EPS

Figure 2 ◆ Groove-joint pliers.

2.3.0 Chisels

In addition to the wood chisel and cold chisel covered in an earlier module, there are also special-purpose chisels, such as those shown in *Figure 3*.

* *Flooring chisel*—This tool is used primarily for ripping up old strip flooring.
* *Brick chisel*—This tool is used to score, cut, and trim bricks. It should be struck with a small sledge or ball-peen hammer, rather than a bricklayer's or nail hammer.

2.4.0 Levels

In addition to the spirit level and torpedo level covered in an earlier module, you will also use the leveling tools and instruments shown in *Figure 4*.

* *Line level*—A line level can be used to level a long span. It consists of a glass tube mounted in a sleeve that has a hook on either end. The hooks are attached at the center of a stretched line, which is moved up or down until the bubble is centered.

* *Water level*—A water level is a simple, accurate tool consisting of a length of clear plastic tubing. It works on the principle that water seeks its own level. The water level is somewhat limited in distance because of the tubing. It can be used effectively for checking the level around obstructions. An example is checking the level from one room to the next when a wall is in the way.
* *Builder's level*—The builder's level is basically a telescope with a spirit level mounted on top. It can be rotated 360°, but cannot be tilted up and down. It is used to check grades and elevations and to set up level points over long distances.
* *Transit level*—A transit level, commonly called a *transit*, is similar to the builder's level. Unlike the builder's level, its telescope can be moved up and down 45°, making more operations possible than with a builder's level. When the telescope is locked in a level position, the transit level can be used to perform all the functions of a builder's level. Movement of the telescope in the up and down positions also allows the transit level to be used to plumb columns, building corners, or any other vertical members.
* *Laser level*—A laser level can be used to perform all of the tasks that can be performed with a conventional transit level. The laser level does not depend on the human eye. It emits a high-intensity light beam, which is detected by an electronic sensor (target) at distances up to 1,000'. Both fixed and rotating laser models are available. Rotating models enable one person, instead of two, to perform any layout operation. When a rotating laser is operated in the sweep mode, the head rotates through 360°, allowing the laser beam to sweep multiple sensors placed at different locations.

 WARNING!
Lasers can cause serious damage to your eyes. NEVER point a laser at anyone or look directly at a laser beam.

 Classroom

Show various types of chisels and levels and discuss their applications. Pass around examples for the trainees to examine.

Safety

Discuss the safety precautions associated with the use of laser levels.

Instructor's Notes:

_____

_____

_____

_____

_____

_____

_____

_____

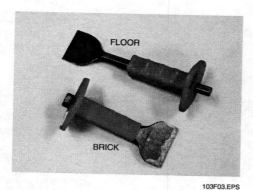

FLOOR

BRICK

103F03.EPS

Figure 3 ◆ Special chisels.

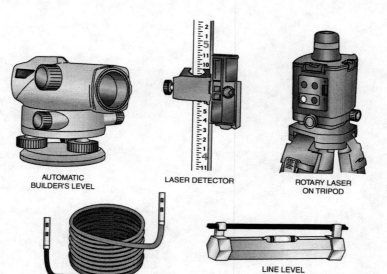

AUTOMATIC
BUILDER'S LEVEL

LASER DETECTOR

ROTARY LASER
ON TRIPOD

WATER LEVEL

LINE LEVEL

103F04.EPS

Figure 4 ◆ Levels.

**Show Transparency 4
(Figure 4).**

 **INSIDE TRACK**

Water Level

A water level is often called a *poor man's transit.* It is very effective for distances where a four- or six-foot level is not long enough, like decks and siding. It is also ideal for checking levels around obstructions, such as a wall that separates two rooms. The water level may be filled with colored water for increased visibility.

Audiovisual

Show Transparency 5 (Figure 5).

Classroom

Explain that terms shown in bold (blue) are defined in the Glossary at the back of this module. Show various types of squares and discuss their applications.

Demonstration

Demonstrate how to use a framing square.

Laboratory

Hand out Job Sheet 27103-1. Under your supervision, have the trainees perform the tasks on the Job Sheet. Note the proficiency of each trainee on his or her Job Sheet and Skill Test Record.

Keeping It Level

Make sure that you are very careful when handling and using a level. Levels are precision instruments and should be treated as such. In order to provide accurate readings, levels need to be calibrated properly. If they aren't, what appears to be a level reading will actually be slightly off. These errors are magnified over long distances, so a small miscalculation can wind up being a very large problem. Avoid this by taking proper care of your instruments.

LASER LEVEL   103P0301.EPS

TRANSIT LEVEL   103P0302.EPS

3.4

2.5.0 Squares

Establishing squareness is an essential part of carpentry; therefore, several types of squares are needed. You have already been introduced to the combination square. *Figure 5* shows a few more. Squares are made from steel, aluminum, and plastic and are available in both standard and metric measurements.

- *Combination square*—A combination square is used to lay out lines at 45° and 90°. The square can be adjusted by sliding the adjustable edge along the length of the blade.
- *Try square*—A try square can be used to lay out a line at a right angle (90°) to an edge or surface and check the squareness of adjoining surfaces and of planed lumber. It is also used for 45° angles.
- *Miter square*—A miter square has a blade fixed at 45° for laying out and marking lines such as for a miter cut. The opposite angle is fixed at 135°.
- *Sliding T-bevel*—A sliding T-bevel is an adjustable gauge for setting, testing, and transferring angles, such as when making a miter or bevel cut. The metal blade pivots and can be locked at any angle.
- *Drywall square*—A drywall square, also called a T-square, is used for laying out square and angled lines on large surfaces such as plywood, paneling, wallboard, etc.
- *Framing square*—A framing square, also called a rafter scale, is used for squaring lines and marking 45° lines across wide boards and checking inside corners for squareness. The two surfaces of the square are imprinted with a number of useful tables and formulas, including a rafter table and an Essex board measure table.
- *Speed square*—A speed square is used to mark and check 90° and 45° angles. It is especially useful when laying out angle cuts for roof rafters and other roof system components. Standard speed squares are 6" triangular tools. The large triangle has a 6" scale on one edge, a full 90° scale on another edge, and a T-bar on the third edge. The small triangle has a 2½" scale on one side. There are also 12" speed squares available for stair layout.

Instructor's Notes:

_____

_____

_____

_____

_____

_____

_____

_____

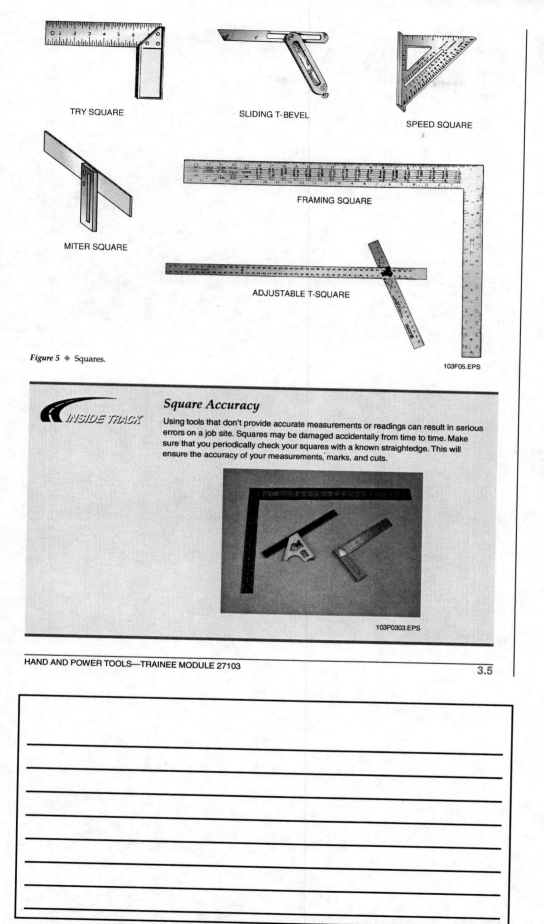

TRY SQUARE

SLIDING T-BEVEL

SPEED SQUARE

MITER SQUARE

FRAMING SQUARE

ADJUSTABLE T-SQUARE

Figure 5 ◆ Squares.

103F05.EPS

Square Accuracy

Using tools that don't provide accurate measurements or readings can result in serious errors on a job site. Squares may be damaged accidentally from time to time. Make sure that you periodically check your squares with a known straightedge. This will ensure the accuracy of your measurements, marks, and cuts.

103P0303.EPS

Show Transparency 6 (Figure 6).

Show various types of planes and wood chisels and discuss their applications.

Demonstrate how to use and hone planes and wood chisels.

Hand out Job Sheet 27103-2. Under your supervision, have the trainees perform the tasks on the Job Sheet. Note the proficiency of each trainee on his or her Job Sheet and Skill Test Record.

2.6.0 Planes

Planes (*Figure 6*) are used to remove excess wood from surfaces. A sharp steel blade protrudes from the flat bottom edge of the plane. On most planes, the depth of the cut can be adjusted by raising or lowering the blade. It is extremely important that the blade be kept sharp.

- *Block plane*—A block plane is used to plane small pieces, edges, and joint surfaces when a small amount of change is needed. Block planes come in 4" to 7" lengths and 1⅜" to 1⅝" widths. Blades are usually adjustable from 12° to 21°. A version of a block plane, called a *low-angle block plane*, is made for shaping fine trimwork and end grain. Its blade rests at a 12° angle for fine cuts on end and cross grains.
- *Jack plane*—A jack plane is a type of bench plane used to plane rough work. Jack planes are 14" long and 2" wide and have an adjustable blade.
- *Smoothing plane*—A smoothing plane is another type of bench plane. Smoothing planes are used to plane small work to a smooth, even surface without removing too much material. Smoothing planes come in lengths of 7" to 10" and widths of 1⅝" to 2".
- *Jointer plane*—A jointer plane is used to plane and true long boards such as the edges of doors. They come in 18" to 30" lengths. The cutting edge is ground straight and is set for a fine cut.

BLOCK PLANE

SMOOTHING PLANE

JACK PLANE

JOINTER PLANE

103F06.EPS

Figure 6 ◆ Planes.

3.6

Instructor's Notes:

_____

_____

_____

_____

_____

_____

_____

_____

2.7.0 Clamps

In addition to the C-clamp, which was introduced in the Core Curriculum, you will use several other types of clamps. Some of the most commonly used clamps are shown in *Figure 7*.

- *Hand-screw clamp*—A hand-screw clamp can be used to hold pieces together while adhesive dries. It is useful for beveled or tapered pieces.
- *Locking C-clamp*—Locking C-clamps come in a variety of sizes up to 24". They are based on the same design as the clamping pliers; when the handles are squeezed together, the jaws lock into place. The jaw opening is adjustable.
- *Spring clamp*—This simple clamp is spring-loaded so that the jaws will lock down on the work when the handles are released.
- *Quick clamp*—Quick clamps are heavy-duty clamps that are designed to apply extreme pressure. The jaws are tightened by pulling on the trigger and loosened by pulling on the release lever.

- *Web (strap) clamp*—Web clamps are designed to secure round, oval, and oddly-shaped work. A ratchet assembly is used to tighten the strap.
- *Pipe clamp*—This heavy-duty clamp can also be used as a spreader. Pipe clamps are made to fit ½" or ¾" pipe. The length of the clamp is determined by the length of the pipe. A sliding jaw operates with a spring-locking device, and a middle jaw is controlled by a screw set in a third, fixed jaw. A reversing pipe clamp has a sliding jaw that can also be used in the reverse direction for spreading by applying pressure away from the clamp instead of between the clamp jaws.

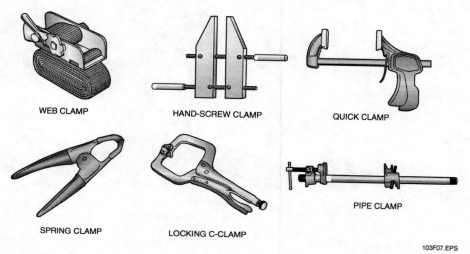

WEB CLAMP   HAND-SCREW CLAMP   QUICK CLAMP

SPRING CLAMP   LOCKING C-CLAMP   PIPE CLAMP

103F07.EPS

Figure 7 ◆ Clamps.

Clamps

As you gain experience, you will learn the various uses for different types of clamps. For example, a quick clamp is helpful in tight spots when you only have one hand available. A hand-screw clamp can be used to hold pieces together when laminating.

Demonstrate how to use a hacksaw and replace its blade.

Hand out Job Sheet 27103-3. Under your supervision, have the trainees perform the tasks on the Job Sheet. Note the proficiency of each trainee on his or her Job Sheet and Skill Test Record.

Assign reading of Sections 3.0.0–3.2.2 for the next class session.

Have each trainee complete to your satisfaction Performance Profile Task 2 on hand tool use. Fill out Performance Profile Sheets for each trainee.

2.8.0 Saws

In an earlier module, you were introduced to the common crosscut and rip saws. Carpenters also use many other types of saws, such as those shown in *Figure 8*. Keep your saws sharpened and properly protected when not in use.

- *Dovetail saw*—This saw, with its very fine teeth, is used to cut very smooth joints, such as mortise and tenon and dovetail joints. *Figure 9* shows a simple mortise and tenon joint. The dovetail saw is normally 10" long and has 17 teeth per inch. The blade is extremely thin, but has a band across the top edge to stiffen it.
- *Backsaw (miter saw)*—This saw has a strong back and very fine teeth. It is commonly used with a miter box to cut trim. Backsaws come in lengths from 12" to 28" and have 11 to 13 teeth per inch.
- *Compass (keyhole) saw*—This saw is used for making rough cuts in drywall, plywood, hardboard, and paneling. It is often used to cut access holes for piping and electrical boxes.
- *Coping saw*—The coping saw has a thin, fine-toothed blade and is ideal for making curved and scroll cuts in moldings and trimwork.
- *Hacksaw*—A hacksaw is used to cut metal. It has a replaceable blade with 18 to 32 teeth per inch. Blade lengths range from 8" to 32".

Miter Box

A backsaw is often used with a miter box. A miter box is either slotted or has holders that guide the saw during a cut. This guided saw movement results in very accurate cuts. Although the miter box was once a necessity for trim work, most carpenters now use a compound miter saw.

103P0305.EPS

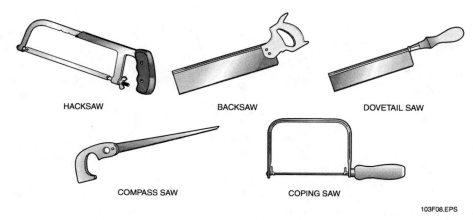

HACKSAW   BACKSAW   DOVETAIL SAW

COMPASS SAW   COPING SAW

103F08.EPS

Figure 8 ◆ Special saws.

Instructor's Notes:

_____

_____

_____

_____

_____

_____

_____

_____

103F09.EPS

Figure 9 ◆ Mortise and tenon joint.

3.0.0 ◆ POWER TOOLS

In this section, we will briefly review some of those tools studied earlier in the Core Curriculum and introduce you to several new ones and their uses. The intent of this section is to familiarize you with each of the tools and the related safety rules that apply when using them. Before you will be allowed to operate a specific power tool, you must be able to show that you know the safety rules associated with it. As your training progresses, you will learn how to operate each of the tools while under the supervision of your instructor and/or supervisor. Specific operating procedures and safety rules for using a tool are provided in the operator's/user's manual supplied by the manufacturer with each tool. Before operating any power tool for the first time, you should always read this manual to familiarize yourself with the tool. If the manual is missing, you or your supervisor should contact the manufacturer for a replacement.

3.1.0 Guidelines for Using All Power Tools

Before proceeding with our descriptions of power tools, it is important to first overview the general safety rules that apply when using all power tools, regardless of type. It is also important to review the general guidelines that should be followed in order to properly care for power tools.

Note

Power tools may be operated by electricity (AC or DC), air, combustion engines, or explosive powder.

3.1.1 Safety Rules Pertaining to All Power Tools

The rules for the safe use of all power tools are:

- Do not attempt to operate any power tool before being checked out by the instructor on that particular tool.
- Always wear eye protection and a hard hat when operating all power tools.
- Wear face and hearing protection when required.
- Wear proper respirator equipment when necessary.
- Wear the appropriate clothing for the job being done. Always wear tight-fitting clothing that cannot become caught in the moving tool. Roll up or button long sleeves, tuck in shirttails, and tie back long hair. Do not wear any jewelry or watches.
- Do not distract others or let anyone distract you while operating a power tool.
- Do not engage in horseplay.
- Do not run or throw objects.
- Consider the safety of others, as well as yourself.
- Do not leave a power tool running while it is unattended.
- Assume a safe and comfortable position before using a power tool.
- Be sure that a power tool is properly grounded and connected to a GFCI circuit before using it.
- Be sure that a power tool is disconnected before performing maintenance or changing accessories.
- Do not use a dull or broken tool or accessory.
- Use a power tool only for its intended use.
- Keep your feet, fingers, and hair away from the blade and/or other moving parts of a power tool.
- Do not use a power tool with guards or safety devices removed or disabled.
- Do not operate a power tool if your hands or feet are wet.
- Keep the work area clean at all times.
- Become familiar with the correct operation and adjustments of a power tool before attempting to use it.
- Keep a firm grip on the power tool at all times.
- Use electric extension cords of sufficient size to service the particular power tool you are using.
- Report unsafe conditions to your instructor or supervisor.

Review the safety guidelines for using power tools.

3.1.2 Guidelines Pertaining to the Care of All Power Tools

Guidelines for the proper care of power tools are:

- Keep all tools clean and in good working order.
- Keep all machine surfaces clean and waxed.
- Follow the manufacturer's maintenance procedures.
- Protect cutting edges.
- Keep all tool accessories (such as blades and bits) sharp.
- Always use the appropriate blade for the arbor size.
- Report any unusual noises, sounds, or vibrations to your instructor or supervisor.
- Regularly inspect all tools and accessories.
- Keep all tools in their proper place when not in use.
- Use the proper blade for the job being done.

3.2.0 Power Saws

There are a wide variety of different power saws that can be used by carpenters. These include:

- Circular saws
- Table saws
- Radial arm saws
- Power miter saws/compound miter saws
- Frame and trim saws
- Reciprocating saws
- Jigsaws
- Band saws

3.2.1 Circular Saws

Circular saws (*Figure 10*) are versatile, portable saws used for the following tasks:

- Ripping (rip cut)
- Crosscutting (crosscut)
- Mitering
- Pocket (plunge) cuts
- Bevel cuts

The size of a circular saw is determined by the diameter of the largest size blade that can be used

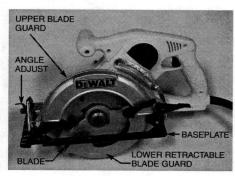

UPPER BLADE GUARD

ANGLE ADJUST

BASEPLATE

BLADE

LOWER RETRACTABLE BLADE GUARD

103F10.EPS

Figure 10 ◆ Circular saw.

with the saw, which determines how thick a material can be cut. Saws using a 7¼" blade are popular. Circular saws have upper and lower guards that surround the blade. The upper guard is fixed; the lower guard is spring-loaded and retracts as the saw cuts into the workpiece. The saw has a baseplate that rests on the material being cut and can be adjusted to change the depth of the cut or to make bevel cuts ranging from 0° to 45°. The saw is started by pressing the trigger in the handle and stopped by releasing it. It will run only while the trigger is pressed. To make ripping of narrower boards easier, most saws come equipped with an adjustable rip fence that can be attached to the saw baseplate in slots in front of the blade.

WARNING!

When using a circular saw to cut on an angle, the blade guard may have the tendency to bind. DO NOT reach below the saw to release the guard. Severe injury, such as lost fingers, could result. If you cannot complete the cut, find an alternative saw to finish the job. Never jeopardize your safety for any reason.

Discuss the various uses of a portable circular saw.

Demonstrate how to make straight, angled, and rip cuts using a circular saw. Emphasize safety precautions.

Instructor's Notes:

Circular Saws

The circular saw is perhaps the most important tool on a job site. If you are going to be a successful professional carpenter, you must master the use of a circular saw and be able to make all of the different cuts with precision. In order to achieve this, you will need to practice. Mastery can only come from extensive practice and experience.

103P0306.EPS

Rules for the safe use of a portable circular saw are:

- Always wear safety glasses.
- Hold the saw firmly.
- Always start the saw before making contact with the material.
- Keep the electric cord clear of the blade to avoid cutting it.
- Check the condition of the blade and be sure it is secure before starting the saw.
- Be sure the blade guards are in place and working properly.
- Set the blade only deep enough to cut through the material you are using.
- Check the stock for nails and any other metal before cutting it.
- Be sure the saw has reached maximum speed before starting the cut.
- Keep your hands clear of the blade.
- If the blade binds in a cut, stop the saw immediately and hold it in position until the blade stops, then back it out of the cut.
- Stop the saw and lay it on its side after finishing the cut.
- Do not hold stock in your hands while ripping. Nail the material to sawhorses.

3.2.2 Table Saws

Table saws (*Figure 11*) are stationary saws used to do the following sawing and related woodworking tasks:

- Ripping
- Crosscutting
- Mitering
- Rabbeting (making rabbet cuts)
- Dadoing (making dados)
- Cutting molding

The table saw consists of a machined flat metal table surface with a slotted table insert in the center through which the circular saw blade extends. The table provides support for the workpiece as it is fed into the rotating blade. The size of a table saw is determined by the diameter of the largest size circular saw blade it can use, with 10" and 12" diameters being common. The depth of the cut is adjusted by moving the blade vertically up or down. The blade can also be tilted to make beveled cuts up to 45°, as indicated on a related degree scale.

For crosscutting, the table is grooved to accept a guide called a *miter gauge*. The workpiece is held against the vertical face of the miter gauge to keep it at a perfect angle with respect to the blade. Pushing the miter gauge forward moves the workpiece through and past the blade so that each cut is flat and square. Typically, the miter gauge can be turned to any angle up to 60°, with positive stops at 0° and 45° right and left.

For ripping, the saw comes equipped with a rip guide or fence. This clamps to the front and rear edges of the table and has rails that allow it to be moved toward or away from the blade while

Hand out Worksheet 27103-1 and Job Sheets 27103-4 and 27103-5. Under your supervision, have the trainees complete the circular saw safety test before performing the related tasks on the Job Sheets. Note the proficiency of each trainee on his or her Job Sheets and Skill Test Record.

Show Transparency 9 (Figure 11).

Demonstration

Demonstrate how to make straight, angled, and rip cuts using a table saw. Emphasize safety precautions.

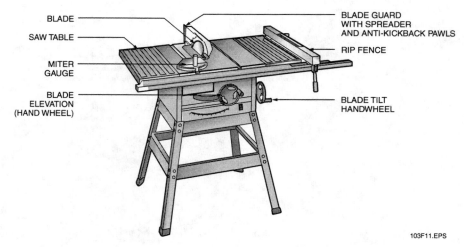

BLADE

SAW TABLE

MITER
GAUGE

BLADE
ELEVATION
(HAND WHEEL)

BLADE GUARD
WITH SPREADER
AND ANTI-KICKBACK PAWLS

RIP FENCE

BLADE TILT
HANDWHEEL

103F11.EPS

Figure 11 ◆ Table saw.

remaining parallel. It is positioned to the width desired for the work, then locked in place. The workpiece is guided against the fence while making the cut as it is being pushed and/or pulled through the blade.

A blade guard protects the user's hands from the rotating blade and prevents flying chips. At the rear of the blade guard assembly are a kerf spreader and an anti-kickback device. The kerf spreader acts to keep the kerf in the workpiece open behind the blade. The kerf is the width of the cut made by the saw blade in the workpiece as it is being cut. If the kerf is allowed to close, the blade can become pinched, causing the workpiece to start moving backward while cutting and resulting in possible injury to the operator. If kickback starts to occur, sharp pawls on the anti-kickback device function to grip the workpiece and hold it in place.

When using a table saw to make dados, the saw blade is removed and a dado set or adjustable dado is used in place of the saw blade for making the cuts. Similarly, when cutting molding, the saw blade is removed and a molding head and molding cutters are used in place of the saw blade. Rules for the safe use of a table saw are:

• Keep the guard over the blade while the saw is being used.

• Do not stand directly in line with the blade.
• Make sure that the blade does not project more than ⅛" above the stock being cut.
• Never reach across the saw blade.
• Use a push stick for ripping all stock less than 4" wide.
• Disconnect power when changing blades or performing maintenance.
• Never adjust the fence or other accessories until the saw has stopped.
• Enlist a helper or use a work support when cutting long or wide stock. The weight of the stock may cause the table to lean.
• Before cutting stock to length using a miter gauge and rip fence, clamp on a step block for clearance.
• Do not rip without a fence, or crosscut without a miter gauge.
• When the blade is tilted, be sure that the blade will clear the stock before turning on the machine.
• Be sure that the stock has a straight edge before ripping.
• When using a dado or molding head, take extra care to hold the stock firmly.
• Never remove scraps from the saw table with your hands or while the saw blade is running.
• Use the proper blade for the job being done.

Instructor's Notes:

_____

_____

_____

_____

_____

_____

_____

_____

Table Saw Kickback

When making cuts on a table saw, particularly
crosscuts and dado cuts, the saw blade may get
pinched in the piece you are cutting. This could cause
the workpiece to jump back, resulting in an injury.
Although most table saws have an anti-kickback
device that is designed to hold the lumber in place, it's
a good idea to use a featherboard to feed the saw.

3.2.3 Radial Arm Saws

Radial arm saws, also called *cutoff saws*, are stationary saws that are used to do the following
sawing and related woodworking tasks:

- Crosscutting
- Mitering
- Dadoing
- Cutting molding

As shown in *Figure 12*, the distinguishing feature of a radial arm saw is its overhead arm, which
is mounted on a rigid vertical column. This arm
can be moved horizontally around the column in
any direction. Suspended from the arm is a yoke
(mounting frame) to hold the motor with the saw
blade attached to its shaft. The yoke rides freely in
a carriage or track along the underside of the arm
to carry the motor back and forth as it is pulled and
pushed by the operator when doing crosscutting
and miter cutting operations. It can also be positioned and locked in place at any location along
the length of the carriage to facilitate ripping operations. The motor/blade can be tilted to any
desired angle for making bevel cuts. It can also be
rotated 180° in a clockwise or counterclockwise
direction from normal (blade aligned with the
arm) for making rip cuts. Regardless of position,
the depth of the cut is controlled by raising or lowering the height of the arm above the fixed table.

> **WARNING!**
>
> Never put your hands or fingers near the blade or
> blade path. The stock could self-feed and
> crosscut very rapidly and cause serious injuries.

Two or more boards and a fence form the saw
table. The fence, which acts as a guide or stop for
the workpiece, is inserted between the boards and
tightened with a table clamp or lock.

The size of a radial arm saw is determined by
the largest diameter circular saw blade it can use,
with 10" and 12" diameters being the most common. A saw guard covering the blade prevents

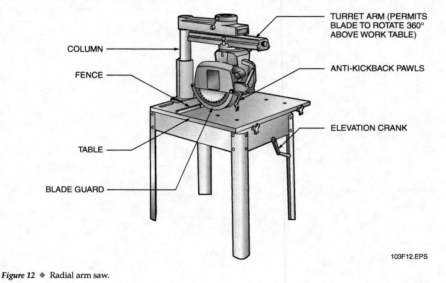

COLUMN

FENCE

TABLE

BLADE GUARD

TURRET ARM (PERMITS BLADE TO ROTATE 360° ABOVE WORK TABLE)

ANTI-KICKBACK PAWLS

ELEVATION CRANK

103F12.EPS

Figure 12 ◆ Radial arm saw.

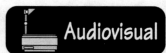

**Show Transparency 10
(Figure 12).**

**Demonstrate how to make
straight, angled, and rip
cuts using a radial arm saw.
Emphasize safety
precautions.**

Laboratory

Hand out Worksheet 27103-1 and Job Sheets 27103-4 and 27103-5. Under your supervision, have the trainees complete the radial arm saw safety test before performing the related tasks on the Job Sheets. Note the proficiency of each trainee on his or her Job Sheets and Skill Test Record.

Audiovisual

Show Transparency 11 (Figure 13).

Demonstration

Demonstrate how to make straight, miter, and compound cuts using a compound miter saw. Emphasize safety precautions.

Laboratory

Hand out Worksheet 27103-1 and Job Sheet 27103-6. Under your supervision, have the trainees complete the compound miter saw safety test before performing the related tasks on the Job Sheet. Note the proficiency of each trainee on his or her Job Sheet and Skill Test Record.

injury. An anti-kickback device attached to the guard is used when making rip cuts. If the workpiece starts to kick back while cutting, pawls on this device grip the workpiece and hold it in place. A kerf spreader located behind the blade keeps the cut in the workpiece open during rip cuts, preventing the blade from jamming the work.

Although the radial arm saw can be used for several cutting tasks, it is used mainly for crosscutting operations. This is because the workpiece being cut remains stationary on the table and aligned against the fence while the saw is pulled across it to make the cut. This eliminates the need to push or pull the workpiece through the blade. Note that for ripping stock, the blade must be positioned so that the workpiece is pushed into and/or pulled through the blade in the same manner as is done with a table saw.

Rules for the safe use of a radial arm saw are:

- Make sure the saw is set to the proper depth before turning it on.
- Use only a sharp and properly-set blade.
- Do not use a radial arm saw for ripping if you have a table saw.
- Do not stand directly in line with the blade.
- Make sure that the saw has reached its maximum speed before starting the cut.
- Do not let the saw feed too fast when crosscutting heavy stock. The saw will have a tendency to feed itself.
- When ripping, follow the same safety rules that you would when using a table saw.
- Always use an anti-kickback guard when ripping.

- Do not feed or pull the stock too fast when ripping because this will jam the saw.
- When ripping, feed the stock against the direction of the blade rotation.
- Do not stop the blade by forcing a scrap of wood into the blade.
- Always lock the saw or hold it in place before starting it.

3.2.4 Power Miter Saws/Compound Miter Saws

Power miter saws/compound miter saws (*Figure 13*) are also commonly called *power miter boxes*. This saw combines a miter box or table with a circular saw, allowing it to be used to make straight and miter cuts. The saw blade pivots horizontally from the rear of the table and locks in position to cut angles from 0° to 45° right and left. Stops are set for common angles. The difference between the power miter saw and the compound miter saw is that the blade on the compound miter saw can be tilted vertically, allowing the saw to be used to make a compound cut (combined bevel and miter cut).

Rules for the safe use of a power miter saw/compound miter saw are:

- Always check the condition of the blade and be sure the blade is secure before starting the saw.
- Keep your fingers clear of the blade.
- Never make adjustments while the saw is running.
- Never leave a saw until the blade stops.

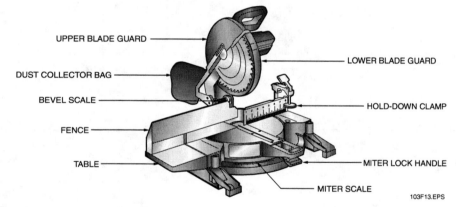

Figure 13 ◆ Compound miter saw.

CARPENTRY LEVEL ONE—TRAINEE MODULE 27103

Instructor's Notes:

_____

_____

_____

_____

_____

_____

_____

CARPENTRY LEVEL ONE—INSTRUCTOR'S GUIDE MODULE 27103

- Be sure the saw is sitting on a firm base and is properly fastened to the base.
- Be sure the saw is locked securely at the correct angle.
- Do not attempt to cut oversized material.
- Turn off the saw immediately after making the cut and use the brake to stop the blade.
- Enlist a helper to support the end of long stock.
- Be sure the blade guards are in place and working properly.
- Be sure the saw blade has reached its maximum speed before starting the cut.
- Hold the workpiece firmly against the fence when making the cut.

3.2.5 Frame and Trim Saws

Frame and trim saws (*Figure 14*), commonly called *Sawbuck™ saws* or *sliding compound miter saws*, combine the features of a radial arm saw and a power compound miter saw. The table rotates with the track arm, and the saw head can be adjusted to make crosscuts, straight miter cuts, bevel cuts, and compound miter cuts. Because the blade assembly slides along supporting rods, wider pieces of stock can be cut with this saw than with a power miter or compound miter saw. This type of saw is ideal for cutting crown molding.

Rules for the safe use of a frame and trim saw are:

- Check the condition of the blade and be sure that it is secure before starting the saw.
- Keep your fingers clear of the blade.
- Never reach around or across the blade.
- Never make adjustments with the saw running.
- Allow the saw blade to reach its maximum speed before starting the cut.
- Turn the saw off immediately after making the cut and use the brake to stop the blade.

- Be sure that the blade guards are in place and working properly.
- Never leave the saw until the blade has stopped.
- Be sure the saw is sitting on a level, firm base.
- Be sure the saw is locked securely at the correct angle.
- Hold the material firmly against the fence while making cuts. Clamp the material to the fence when possible.

3.2.6 Circular Saw Blades

There are a wide variety of saw blades (*Figure 15*) available for use with circular, table, radial arm, power miter, and similar saws. Each blade is designed to make an optimum cut in a different type and/or density of material. Generally, blades are standard high-speed steel or tipped with carbide. Carbide-tipped blades stay sharper longer, but they are more brittle and can be damaged if improperly handled. The number of teeth on a blade, the grind of each tooth, and the space between the teeth (gullet depth) determines the smoothness and speed of the cut.

Crosscutting requires the severing of wood fibers, whereas during ripping, the teeth must gouge out lengths of wood fiber. Blades made specifically for crosscutting have small teeth that are alternately beveled across their front edges. The gullets are small because only fine sawdust is produced. Blades made specifically for ripping have large, scraper-like teeth with deep, curved gullets. When both types are combined in a combination blade, the scraper, called a *raker*, has a much deeper gullet than the crosscut teeth. In chisel-teeth combination blades, the teeth and gullets are a compromise between rip and crosscut designs. They can be used for both types of cutting, but they cut less smoothly than blades designed to do just one job. Blades designed for

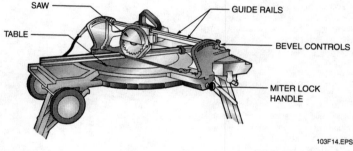

SAW · · · · · · · · · GUIDE RAILS

TABLE · · · · · · · · · BEVEL CONTROLS

· · · · · · · · · MITER LOCK HANDLE

103F14.EPS

Figure 14 ◆ Frame and trim saw.

Show Transparency 12 (Figure 14).

Demonstrate how to make crosscuts, straight miter cuts, bevel cuts, and compound cuts using a frame and trim saw. Emphasize safety precautions.

Hand out Worksheet 27103-1 and Job Sheet 27103-6. Under your supervision, have the trainees complete the frame and trim saw (sawbuck) safety test before performing the related tasks on the Job Sheet. Note the proficiency of each trainee on his or her Job Sheet and Skill Test Record.

Show Transparency 13 (Figure 15).

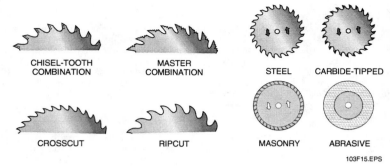

Figure 15 ◆ Circular saw blades.

Identify the applications for various types of saw blades.

Assign reading of Sections 3.2.7–3.3.0 for the next class session.

Show Transparency 14 (Figure 16).

cutting plywood and other easily splintered or chipped materials have small, scissor-like teeth. The smoothness of the cut is determined by the number of teeth and their sharpness. Use a blade recommended by the blade manufacturer for the type of material being cut. Always make sure that the blade diameter, arbor hole size, and maximum rotation speed are compatible with the saw on which it is to be used. Manufacturers have different names for similar circular saw blades. Some types of circular saw blades widely used for woodworking are:

- *Master combination blade (48 to 60 teeth)*—This is an all-purpose blade used for smooth, fast cutting. It can be used for crosscutting, ripping, mitering, and general-purpose work.
- *Chisel-tooth combination blade (24 to 30 teeth)*—This is an all-purpose blade that is free-cutting in any direction. It can be used for crosscutting, ripping, mitering, and general-purpose work.
- *Combination rip blade (36 to 44 teeth)*—This is an all-purpose blade used for crosscutting, ripping, mitering, and general-purpose work.
- *Carbide-tipped blade*—Most of the combination blades described above can be purchased with carbide-tipped teeth. These blades are popular because of their economy and quality.

- *Hollow-ground plane (miter) blade (48 to 60 teeth)*—Hollow-ground blades are used to make smooth finish cuts on all solid woods and are ideal for cabinet work. They can be used for crosscutting, ripping, mitering, and finish work.

3.2.7 Reciprocating Saws

Reciprocating saws (*Figure 16*) are heavy-duty saws with a horizontal back and forth movement of the blade. They are an all-purpose saw used for making heavy-duty cuts such as:

- Plunge cuts
- Crosscuts
- Relief cuts
- Inside curves and corner cuts
- Pocket cuts

Figure 16 ◆ Reciprocating saw.

3.16

Instructor's Notes:

_____
_____
_____
_____
_____
_____
_____
_____

Reciprocating saw models with variable speeds ranging from 0 to 2,400 strokes per minute (SPM) are common. Speed selection is made with a variable control at the trigger on-off switch. Greater horsepower and slower speeds are generally needed when cutting through metals or when cutting along a curved or angled line. The typical length of the horizontal sawing stroke is 1⅛". A multi-positioned foot at the front of the saw can be put in three different positions for use in flush cutting, ripping, and crosscutting.

A wide variety of blades are made for use with reciprocating saws. Each type of blade is designed to make an optimum cut in a different kind of material. The blade length determines the thickness of the material that can be cut. Use the shortest blade that will do the job. Reciprocating saw blade lengths range from 3½" to 12". The number of teeth range from 3½ to 32 per inch. Blades with 3½ to 6 teeth per inch are generally used for sawing wood, while blades with 6 to 10 teeth per inch are used for general-purpose sawing. Blades with 10 to 18 teeth per inch are used for cutting metal. Always use the blade recommended by the blade manufacturer for the type of material being cut.

A reciprocating saw is an ideal tool for renovations and remodeling, such as removing existing framing.

Rules for the safe use of a reciprocating saw are:

- Keep your hands clear of the blade.
- Hold the saw firmly.

- Be sure the blade has clearance on the opposite side of the material.
- Keep the electric cord clear of the blade.
- Do not force the blade into the material.
- Use the correct blade for the job being done.
- Turn off the saw immediately after finishing the cut.
- Be sure the material being cut is properly supported.
- Be careful not to cut live electrical wires or plumbing in walls and floors.
- Be sure the saw foot is kept firmly against the work during the entire cut.

3.2.8 Jigsaws

Jigsaws (*Figure 17*), also commonly called *saber saws*, are lighter-duty saws than reciprocating saws. Unlike the reciprocating saw, they have a vertical back and forth blade movement. Jigsaws are an all-purpose tool used to make cuts such as:

- Curves and scrolls
- Inside curves and corner cuts
- Pocket cuts
- Ripping
- Crosscuts
- Relief cuts

With the proper blade, jigsaws can cut wood, metal, plastic, and other materials. Variable speed models with speeds ranging from 0 to 3,200 SPM

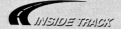
Cordless Power Tools

Today's cordless tools offer both substantial power capability and convenience. Once limited to only a few types, they are now available to serve a wide variety of applications, including the cordless drill and reciprocating saw shown here.

103P0307.EPS

Demonstrate how to make different cuts using a reciprocating saw. Emphasize safety precautions.

Hand out Worksheet 27103-1. Under your supervision, have the trainees complete the reciprocating saw safety test before practicing with this tool. Note the proficiency of each trainee.

Show a portable jigsaw with various blades.

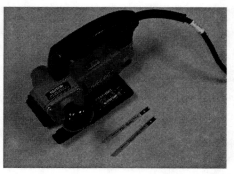

103F17.EPS

Figure 17 ◆ Jigsaw.

are common. The typical length of the vertical sawing stroke is 1". Jigsaws have a baseplate or shoe. Its broad surface helps keep the blade aligned. It also helps prevent the work from vibrating and allows the teeth to bite into the material. The baseplate can be tilted for making bevel cuts. Many jigsaws have a large scrolling knob that can be unlocked from a stationary position, then used to rotate the blade while sawing the material. This makes it easier to cut tight curves, corners, and patterns. A pocket cut is a typical cut that you would make with a jigsaw.

Wood-cutting, metal-cutting, and special-purpose blades are available. Use the shortest blade that will do the job. Blades for cutting wood have as few as 6 teeth per inch for fast, coarse cutting and as many as 14 teeth per inch for fine work. Tapered-ground blades are made that produce splinter-free cuts in plywood. Blades made for cutting both metals and plastic laminates typically have between 12 and 32 teeth per inch. Always use the blade recommended by the blade manufacturer for the type of material being cut.

Rules for the safe use of a jigsaw are:

- Keep your hands clear of the blade.
- Hold the saw firmly.
- Be sure the blade has clearance on the opposite side of the material.
- Keep the electric cord clear of the blade to avoid cutting it.
- Do not force the blade into the material.
- Use the correct blade for the job.
- Turn off the saw immediately after finishing the cut.
- Be sure the material being cut is properly supported.

3.18

3.2.9 Band Saws

Band saws are made in both stationary table and portable models. Table band saws (*Figure 18*) are used for special cuts such as creating designs in wood and making molding. These include:

- Irregular cuts
- Circular cuts
- Resawing
- Relief cuts

The table band saw consists of an endless blade looped around a set of wheels protected by hinged upper and lower wheel guards. Some models have three wheels. A motor-driven wheel drives the blade. Blades come in various widths with different numbers of teeth. The width of the blades can range between ⅛" and ¾", with ⅛" and ¼" blades being common. The more teeth per inch on the blade, the finer the cut; fewer teeth allow a faster but coarser cut. Blade materials and the number and pitch of the teeth are designed to make an optimum cut in different kinds of materials. Always use the blade recommended by the blade manufacturer for the type of material being cut.

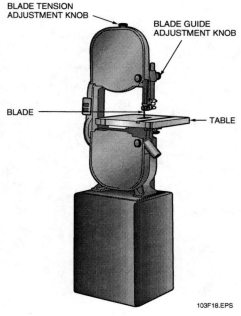

BLADE TENSION ADJUSTMENT KNOB

BLADE GUIDE ADJUSTMENT KNOB

BLADE

TABLE

103F18.EPS

Figure 18 ◆ Table band saw.

Instructor's Notes:

_____

_____

_____

_____

_____

_____

_____

_____

An upper blade guide and guard move up and down to accommodate work of various thicknesses. The upper blade guide and another one under the table keep the blade aligned. The maximum cutting depth of a particular saw is determined by the distance between the saw's table and the upper blade guard when it is set at its highest position. In smaller saws, the cutting depth runs up to 4"; in larger models, it can be greater than 7½". The throat capacity (the distance between the blade and the vertical frame) determines the widest board that can be cut.

A blade tension adjustment that moves one of the wheels can be adjusted in order to obtain the proper amount of tension between the wheels as necessary to keep the blade running true. The tracking of the blade can also be adjusted so that the blade rides centered on the wheels. Some band saw tables have a tilt/scale mechanism that allows the table to be adjusted for making bevel cuts up to 45°. For other models, called *tilt-head saws*, the angle of the saw itself relative to the table is adjusted for making beveled cuts. Like the table saw, the band saw table is grooved to accommodate a miter guide used when making straight and angle cuts. A rip fence can also be used with table band saws to aid in ripping operations.

Rules for the safe use of a table band saw are:

- To avoid breakage, use the proper blade size.
- Do not feed the stock too fast.
- Use relief cuts when necessary.
- Adjust the saw guide before each cut.
- Keep your hands and arms clear of the blade.
- Do not try to cut curves smaller than the recommended size for the blade you are using.
- Avoid long backouts.
- Never leave the saw until the blade has stopped.
- Never make adjustments with the saw running.
- Do not try to back out of a cut if the blade binds or pinches in the wood.
- If the blade breaks, shut off the power immediately and step back until the machine stops.

The portable band saw (*Figure 19*) can be used to cut workpieces that are too large to fit on a table band saw's work table. It is also useful for cutting metal studs, angle iron, steel rods, and pipe. Typical cutting capacities are up to 3½" for round materials and 3½" × 4½" for rectangular materials. The portable band saw has a continuous loop blade that runs in one direction around guides located at either end of the saw. Two-speed and variable speed models are common. A portable band saw is an ideal tool for cutting conduit.

103F19.EPS

Figure 19 ◆ Portable band saw.

Rules for the safe use of a portable band saw are:

- Keep your hands clear of the blade.
- Hold the saw firmly.
- Keep the electric cord clear of the blade.
- Be sure the blade guards are in place.
- Use the proper size and type of blade for the job being done.
- Do not force the blade into the material.
- Be sure the material being cut is properly supported.

3.3.0 Jointers

Jointers (*Figure 20*), commonly called *jointer-planers*, are used for the following cutting operations:

- Straightening
- Edge planing
- Beveling
- Chamfering
- Tapering

Jointers consist of an infeed table, a cutter head with knives, an outfeed table, and a fence. The knife length determines the size of the jointer, with 6" being the most common. Typically, the knives rotate at a speed of 4,500 RPM. A spring-loaded cutter guard covers the cutter head knives, except when pushed aside by the workpiece as it is being fed over the knives. The height of the infeed table is adjustable to change the depth of the cut. The outfeed table is also adjustable and must be adjusted to be level with the knife edges at their highest point of rotation. In some models, the infeed and outfeed adjustment is controlled by a single knob. A fence is provided to guide the stock over the table and knives. It can be adjusted up to 45° left and right for making bevel and

Demonstrate how to make straight, angled, and rip cuts using both table and portable band saws. Emphasize safety precautions.

Laboratory

Hand out Worksheet 27103-1 and Job Sheets 27103-4 and 27103-5. Under your supervision, have the trainees complete the portable and table band saw safety tests before performing the related tasks on the Job Sheets. Note the proficiency of each trainee on his or her Job Sheets and Skill Test Record.

Audiovisual

Show Transparency 16 (Figure 20).

Demonstration

Demonstrate facing, edge jointing, beveling, rabbeting, and tapering using a jointer. Emphasize safety precautions.

103F20.EPS

Figure 20 ◆ Jointer-planer.

Jointer Safety

Always use a push stick or push block to feed the stock into the jointer, as shown here.

103P0308.EPS

Instructor's Notes:

chamfer cuts. Most fences have positive stops at 90° and 45°.

Rules for the safe use of a jointer are:

- Keep the guard over the knives at all times.
- Adjust the depth of the cut before turning on the power.
- Do not use stock less than 12" long.
- Do not use strips less than 1" wide or ¼" thick.
- Always use a push stick or push block when cutting stock.
- Secure the clamping screws tightly on the fence so that the fence will not slip when in use.
- Do not start the cut until the jointer has gained its maximum speed.
- Keep your fingers as high as possible on the stock; never drag your thumb at the back of the stock.
- Avoid making cuts deeper than ⅛"; too deep a cut can cause a kickback.
- When passing work over a jointer, change the position of your hands so that they will never be directly over the knives.
- Follow through with the stock and allow the guard to return to the closed position.
- Do not attempt to plane discs or oddly-shaped pieces that cannot be held securely against the fence.
- Never leave the jointer before the cutter head stops turning.
- Stand to one side when turning on the jointer.
- Have your instructor or supervisor check all special setups before using the jointer.
- Use a roller support or have someone help you when joining pieces.
- Cut with the grain of the wood to avoid kickback.

3.4.0 Thickness Planer

A thickness planer (*Figure 21*) is used to size stock to a desired thickness after one surface has been straightened or flattened by a jointer. It is also used for face and end planing of stock. The workpiece is placed on the table and automatically advanced on rollers past the rotating blades in the cutterhead located at the top of the machine. Benchtop models typically accept stock from 4" to 12" wide and up to 6" thick. Floor models accept larger stock, typically from 12" to 36" wide and up to 8" thick.

WARNING!

Wear proper hearing protection every time you are using a thickness planer.

103F21.EPS

Figure 21 ◆ Thickness planer.

Rules for the safe use of a planer are:

- Keep your hands clear of the blades.
- Allow the automatic feed mechanism to operate.
- Check for the correct depth of cut before starting.
- Do not attempt to remove too much material with one cut.
- Make adjustments only after the power has been turned off. Make speed adjustments according to the manufacturer's instructions.
- Check the stock to be sure that it is free of nails or other objects that may damage the machine or be dangerous to the operator.
- Do not use stock less than 12" long. The stock should be long enough to reach the outfeed rollers before it leaves the infeed rollers.
- Never bend over and look at the stock while it is being planed.
- Never push material with your body.
- Feed the thick end of the material first.
- If more than one board is being fed, only boards of the same size may be fed at the same time.

3.5.0 Shapers

Shapers (*Figure 22*) are used for the following wood-shaping tasks:

- Shaping edges
- Shaping molding
- Grooving
- Fluting
- Reeding

Shapers consist of a work table, a motor-driven adjustable-height spindle, and an adjustable fence. A high-speed molding cutter is mounted on the spindle, which extends up through the work table. Molding cutters typically have three blades

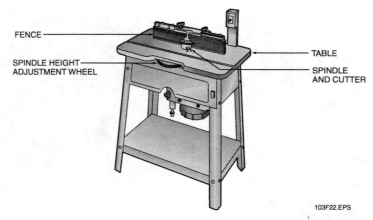

FENCE

SPINDLE HEIGHT ADJUSTMENT WHEEL

TABLE

SPINDLE AND CUTTER

103F22.EPS

Figure 22 ◆ Shaper.

Demonstrate edge shaping using a shaper. Emphasize safety precautions.

Hand out Worksheet 27103-1 and Job Sheet 27103-9. Under your supervision, have the trainees complete the shaper safety test prior to performing the related tasks on the Job Sheet. Note the proficiency of each trainee on his or her Job Sheet and Skill Test Record.

Show Transparency 19 (Figure 23).

and are available in several designs, each made to produce a different edge shape. The adjustment of spindle elevation and cutter height affects the resulting edge profile made by the cutter on the workpiece. The adjustment of the fence affects the depth of the edge cut. Some shapers are equipped with a reversing switch that allows the spindle to rotate in either a clockwise or counterclockwise direction.

Rules for the safe use of a shaper are:

- Feed the thick end of the material first.
- Whenever possible, install the cutter so that the bottom of the stock is shaped; in this way, the stock will cover most of the cutter and act as a guard.
- Never push material with your body.
- Be sure the cutter is locked securely to the spindle.
- Always position fences so that they will support the work that has passed the cutters.
- Adjust the spindle for the correct height and lock it in position. Rotate the spindle by hand to be sure that it clears all guards and fences.
- Check the direction of rotation by snapping the switch on and off; watch as the cutter comes to rest.
- Always feed the stock against the cutting edge.
- Examine the stock carefully for defects before cutting; never cut through a loose knot or through stock that is cracked or split.
- Hold the stock down and against the fence with your hands on top of the material and out of range of the cutters.
- Use all guards, jigs, and clamping devices when appropriate.

- Do not set spring hold-down clips too tightly against the work; use just enough tension to hold the work against the fence.
- Always use a depth collar when shaping irregular work. Use a guide pin in the table to start the cutting.
- Swing the work into the cutters when depth collars and a guide pin are used for contour work. It is also a good idea to keep the stock in motion in the direction of the feed.
- Never shape a piece of stock that is less than 10" long.
- If more than one board is being fed, only boards of the same size may be fed at the same time.

3.6.0 Sanders and Grinders

Power sanders and grinders fitted with the appropriate abrasive papers are used to shape workpieces, remove imperfections in wood, and create smooth surfaces before finishing. Common sanders include the stationary belt-disc sander and a variety of portable sanders/grinders.

3.6.1 Belt-Disc Sanders

Belt-disc sanders (*Figure 23*) are stationary sanders used for a large variety of sanding operations, including:

- Rough sanding
- Fine sanding
- Shaping

3.22

Instructor's Notes:

_____

_____

_____

_____

_____

_____

_____

_____

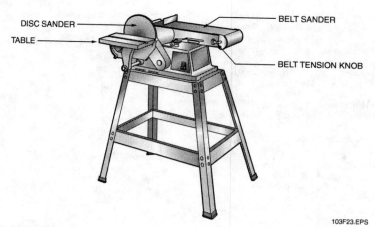

DISC SANDER
TABLE
BELT SANDER
BELT TENSION KNOB

103F23.EPS

Figure 23 ◆ Belt-disc sander.

Belt-disc sanders combine a motor-driven belt sander and a disc sander in one machine. They are made in a variety of different belt and disc sizes. A typical machine is equipped with a 6" wide sanding belt and a 12" diameter sanding disc. Normally, the belt sander unit can be tilted to operate in the vertical position for edge or end sanding, the horizontal position for surface sanding, or at any position in between. The disc table can be tilted up to 45° up or down. The belt sander mechanism has controls that are used to adjust the tension and tracking of the sander belt.

The sander belt is a continuous loop with a cloth backing. It must be of the proper width and length for the sander being used. The abrasive is typically aluminum oxide and is made in fine, medium, coarse, very coarse, and extra-coarse grits. The sanding disc is a precut and center-punched disc with a diameter to fit the tool being used. Sanding discs are also made in a variety of grit sizes. They are typically made of aluminum oxide abrasive on a paper backing coated with a pressure-sensitive adhesive that sticks to the surface of the disc wheel.

WARNING!

When using a belt-disc sander, note the direction that the disc is spinning. Always move stock on the downspin side of the disc so that the motion of the disc keeps the stock on the table. Otherwise, the stock could become airborne and cause injury.

Rules for the safe use of a belt-disc sander are:

- Never leave a sander until it has stopped.
- Never make adjustments while the sander is running.
- Keep your fingers clear of the sanding surfaces.
- Hold the stock firmly with both hands while feeding it into the sander.
- Do not force the stock against the abrasive surface.
- Do not use plastics or other materials that may damage the sanding surfaces.
- Do not attempt to sharpen tools on a sander.
- Never wear gloves while using a belt-disc sander; they may become caught in the rotating mechanism and cause severe injury.

3.6.2 Portable Sanders/Grinders

There are three basic types of portable sanders/grinders commonly used by carpenters: pad sanders, belt sanders, and rotary disc sander/grinders. Half-sheet and palm sanders (*Figure 24*) are finish sanders used for lighter and finer sanding. They are vibrating pad sanders with either a back-and-forth (oscillating) motion or a circular (orbiting) motion. An oscillating motion sander is good for fine, with-the-grain finish sanding. The orbital motion sander circles across the grain for faster stock removal. Half-sheet sanders have flat, rectangular pads to which a half sheet of sandpaper is attached. They work well on large, flat surfaces. Palm sanders, so named because they fit comfortably in the palm of the hand, are good for finish sanding on both flat and rounded surfaces.

Demonstrate rough and fine sanding and shaping using a belt-disc sander. Emphasize safety precautions.

Hand out Worksheet 27103-1 and Job Sheet 27103-10. Under your supervision, have the trainees complete the belt-disc sander safety test before performing the related tasks on the Job Sheet. Note the proficiency of each trainee on his or her Job Sheet and Skill Test Record.

Show portable sanders with various sanding sheets.

Portable belt sanders (*Figure 25*) are high-speed, flat surface sanders used mainly for heavy-duty rough sanding such as when leveling and smoothing large, rough-sawed knotty or warped boards. Portable belt sanders use a continuous-loop abrasive belt installed over two drum-like rollers that control the belt travel. The back roller is powered by the motor; the front roller is spring-loaded to the correct belt tension. The size of a belt sander is usually identified by the width of its sanding belt (typically 3" or 4").

Rules for the safe use of a portable belt sander are:

- Don't wear loose clothing or jewelry.
- Always wear a face shield and safety glasses.
- Make sure the power switch is turned off before plugging in the sander.
- Hold the sander firmly with both hands.
- Always start the sander before making contact with the surface of the material.

- Be sure the material to be sanded is properly secured.
- Keep the power cord away from the sander belt.
- Be sure the sander has stopped running before putting it down.

A rotary grinder-sander (*Figure 26*), also called an *angle grinder,* is a high-speed tool used for rapid shaping and/or sanding of wood surfaces and board edges or ends. With the proper grinding wheel, wire brush, etc., installed on the tool, it can also be used for:

- Grinding and cutting metal
- Grinding and cutting concrete
- Polishing and buffing
- Removing rust and scale from steel
- Wire brushing

PALM SANDER   103F24A.EPS

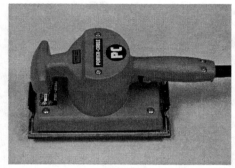

HALF-SHEET SANDER   103F24B.EPS

Figure 24 ◆ Palm and half-sheet sanders.

DUST BAG

BELT TRACKING ADJUSTMENT KNOB

BELT GUARD

103F25.EPS

Figure 25 ◆ Belt sander.

Instructor's Notes:

_____

_____

_____

_____

_____

_____

_____

_____

Sandpaper

There isn't really any sand in sandpaper. Instead, a variety of different types of natural and synthetic abrasives are adhered to various backing materials. In ancient times, sharkskin was used. Today, the mineral garnet is commonly used as a natural abrasive. Aluminum oxide is a man-made abrasive that is perhaps the most popular for wood sanding. Just as the construction of sandpaper varies, so does the actual size of the abrasive particles, referred to as *grit*. Sandpaper is identified by its grit number. Larger particles make a very coarse, or rough, sandpaper and smaller particles will produce a fine surface. The smaller the number, the more coarse the sandpaper. Common grit numbers used for rough sanding are 50, 60, and 80, while 120 and 150 are primarily used for basic finish work.

Assign reading of Sections 3.7.0–3.9.0 for the next class session.

103F26.EPS

Figure 26 ◆ Rotary grinder-sander.

Rules for the safe use of a rotary sander/angle grinder are:

- Be sure the grinding head is in good condition and has been properly secured.
- Wear gloves when appropriate.
- Use the proper grinding head for the material being ground.
- Be sure the grinder head has stopped turning before putting the grinder down.
- Never grind and weld on the same material at the same time.

The sanding sheets, belts, and discs used with portable power sanders are made of flint, garnet, silicon carbide, or aluminum oxide materials. Silicon carbide and aluminum oxide are recommended for most sanding operations. These abrasive materials are divided into fine, medium, and coarse grades. These grades are further divided into varying grit numbers, which are stamped on the back of the sheet, belt, or disc. The lower the number, the rougher the sandpaper will be.

3.7.0 Drill Press

Drill presses (*Figure 27*) are versatile tools used for various tasks, including:

- Countersinking
- Drilling
- Sanding
- Mortising
- Cutting holes with a hole saw

Drill presses consist of an electric drill and work table combined into one unit. Both floor and benchtop models are available. The throat capacity of the drill press (the distance between its rear

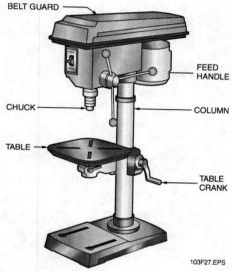

BELT GUARD

FEED HANDLE

CHUCK

COLUMN

TABLE

TABLE CRANK

103F27.EPS

Figure 27 ◆ Drill press.

Show Transparency 20 (Figure 27).

Demonstrate drilling and boring using a drill press. Emphasize safety precautions.

Hand out Worksheet 27103-1 and Job Sheet 27103-11. Under your supervision, have the trainees complete the drill press safety test prior to performing the related tasks on the Job Sheet. Note the proficiency of each trainee on his or her Job Sheet and Skill Test Record.

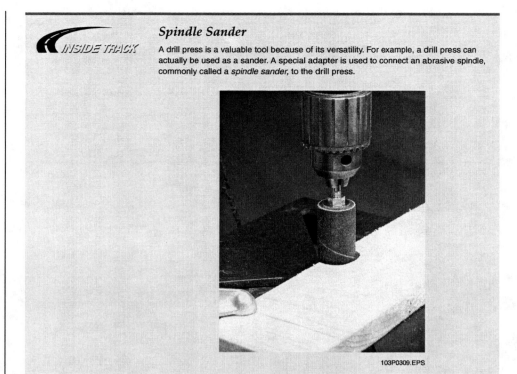

INSIDE TRACK

Spindle Sander

A drill press is a valuable tool because of its versatility. For example, a drill press can actually be used as a sander. A special adapter is used to connect an abrasive spindle, commonly called a *spindle sander*, to the drill press.

103P0309.EPS

post and the center of the bit) determines the maximum size of the workpiece it can accept. The drill table can be raised and lowered, and on some models it can also be tilted for drilling angled holes. The motor mounts on an adjustable bracket and is controlled by a belt and pulley drive. The belt is shifted between different grooves on the pulley to vary the speed of drilling. A chart is usually mounted on the drill press that gives guidelines for the approximate speed to use for different drilling operations.

Drill bits are installed or removed using a chuck key to tighten or loosen the chuck. For drilling, the drill is lowered into the stock by means of a feed handle, which returns to its original position when released. Operations such as sanding and mortising with a drill press require the use of special accessories made for that purpose.

Rules for the safe use of a drill press are:

- Never leave a drill press until the chuck has stopped.

- Always remove the chuck key before starting the drill.

- Do not make adjustments while the drill is running.

- Secure the lock before drilling.

- Use the correct speed for the job being performed.

- Do not force the drill into the stock.

- Never wear gloves when operating a drill press.

- Be sure the drill or tool is secure in the chuck before starting the drill.

- Even if the power has already been turned off, do not attempt to stop the drill by grabbing the chuck; it may still have enough momentum to cause serious injury. Always wait for all moving parts to come to a complete stop before touching them.

Instructor's Notes:

_____

_____

_____

_____

_____

_____

_____

_____

3.8.0 Bench Grinders

Bench grinders (*Figure 28*) are used for sharpening tools, grinding and shaping steel, and grinding burrs and rough places from steel. They normally come equipped with two grinding wheels (usually of different grits), adjustable tool rests, guards, and safety eye shields. Models differ mainly in the diameter of the grinding wheel size, with 8" wheels being the most common. Bench grinder wheels vary in materials. Stone wheels made in different grits are used for sharpening tools and grinding and shaping steel. Wire wheels are used to strip and remove rust from metal objects. Cloth buffing wheels are used for applying a polished finish to metal surfaces.

Rules for the safe use of a bench grinder are:

- Always use the tool rest.
- Be sure the tool rest is adjusted properly.
- Never use grinders without safety glasses.
- Never leave the grinder until the wheels have stopped turning.
- Do not use a grinder for woods or plastics.
- Never wear gloves when using a bench grinder.
- Never use a grinder with a cracked or damaged wheel.
- Never force the stock into the grinding wheel.
- Never use a grinding wheel to remove paint.
- Always set the stock on the tool rest.
- Never grind on the side of a grinding wheel.

3.9.0 Routers/Laminate Trimmers

A standard power router (*Figure 29*) is a portable tool used to cut joints and patterns in wood. It has a high-speed motor that allows the tool to make clean, smooth cuts. The motor shaft powers a chuck (collet) in which a wide variety of specially-designed router bits and cutters can be installed.

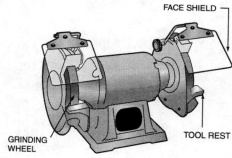

Figure 28 ◆ Bench grinder.

FACE SHIELD

GRINDING WHEEL

TOOL REST

103F28.EPS

Figure 29 ◆ Power router.

103F29.EPS

The motor's vertical position adjusts up or down in order to change the depth of the cut.

Router bits and cutters are made of carbide-tipped steel or high-speed steel. High-speed steel bits are suited for most uses in hard and soft woods. Carbide-tipped bits are good for plastic laminates and plywood. With different bits and available accessories, the router can be used to make many different types of cuts, including:

- Shaping edges
- Rabbeting
- Beveling
- Dovetailing
- Dadoing, fluting, and reeding
- Mortising

A laminate trimmer (*Figure 30*) is a specialized type of router used to trim and shape the edges of plastic laminate materials, such as those used for countertops. It can be used for bevel trimming, flush trimming, and cutting obtuse and acute angles.

Rules for the safe use of a router or laminate trimmer are:

- Keep your hands clear of router bits.
- Hold the tool firmly.
- Be sure that the material is secured properly before routing.
- Approach the workpiece gently; never gouge the bit into the stock.
- Always use jigs and guides with the router; never route or plow freehand.
- Turn off the tool immediately after making the cut, then wait for the bit to stop before putting the tool down.
- Avoid making cuts too deep. Note that a deep cut can cause kickbacks.

Show Transparency 21 (Figure 28).

Demonstrate steel grinding and shaping using a bench grinder. Emphasize safety precautions.

Hand out Worksheet 27103-1. Under your supervision, have the trainees complete the bench grinder safety test before practicing with this tool. Note the proficiency of each trainee.

Show a router and laminate trimmer.

Demonstrate how to use a router and laminate trimmer. Emphasize safety precautions.

Hand out Worksheet 27103-1. Under your supervision, have the trainees complete the router and laminate trimmer safety test before practicing with these tools. Note the proficiency of each trainee.

Bits

Using a router on certain materials will cause excessive bit wear if the proper technique is not used. Particleboard, for example, will quickly wear out a router bit if you attempt to make a cut in one pass. To avoid this type of wear, make your cut in several passes. You will also run into problems if you haven't selected the correct bit for the type of cut you are making. Double-check to make sure you are using the correct bit for the application. The router shown here is being used to make a dado cut.

103P0310.EPS

Routing with a Straightedge

There are several ways to limit the amount of lateral (sideways) motion while using a router. One very effective way is to use a straightedge. A clamp holds the straightedge to the stock and provides a solid guide for the carpenter. When making the cut, keep the base of the router firmly in contact with the straightedge. The result will be a cut that is accurate and straight.

103F30.EPS

Figure 30 ◆ Laminate trimmer.

3.10.0 Portable Power Planes

Portable power planes (*Figure 31*) are excellent for fitting trim and framing members that have been nailed together. They are commonly used for:

- Straightening material
- Edge planing
- Chamfering
- Beveling

There are small block planes used for finish work and jointer planes for heavier work. Jointer planes can be equipped with adjustable fences in order to cut bevels or chamfers.

Rules for the safe use of a power plane are:

- Hold the plane firmly with both hands.
- Be sure the material is secure before planing.
- Always use a door jack when planing door edges.
- Do not try to remove too much material at one time.

3.28

Instructor's Notes:

_____

_____

_____

_____

_____

_____

_____

_____

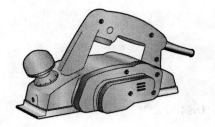

BEVEL

CHAMFER

103F31.EPS

Figure 31 ◆ Power plane.

3.11.0 Portable Drills and Screwguns

Portable drills (*Figure 32*) are made in a great number of styles and sizes, many of them cordless. Light-duty drills generally have a pistol grip. Heavy-duty drills may have a spade-shaped or D-shaped handle and a side handle to provide a secure grip for better control on large drilling jobs. Drill sizes are based on the diameter of the largest drill shank that will fit into the chuck of the drill, with ¼", ⅜", and ½" being common. Most drills have variable speed and reversible controls. Twist bits are used in electric drills to make holes in wood and metal. For boring larger holes in wood, spade and power bore bits are commonly used. Drills can be used for many operations, including:

- Boring and drilling
- Cutting holes with hole saws
- Mixing materials
- Driving screws

Power-driven or cordless screwdrivers (*Figure 33*), also called *screwguns*, are used for the rapid and efficient driving or removal of all types of screws, including wood, machine, and thread-cutting screws. Heavy-duty types can be used for driving and removing lag bolts, flooring screws, etc. Electric screwdrivers normally have an adjustable depth control to prevent overdriving the screws. Many have a clutch mechanism that disengages when the screw has been driven to a preset depth. Some power screwdrivers are designed to perform specific fastening jobs, such as fastening drywall to walls and ceilings.

Hammer drills (*Figure 34*) and rotary hammers are types of power drills. Equipped with a carbide-tipped percussion bit, they are used mainly to drill holes in concrete and other masonry materials. With different bits or cutters, they can also be used for:

- Drilling holes in masonry
- Setting anchors
- Performing light chipping work

The hammer drill is a lighter-duty tool used for drilling smaller holes; the rotary hammer is a

Show a portable drill, hammer drill, and screwgun and discuss their applications.

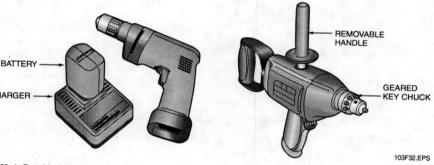

BATTERY

CHARGER

REMOVABLE HANDLE

GEARED KEY CHUCK

103F32.EPS

Figure 32 ◆ Portable drills.

heavier-duty tool used to drill larger holes. Both the hammer drill and rotary hammer operate with a dual action. They rotate and hammer at the same time, enabling them to drill holes much faster than can be done with a standard drill equipped with a masonry bit. Most models are reversible and can be easily switched to a standard rotary-action drill. They also have a depth gauge that can be set to control the depth of the hole being drilled. A concrete bit can be used with some models of

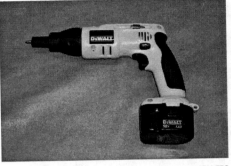

103F33.EPS

Figure 33 ◆ Power-driven screwdriver (drywall gun).

103F34.EPS

Figure 34 ◆ Hammer drill.

rotary hammers for chipping and edging concrete. Note that chipping hammers, similar to rotary hammers, are also made specifically for use in chipping masonry and concrete and for bush-hammering concrete.

Rules for the safe use of portable drills and hammer drills are:

* Hold the tool firmly.
* Always remove the chuck key before starting the drill.
* Be sure the drill or tool is secure in the chuck before starting the drill.
* Never attempt to stop the drill by taking hold of the chuck.
* Do not force the drill into the material; let the hammering action of the tool do the work.
* Be sure the material is properly secured before drilling.
* Never point the drill at anyone.

Rules for the safe use of an electric screwdriver (screwgun) are:

* Hold the tool firmly.
* Use the correct type and size bit for the screws being used.
* Set the gun for the proper depth.
* Never place the screw point against any part of your body.
* Never hold your hand behind the material into which you are driving the screw.

Rules for the safe use of rotary hammers and chipping hammers are:

* Do not force or overload the hammers.
* Use the correct tool for the job being done.
* Keep your hands and feet clear of the tool.
* Never point the hammer at anyone while it is running.
* Do not use the point or chisel for prying.
* Be sure the tools are properly secured in the hammers.

3.12.0 Power Metal Shears

Power metal shears (*Figure 35*) are used to make burr-free cuts in sheet metal, metal strips, special lightweight materials, and metal studs. They can cut straight lines, tight right and left curves, and

Hammer Drill Safety

Make sure that you have a firm grip on the side handle when using a hammer drill. You should never hold on to just the main handle. Use both hands to equalize the rotation of the drill. Most hammer drills have enough torque to break your wrist.

Instructor's Notes:

_____

_____

_____

_____

_____

_____

_____

Power Metal Shears

Power metal shears are commonly used for cutting light-gauge material such as chimney and roof flashing, making openings in ductwork, trimming metal roof panels, and clipping strapping and stud hangers. They are not appropriate for cutting heavier gauge metal like studs. Power metal shears will also leave rough edges and burrs so you should avoid using them for finish work.

103F35.EPS

Figure 35 ◆ Power shears.

round, square, and irregularly-shaped holes. A trigger-operated control allows variable speed operation. Some models have a head that can swivel 360°.

Rules for the safe use of power metal shears are:

- Keep your fingers away from the sharp edges of the tool and stock being cut.
- Never force the tool into the material.
- Do not attempt to cut material that is too thick.
- Keep the electric cord clear of the blade.

3.13.0 Pneumatic/Cordless Nailers and Staplers

 WARNING!
Pneumatic nailers and staplers are extremely dangerous. They are just as dangerous and deadly as a firearm. You should NEVER point these tools at anyone or in anyone's direction. When nailing or stapling, keep your hands away from the contact area.

Pneumatic nailers and staplers are fastening tools (*Figure 36*) powered by compressed air, which is fed to the tool through an air hose connected to an air compressor. These tools, known as *guns*, are widely used for quick, efficient fastening of framing, subflooring, sheathing, etc. Nailers and staplers are made in a variety of sizes to serve different purposes. Under some conditions, staples have some advantages over nails. For example, staples do not split wood as easily as a nail when driven near the end of a board. Staples are also excellent for fastening sheathing, shingles, building paper, and other materials because their two-legged design covers more surface area. However, both tools are sometimes used to accomplish the same fastening jobs. Nailers are typically used for:

- Applying sheathing
- Applying decking
- Applying roofing
- Installing framing
- Installing finish work
- Constructing cabinets

Staplers are typically used for:

- Applying sheathing
- Applying decking
- Applying roofing
- Installing insulation
- Installing ceiling tile
- Installing paneling

For some models of fasteners, the nails or staples come in strips and are loaded into a magazine, which typically holds 100 or more fasteners. Some tools have an angled magazine, which makes it easier for the tool to fit into tight places. Coil-fed models typically use a coil of 300 nails loaded into a circular magazine. Lightweight nailing guns can handle tiny finishing nails. Larger framing nailers can shoot smooth-shank nails of up to 3¼" in length.

Demonstrate how to use power metal shears. Emphasize safety precautions.

Hand out Worksheet 27103-1. Under your supervision, have the trainees complete the power metal shears safety test before practicing with this tool. Note the proficiency of each trainee.

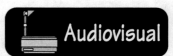

Show Transparency 23 (Figure 36).

Demonstrate how to use pneumatic nailers and staplers. Emphasize safety precautions.

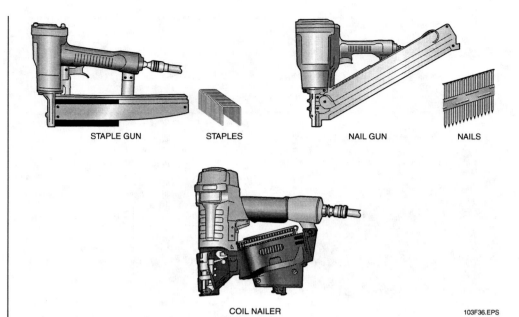

STAPLE GUN   STAPLES   NAIL GUN   NAILS

COIL NAILER

103F36.EPS

Figure 36 ◆ Pneumatic nailers and stapler.

Rules for the safe use of pneumatic fasteners are:

• Be sure all safety devices are in good shape and are functioning properly.
• Use the pressure specified by the manufacturer.
• NEVER disengage the firing safety mechanism.
• Always assume that the fasteners are loaded.
• Never point a pneumatic fastener at yourself or anyone else.
• If you see someone using these tools improperly, report it to your instructor or supervisor.
• Be sure the fastener is disconnected from the power source before making adjustments or repairs.
• Use caution when attaching the fastener to the air supply because the fastener may discharge.
• Never leave a pneumatic fastener unattended while it is still connected to the power source.
• Use nailers and staplers only for the type of work for which they were intended.
• Use only nails and staples designed for the fastener being used.
• Never use fasteners on soft or thin materials that nails may completely penetrate.

Nailers and staplers are also made in cordless models (*Figure 37*). These use a tiny internal combustion engine powered by a disposable fuel cell and a rechargeable battery. The action of the piston drives the fastener. A cordless stapler can drive about 2,500 staples with one fuel cell. A cordless framing nailer can drive about 1,200 nails

103F37.EPS

Figure 37 ◆ Impulse cordless nail gun.

Instructor's Notes:

_____

_____

_____

_____

_____

_____

_____

_____

on one fuel cell. The battery on a cordless tool must be periodically recharged. It pays to have a spare battery to use while one is being charged. Rules for the safe operation of a cordless nailer or stapler are basically the same as those described previously for pneumatic nailers and staplers.

3.14.0 Powder-Actuated Fastening Tools

A powder-actuated fastening tool (*Figure 38*) is a low-velocity fastening system powered by gunpowder cartridges, commonly called boosters.

Powder-actuated tools are used to drive specially-designed fasteners into masonry and steel. *Figure 39* shows the major components of a typical powder-actuated fastening gun.

Manufacturers use color-coding schemes to identify the strength of a powder load charge. It is extremely important to select the right charge for the job, so learn the color-coding system that applies to the tool you are using. *Table 1* shows an example of a color-coding system.

Assign reading of Section 3.14.0 for the next class session.

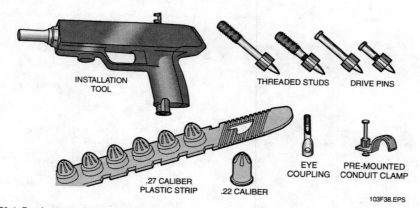

Figure 38 ◆ Powder-actuated fastening tool.

Show Transparencies 24 and 25 (Figures 38 and 39).

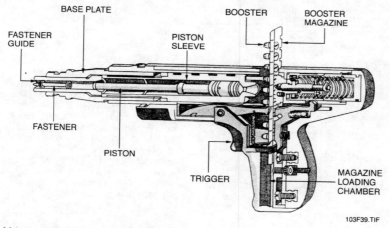

Figure 39 ◆ Major parts of a powder-actuated fastening tool.

Teaching Tip

Review the Case History on powder-actuated tools.

Safety

Review and emphasize the safety guidelines for powder-actuated tools and charges.

Demonstration

Demonstrate how to use a powder-actuated tool.

Laboratory

Hand out Worksheet 27103-1 and Job Sheet 27103-13. Under the supervision of a certified instructor, have the trainees complete the powder-actuated fasteners safety test before performing the related tasks on the Job Sheet. Note the proficiency of each trainee on his or her Job Sheet and Skill Test Record.

Powder-Actuated Tools

CASE HISTORY

A 22-year-old carpenter apprentice was killed when he was struck in the head by a nail fired from a powder-actuated tool in an adjacent room. The tool operator was attempting to anchor plywood to a hollow wall and fired the gun, causing the nail to pass through the wall, where it traveled nearly thirty feet before striking the victim. The tool operator had never received training in the proper use of the tool, and none of the employees in the area were wearing personal protective equipment.

The Bottom Line: Never use a powder-actuated tool to secure fasteners into easily penetrated materials; these tools are designed primarily for installing fasteners into masonry. The use of powder-actuated tools requires special training and certification. In addition, all personnel in the area must be aware that the tool is in use and should be wearing appropriate personal protective equipment.

Table 1   Powder Charge Color-Coding System

Power Level*	Color
1	Gray
2	Brown
3	Green
4	Yellow
5	Red
6	Purple

*From the least powerful (1) to the most powerful (6).

WARNING!

OSHA requires that all operators of powder-actuated tools be qualified and certified by the manufacturer of the tool. Certification cards must be carried whenever using the tool.

WARNING!

If the gun does not fire, hold it against the work surface for at least 30 seconds. Follow the manufacturer's instructions for removing the cartridge. Do not try to pry it out, because some cartridges are rim-fired and could explode.

Other rules for safely operating a powder-actuated tool are:

- Do not use a powder-actuated tool until you are certified.

- Follow all safety precautions in the manufacturer's instruction manual.
- Always wear safety goggles and a hard hat when operating a powder-actuated tool.
- Use the proper size pin for the job you are performing.
- When loading the driver, put the pin in before the charge.
- Use the correct booster (powder load) according to the manufacturer's instructions.
- Never hold the end of the barrel against any part of your body or cock the tool against your hand.
- Never hold your hand behind or near the material you are fastening.
- Do not shoot close to the edge of concrete.
- Never attempt to pry the booster out of the magazine with a sharp instrument.
- Always wear ear protection.
- Always hold the muzzle perpendicular (90°) to the work.

Summary

Knowing how to choose the proper tool for the job, use it skillfully, and keep it in good condition are essential carpentry skills. The use of power tools allows more work to be done in a shorter period of time. However, they can cause serious injury if operated incorrectly and/or unsafely. The carpenter must always be alert to the dangers of operating power tools. Always follow the safety rules for tool use given in this module, by your instructor or supervisor, and by the tool manufacturers in the operating/user's manual(s) for any tool you are using.

Performance Profile Test

Have each trainee complete to your satisfaction Performance Profile Tasks 1 on hand and power tool identification, 3 on portable power tool use, and 4 on stationary power tool use. Fill out Performance Profile Sheets for each trainee.

Instructor's Notes:

_____
_____
_____
_____
_____
_____
_____
_____

Review Questions

1. A tool that consists of a glass tube in a sleeve with hooks on either end is called a _____ level.
   a. builder's
   b. line
   c. laser
   d. water

2. A large hand plane used for rough work is the _____ plane.
   a. jack
   b. block
   c. smoothing
   d. jointer

3. The handsaw that would be used to make a scroll cut on a trim piece is the _____ saw.
   a. dovetail
   b. back
   c. coping
   d. compass

4. When using any power tool, be sure that it is _____ before performing maintenance or changing accessories.
   a. grounded
   b. turned off
   c. checked by your supervisor
   d. disconnected

5. In general, to properly care for power tools, you should _____.
   a. keep the machined surfaces cleaned and oiled
   b. inspect tools and accessories yearly
   c. report any unusual noises, sounds, or vibrations to your supervisor or instructor
   d. keep tools in your trunk or in the back of a pickup truck when not in use

6. The size of saw blade that is used most often with portable circular saws is _____.
   a. 7¼"
   b. 9"
   c. 10¼"
   d. 12"

7. Which of the following operations is not done with a table saw?
   a. Dadoing
   b. Crosscutting
   c. Making irregular cuts
   d. Mitering

8. When ripping with a table saw, make sure that the blade does not project more than _____ above the stock being cut.
   a. ⅛₆"
   b. ⅛"
   c. ¼"
   d. ⅜"

9. Which of the following operations is not done with a radial arm saw?
   a. Cutting molding
   b. Crosscutting
   c. Rabbeting
   d. Making pocket cuts

10. You should not use a radial arm saw for _____ if you have a table saw.
    a. crosscutting
    b. mitering
    c. ripping
    d. dadoing

11. To make pocket cuts in a material, you would use a _____ .
    a. circular saw
    b. reciprocating saw
    c. jigsaw
    d. all of the above

12. Which of the following operations is not done with a jointer?
    a. Mitering
    b. Chamfering
    c. Tapering
    d. Straightening

13. Making too deep a cut when using a jointer can cause _____.
    a. unwanted gouges in the stock
    b. a kickback of the stock
    c. the jointer cutting heads to speed up
    d. the jointer cutting heads to become dull

14. When using a shaper, never cut through _____ in the stock.
    a. loose knots
    b. cracks or splits
    c. warps
    d. both a and b

15. Which of the following operations is not done with a belt-disc sander?
    a. Fine sanding
    b. Shaping
    c. Cleaning paint from trim
    d. Rough sanding

Have the trainees complete the Review Questions and go over the answers prior to administering the Module Examination. Answer any questions the trainees may have.

Administer the Module Examination. Record the results on Craft Training Report Form 200 and submit the results to the Training Program Sponsor.

Performance Profile Test

Ensure that all Performance Profile Tests have been completed and Performance Profile Sheets for each trainee are filled out. Be sure to record the results of the Testing on Craft Training Report Form 200 and submit the results to the Training Program Sponsor.

16. Before starting a drill press, always _____.
    a. remove the chuck key
    b. raise the feed handle
    c. insert the chuck key
    d. check the position of the belt on the pulley

17. A laminate trimmer is a specialized type of _____.
    a. router
    b. planer
    c. saw
    d. shaper

18. Which of the following operations is not done with a pneumatic nailer?
    a. Applying roofing
    b. Installing paneling
    c. Applying sheathing
    d. Installing framing

19. When using a pneumatic fastener, use only nails and staples _____ for the fastener being used.
    a. long enough
    b. short enough
    c. designed
    d. color-coded

20. The device that holds the charge in a powder-actuated fastener gun is called the _____.
    a. piston
    b. base plate
    c. piston sleeve
    d. booster magazine

Instructor's Notes:

_____
_____
_____
_____
_____
_____
_____

John Hoerlein
Director of Education
Allied Construction Industries

Allied Construction Industries (ACI) is a construction trade association with over 460 member companies serving the Greater Cincinnati area. John Hoerlein began his career with ACI in 1996 as director of a commercial carpentry apprenticeship program. Today, John is the director of education for ACI, providing member companies with training in carpentry apprenticeship as well as construction business management, contractor marketing, superintendent training, and many other subjects. John also served as a subject matter expert for the NCCER Carpentry curriculum.

You chose college over the skilled trades as a route to construction. Why?

I just didn't know about all of the opportunities in the skilled trades. While I was growing up, too many people back then gave me the impression that I had to go to college in order to be successful. Had I known of all the opportunities for high pay, potential for advancement, and the level of satisfaction of the work back then, I might have chosen the skilled trades as a route into construction. In 1988, I completed my associate's degree in construction management from Cincinnati State College, and I completed my bachelor's degree in construction management in 1994. I went to school another three years, part-time, and graduated with a master's degree in business administration with a concentration in construction management in 1997.

What was the single greatest factor that contributed to your success?

Motivation. I found out that you can have all of the education in the world, but if you are not a motivated individual, you are only half as successful as you can be. I am never satisfied with being average, and I never wait around until someone tells me what to do. I always find things to do on my own, and I end up defining my own job description. The less an employer has to look over you, the more the employer wants you to work for them and the better they take care of you.

What advice do you have for new trainees?

Plan for the future. Set goals. Make sure everything you do is consistent with those goals. Don't look at your wallet today, look at it three or four years from now, ten years from now, and beyond. Those who are only interested in the highest pay they can make right now tend to make bad career choices. Determine what you want to do in life and always work towards it. Never give up.

When working with a particular company, I have found that you must look out for the company first. Be productive and work on making money for your employer, not just yourself. Those who look out for the company they work for as if it were their own are much more successful with that company. Learn to build trust by showing up to work every day, being on time, and striving to be better than average. Average workers make average pay and have average opportunities for advancement.

Finally, complete your carpentry training. It will be the stepping stone to much better things in the future.

Trade Terms Introduced in This Module

Bevel cut: A cut made across the sloping edge or side of a workpiece at an angle of less than 90°.

Boosters: Powder loads that are used to propel fasteners in powder-actuated tools.

Chamfering: Edging or end beveling that does not go all the way across the edge or end of the workpiece.

Compound cut: A simultaneous bevel and miter cut.

Crosscut: A cut made across the grain in lumber.

Dado: A rectangular groove or rabbet cut that is made partway through and across the grain in lumber.

Door jack: A holder or stand used to hold a door on edge while planing, routing, or installing hinges.

Dovetail joint: An interlocking joint with a triangular shape like that of a dove's tail.

Fluting: Decorative, closely spaced, concave channels or grooves in lumber such as on a column, post, or pilaster.

Kerf: The width of the cut made by a saw blade. It is also the amount of material removed by the blade in a through (complete) cut or the slot produced by the blade in a partial cut.

Kickback: A sharp, uncontrolled grabbing and throwing of the workpiece by a tool as it rejects material being forced into it.

Miter box: A device that is used to cut lumber at precise angles.

Miter cut: A cut made at the end of a piece of lumber at any angle other than 90°.

Mortise: A rectangular cutout or depression made in a piece of wood to receive something such as a hinge, lock, or tenon.

Pocket (plunge) cut: A cut made to remove an interior section of a workpiece or stock (such as in a countertop for a sink) or to make square or rectangular openings in floors and walls.

Push stick: A safety device used to push material through cutters to protect fingers from being cut.

Rabbet cut: A rectangular cut made along the edge or end of a board to receive another board that has been similarly cut.

Reeding: The process of cutting or routing a type of molding with closely-spaced, parallel, half-round, convex profiles.

Relief cut: A cut in stock that keeps a blade from binding.

Resawing: Cutting thick pieces of stock in the direction of the grain to make two or more thinner pieces.

Rip cut: A cut made in the direction of the grain in lumber.

Sheathing: Boards or sheet material attached to framing members of floors, walls, and roofs, and used as the nailing base for finish materials.

Tenon: A tongue that is cut on the end of a piece of wood and shaped to fit into a mortise.

True: Accurately shaped or fitted.

Instructor's Notes:

_____

_____

_____

_____

_____

_____

_____

Answers to Review Questions

Answer	Section
1. b	2.4.0
2. a	2.6.0
3. c	2.8.0
4. d	3.1.1
5. c	3.1.2
6. a	3.2.1
7. c	3.2.2
8. b	3.2.2
9. d	3.2.3
10. c	3.2.3
11. d	3.2.1, 3.2.7, 3.2.8
12. a	3.3.0
13. b	3.3.0
14. d	3.5.0
15. c	3.6.1
16. a	3.7.0
17. a	3.9.0
18. b	3.13.0
19. c	3.13.0
20. d	3.14.0, Figure 39

Additional Resources

This module is intended to present thorough resources for task training. The following reference works are suggested for further study. These are optional materials for continuing education rather than for task training.

Carpentry. Homewood, IL: American Technical Publishers.

Carpentry. New York, NY: Delmar Publishers.

Modern Carpentry. Tinley Park, IL: The Goodheart-Willcox Company, Inc.

Popular Science Complete Book of Power Tools. New York, NY: Workman Publishing Company.

Table Saw and Bench Power Tool Know How. Sears, Roebuck, and Company. Available at most retail stores.

Instructor's Notes:

_____

_____

_____

_____

_____

_____

_____

_____

The NCCER makes every effort to keep these textbooks up-to-date and free of technical errors. We appreciate your help in this process. If you have an idea for improving this textbook, or if you find an error, a typographical mistake, or an inaccuracy in the NCCER's Craft Training textbooks, please write us, using this form or a photocopy. Be sure to include the exact module number, page number, a detailed description, and the correction, if applicable. Your input will be brought to the attention of the Technical Review Committee. Thank you for your assistance.

Instructors – If you found that additional materials were necessary in order to teach this module effectively, please let us know so that we may include them in the Equipment/Materials list in the Instructor's Guide.

Write: Curriculum Revision and Development Department
National Center for Construction Education and Research
P.O. Box 141104, Gainesville, FL 32614-1104

Fax: 352-334-0932

E-mail: curriculum@nccer.org

Craft	Module Name	
Copyright Date	Module Number	Page Number(s)
Description		

(Optional) Correction

(Optional) Your Name and Address

Floor Systems

27104-01

MODULE OVERVIEW

This module introduces the carpentry trainee to residential floor systems. It covers the materials and general methods used to construct floor systems, with emphasis placed on the platform method of floor framing.

PREREQUISITES

Please refer to the Course Map in the Trainee Module. Prior to training with this module, it is suggested that the trainee shall have successfully completed the following modules:

Core Curriculum; Carpentry Level One, Modules 27101 through 27103

LEARNING OBJECTIVES

Upon completion of this module, the trainee will be able to:

1. Identify the different types of framing systems.
2. Read and understand drawings and specifications to determine floor system requirements.
3. Identify floor and sill framing and support members.
4. Name the methods used to fasten sills to the foundation.
5. Given specific floor load and span data, select the proper girder/beam size from a list of available girders/beams.
6. List and recognize different types of floor joists.
7. Given specific floor load and span data, select the proper joist size from a list of available joists.
8. List and recognize different types of bridging.
9. List and recognize different types of flooring materials.
10. Explain the purposes of subflooring and underlayment.
11. Match selected fasteners used in floor framing to their correct uses.
12. Estimate the amount of material needed to frame a floor assembly.
13. Demonstrate the ability to:
    - Lay out and construct a floor assembly
    - Install bridging
    - Install joists for a cantilever floor
    - Install a subfloor using butt-joint plywood/OSB panels
    - Install a single floor system using tongue-and-groove plywood/OSB panels

PERFORMANCE OBJECTIVES

Under the supervision of the instructor, the trainee should be able to:

1. Identify the different types of framing systems.
2. Read and understand drawings and specifications to determine floor system requirements.
3. Identify floor and sill framing and support members.
4. Lay out and construct a floor assembly.
5. Install bridging.
6. Install joists for a cantilever floor.
7. Install or describe how to install a subfloor using butt-joint plywood/OSB panels.
8. Install a single floor system using tongue-and-groove plywood/OSB panels.
9. Estimate the amount of materials needed to frame a floor assembly.
10. Given specific floor plan, load, and span data, select the proper girder/beam and joist size from a list of available girders/beams and joists.

NCCER STANDARDIZED CRAFT TRAINING PROGRAM

The National Center for Construction Education and Research (NCCER) provides a standardized national program of accredited craft training. Key features of the program include instructor certification, competency-based training, and performance testing. The program provides trainees, instructors, and companies with a standard form of recognition through a National Craft Training Registry. The program is described in full in the *Guidelines for Accreditation,* published by the NCCER. For more information on standardized craft training, contact the NCCER at P.O. Box 141104, Gainesville, FL 32614-1104, 352-334-0911, visit our Web site at www.NCCER.org, or e-mail info@NCCER.org.

HOW TO USE THIS ANNOTATED INSTRUCTOR'S GUIDE

Each page presents two sections of information. The larger section displays each page exactly as it appears in the Trainee Module. The narrow column ties suggested trainee and instructor actions to each page and provides icons to call your attention to material, safety, audiovisual, or testing requirements. The bottom of each page includes space for your notes.

 If you see the Teaching Tip icon, that means there is a teaching tip associated with this section. Also refer to any suggested teaching tips at the end of the module.

SAFETY CONSIDERATIONS

Ensure that the trainees are equipped with appropriate personal protective equipment.

PREPARATION

Before teaching this module, you should review the Module Outline, the Learning and Performance Objectives, and the Materials and Equipment List. Be sure to allow ample time to prepare your own training or lesson plan and gather all required equipment and materials.

MATERIALS AND EQUIPMENT LIST

Materials:
Transparencies
Markers/chalk
Floor adhesive (optional)
Beam material
Grout
Plywood or OSB butt-joint panels to cover floor area
Plywood or OSB (tongue-and-groove, 1¼") to cover floor area
Shim materials
Sill sealer
Steel bridging and instructions
Termite shield
2 × 6s for sills
2 × 10s for joists and headers
1 × 4s or 2 × 10s for bridging
8d box nails for bridging
8d box, screw, or ring shank nails for flooring
16d box nails for joists and headers
8d doublehead box nails
Copies of Worksheets 1 through 3*
Copies of Job Sheets 1 through 5*
Module Examinations*
Performance Profile Sheets*

Pictures, photographs, etc., showing braced, balloon, platform, and post-and-beam framing
Sets of building working drawings and specifications
Examples of several floor plans and specifications
Videotape (optional) *Framing Floors and Sills*
Pictures/photos of building damage that resulted from defective floor and sill framing (optional)

Equipment:
Overhead projector and screen
Whiteboard/chalkboard
Appropriate personal protective equipment
Tool box consisting of standard carpenter's hand tools
Chalkline
Electric drill and assorted drill and flat bits
Framing square
Level
100' tape
Power circular saw and extension cord
Reciprocating saw
Tin snips
Videocassette recorder (VCR)/TV set (optional)

*Packaged with this Annotated Instructor's Guide.

ADDITIONAL RESOURCES

This module is intended to present thorough resources for task training. The following reference works are suggested for both instructors and motivated trainees interested in further study. These are optional materials for continued education rather than for task training.

Carpentry, Leonard Koel. Homewood, IL: American Technical Publishers, 1997.

Carpentry, Gasper J. Lewis. Albany, NY: Delmar Publishers, 2000.

Modern Carpentry, Willis H. Wagner and Howard Bud Smith. Tinley Park, IL: The Goodheart-Willcox Company, Inc., 2000.

Framing Floors and Sills, videotape. Gainesville, FL: The National Center for Construction Education and Research.

TEACHING TIME FOR THIS MODULE

An outline for use in developing your lesson plan is presented below. Note that each Roman numeral in the outline equates to one session of instruction. Each session has a suggested time period of 2½ hours. This includes 10 minutes at the beginning of each session for administrative tasks and one 10-minute break during the session. Approximately 25 hours are suggested to cover *Floor Systems*. You will need to adjust the time required for hands-on activity and testing based on your class size and resources.

Topic   **Planned Time**

**Session I. Introduction; Methods of Framing Houses;
         Building Working Drawings and Specifications**

  A. Introduction

  B. Methods of Framing Houses   _____

    1. Platform Frame   _____

    2. Braced Frame   _____

    3. Balloon Frame   _____

    4. Post-and-Beam Frame   _____

  C. Building Working Drawings and Specifications   _____

    1. Architectural Drawings   _____

      a. Foundation Plan   _____

      b. Floor Plan   _____

      c. Section and Detail Drawings   _____

      d. Structural Drawings   _____

    2. Plumbing, Mechanical, and Electrical Plans   _____

    3. Reading Blueprints   _____

    4. Specifications   _____

Session II. The Floor System

  A. The Floor System

    1. Sills   _____

    2. Beams/Girders and Supports   _____

      a. Solid Lumber Girders   _____

      b. Built-Up Lumber Girders   _____

      c. Engineered Lumber Girders   _____

      d. Steel I-Beam Girders   _____

      e. Beam/Girder Supports   _____

    3. Floor Joists

      a. Notching and Drilling of Wooden Joists   _____

      b. Wood I-Beams   _____

      c. Trusses   _____

      d. Notching and Drilling of Wooden Joists   _____

  4. Bridging   _____

  5. Subflooring   _____

      a. Plywood Subfloors   _____

      b. Manufactured Board Panel Subfloors   _____

      c. Board Subfloors   _____

Session III. Laying Out and Constructing a Platform Floor Assembly; Laboratory

  A. Laying Out and Constructing a Platform Floor Assembly   _____

    1. Checking the Foundation for Squareness   _____

    2. Installing the Sill   _____

    3. Installing a Beam/Girder   _____

    4. Laying Out Sills and Girders for Floor Joists   _____

    5. Laying Out Joist Locations for the Partition and Floor Openings   _____

    6. Cutting and Installing Joist Headers   _____

    7. Installing Floor Joists   _____

    8. Framing Opening(s) in the Floor   _____

    9. Installing Bridging   _____

   10. Installing Subflooring   _____

  B. Laboratory   _____

  Hand out Worksheets 27104-1 and -2. Have the trainees complete the tasks on the Worksheets. Note the proficiency of each trainee.

Session IV. Laboratory

  A. Laboratory   _____

  Hand out Job Sheet 27104-1. Under your supervision, have the trainees perform the tasks on the Job Sheet. Note the proficiency of each trainee on his or her Job Sheet and Skill Test Record.

Session V. Laboratory

  A. Laboratory   _____

  Hand out Job Sheet 27104-2. Under your supervision, have the trainees perform the tasks on the Job Sheet. Note the proficiency of each trainee on his or her Job Sheet and Skill Test Record.

Session VI. Laboratory

  A. Laboratory   _____

  Hand out Job Sheet 27104-3. Under your supervision, have the trainees perform the tasks on the Job Sheet. Note the proficiency of each trainee on his or her Job Sheet and Skill Test Record.

Session VII. Laboratory

  A. Laboratory   _____

  Hand out Job Sheet 27104-4. Under your supervision, have the trainees perform the tasks on the Job Sheet. Note the proficiency of each trainee on his or her Job Sheet and Skill Test Record.

Session VIII. Installing Joists for Projections and Cantilever Floors; Laboratory

A. Installing Joists for Projections and Cantilever Floors

B. Laboratory

Hand out Job Sheet 27104-5. Under your supervision, have the trainees perform the tasks on the Job Sheet. Note the proficiency of each trainee on his or her Job Sheet and Skill Test Record.

Session IX. Estimating the Quantity of Floor Materials; Laboratory

A. Estimating the Quantity of Floor Materials

1. Sill, Sill Sealer, and Termite Shield
2. Beams/Girders
3. Joists and Joist Headers
4. Bridging
5. Flooring

B. Laboratory

Hand out Worksheet 27104-3. Have the trainees complete the tasks on the Worksheet. Record the proficiency of each trainee.

Session X. Guidelines for Determining Proper Girder and Joist Sizes; Laboratory; Module Examination and Performance Testing

A. Guidelines for Determining Proper Girder and Joist Sizes

1. Sizing Girders
2. Sizing Joists

B. Laboratory

Have the trainees select the proper girder/beam and joist size from the tables in the Trainee Module for various floor plans, floor loads, and span data.

C. Summary

1. Summarize module.
2. Answer answers.

D. Module Examination

1. Trainees must score 70% or higher to receive recognition from the NCCER.
2. Record the testing results on Craft Training Report Form 200 and submit the results to the Training Program Sponsor.

E. Performance Testing

1. Trainees must perform each task to the satisfaction of the instructor to receive recognition from the NCCER.
2. Record the testing results on Craft Training Report Form 200 and submit the results to the Training Program Sponsor.

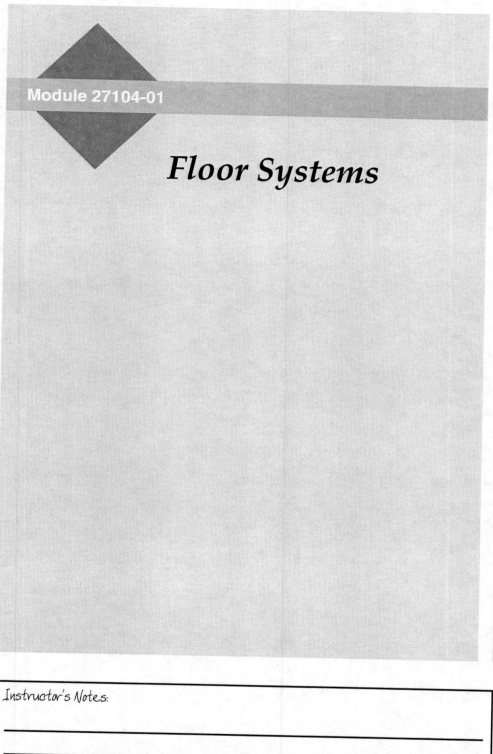

Module 27104-01

Floor Systems

Instructor's Notes:

_____

_____

_____

_____

_____

_____

_____

_____

Course Map

This course map shows all of the modules in the first level of the Carpentry curriculum. The suggested training order begins at the bottom and proceeds up. Skill levels increase as a trainee advances on the course map. The training order may be adjusted by the local Training Program Sponsor.

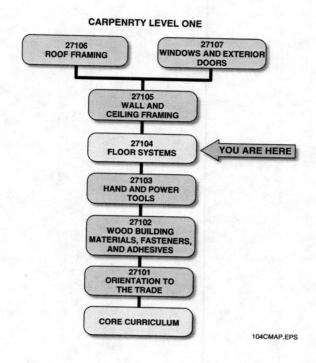

CARPENRTY LEVEL ONE

104CMAP.EPS

Assign reading of Module 27104.

1.0.0   INTRODUCTION 4.1
2.0.0   METHODS OF FRAMING HOUSES 4.1
    2.1.0   Platform Frame 4.2
    2.2.0   Braced Frame 4.3
    2.3.0   Balloon Frame 4.3
    2.4.0   Post-and-Beam Frame 4.4
3.0.0   BUILDING WORKING DRAWINGS AND SPECIFICATIONS 4.6
    3.1.0   Architectural Drawings 4.7
    3.1.1   Foundation Plan 4.7
    3.1.2   Floor Plan 4.7
    3.1.3   Section and Detail Drawings .4.7
    3.1.4   Structural Drawings 4.9
    3.2.0   Plumbing, Mechanical, and Electrical Plans 4.10
    3.3.0   Reading Blueprints 4.10
    3.4.0   Specifications 4.11
4.0.0   THE FLOOR SYSTEM 4.11
    4.1.0   Sills 4.11
    4.2.0   Beams/Girders and Supports 4.14
    4.2.1   Solid Lumber Girders 4.14
    4.2.2   Built-Up Lumber Girders 4.14
    4.2.3   Engineered Lumber Girders .4.14
    4.2.4   Steel I-Beam Girders 4.15
    4.2.5   Beam/Girder Supports 4.15
    4.3.0   Floor Joists 4.18
    4.3.1   Notching and Drilling of Wooden Joists 4.21
    4.3.2   Wood I-Beams 4.21
    4.3.3   Trusses 4.22
    4.4.0   Bridging 4.23
    4.5.0   Subflooring 4.23
    4.5.1   Plywood Subfloors 4.25
    4.5.2   Manufactured Board Panel Subfloors 4.25
    4.5.3   Board Subfloors 4.25
5.0.0   LAYING OUT AND CONSTRUCTING A PLATFORM
        FLOOR ASSEMBLY 4.27
    5.1.0   Checking the Foundation for Squareness 4.27
    5.2.0   Installing the Sill 4.27
    5.3.0   Installing a Beam/Girder 4.29
    5.4.0   Laying Out Sills and Girders for Floor Joists 4.31
    5.5.0   Laying Out Joist Locations for the Partition
        and Floor Openings 4.32

5.6.0	Cutting and Installing Joist Headers	4.32
5.7.0	Installing Floor Joists	4.33
5.8.0	Framing Opening(s) in the Floor	4.33
5.9.0	Installing Bridging	4.35
5.10.0	Installing Subflooring	4.35

6.0.0 INSTALLING JOISTS FOR PROJECTIONS AND CANTILEVERED FLOORS 4.38

7.0.0 ESTIMATING THE QUANTITY OF FLOOR MATERIALS ... 4.38

7.1.0	Sill, Sill Sealer, and Termite Shield	4.38
7.2.0	Beams/Girders	4.38
7.3.0	Joists and Joist Headers	4.38
7.4.0	Bridging	4.38
7.5.0	Flooring	4.40

8.0.0 GUIDELINES FOR DETERMINING PROPER GIRDER AND JOIST SIZES 4.40

| 8.1.0 | Sizing Girders | 4.41 |
| 8.2.0 | Sizing Joists | 4.41 |

SUMMARY 4.43

REVIEW QUESTIONS 4.44

GLOSSARY 4.47

ANSWERS TO REVIEW QUESTIONS 4.48

ADDITIONAL RESOURCES 4.49

Figures

Figure 1	Platform framing	4.2
Figure 2	Balloon framing	4.4
Figure 3	Post-and-beam framing	4.5
Figure 4	Typical format of a working drawing set	4.7
Figure 5	Typical section drawing	4.8
Figure 6	Typical platform frame floor system	4.12
Figure 7	Typical sill installation	4.12
Figure 8	Typical sill anchor strap	4.13
Figure 9	Types of girders	4.14
Figure 10	Typical methods of supporting girders	4.16
Figure 11	Typical post anchors and caps	4.16
Figure 12	Example of column spacing	4.17
Figure 13	Post or pier support of a girder at the foundation wall	4.17
Figure 14	Girder pocket and girder hanger girder supports	4.18
Figure 15	Floor joists	4.19
Figure 16	Methods of joist framing at a girder	4.19
Figure 17	Typical types of joist hangers	4.20

Instructor's Notes:

_____

_____

_____

_____

_____

_____

_____

_____

Figure 18   Notching and drilling of wooden joists 4.21
Figure 19   I-beams 4.22
Figure 20   Typical floor system constructed with engineered I-beams
             (second floor shown) 4.22
Figure 21   Typical trusses 4.23
Figure 22   Typical floor system constructed with trusses 4.24
Figure 23   Types of bridging 4.24
Figure 24   Subflooring installation 4.24
Figure 25   Checking the foundation for squareness 4.27
Figure 26   Inside edges of sill plates marked on the top of the
             foundation wall 4.28
Figure 27   Square lines across the sill to locate the anchor bolt hole 4.29
Figure 28   Example girder and support column data 4.30
Figure 29   Marking the sill for joist locations 4.31
Figure 30   Double joists at a partition that runs parallel to the joists 4.32
Figure 31   Box sill installed on foundation 4.33
Figure 32   Installing joists 4.34
Figure 33   Floor opening construction 4.34
Figure 34   Framing square used to lay out bridging 4.35
Figure 35   Installing butt-joint floor panels 4.36
Figure 36   Installing tongue-and-groove floor panels .4.37
Figure 37   Cantilevered joists 4.39
Figure 38   Determining floor system materials 4.40
Figure 39   Sizing girders 4.41
Figure 40   Floor loads 4.42
Figure 41   Sizing joists 4.42

Tables

Table 1   Guide to APA Performance-Rated Plywood Panels 4.26
Table 2   Wood Cross Bridging Multiplication Factor 4.40
Table 3   Safe Girder Loads 4.42
Table 4   Safe Joist Spans 4.43

Floor Systems

Materials

Ensure you have everything required to teach the course. Check the Materials and Equipment List at the front of this Instructor's Guide.

Objectives

Upon completion of this module, the trainee will be able to:

1. Identify the different types of framing systems.
2. Read and understand drawings and specifications to determine floor system requirements.
3. Identify floor and sill framing and support members.
4. Name the methods used to fasten sills to the foundation.
5. Given specific floor load and span data, select the proper girder/beam size from a list of available girders/beams.
6. List and recognize different types of floor joists.
7. Given specific floor load and span data, select the proper joist size from a list of available joists.
8. List and recognize different types of bridging.
9. List and recognize different types of flooring materials.
10. Explain the purposes of subflooring and underlayment.
11. Match selected fasteners used in floor framing to their correct uses.
12. Estimate the amount of material needed to frame a floor assembly.
13. Demonstrate the ability to:
    - Lay out and construct a floor assembly
    - Install bridging
    - Install joists for a cantilever floor
    - Install a subfloor using butt-joint plywood/OSB panels
    - Install a single floor system using tongue-and-groove plywood/OSB panels

Prerequisites

Successful completion of the following Task Modules is recommended before beginning study of this Task Module: Core Curriculum; Carpentry Level One, Modules 27101 through 27103.

Required Trainee Materials

1. Trainee Task Module
2. Appropriate personal protective equipment

1.0.0 ◆ INTRODUCTION

This module briefly introduces the different methods used for framing buildings. Some of the types of framing described are rarely used today. They are seen in existing installations, however, so knowledge about them is helpful, especially when remodeling. The remainder of the module describes floor framing with an emphasis on the platform method of framing. Included are descriptions of the materials and general methods used to construct floors. Proper construction of floors is essential because, no matter how well the foundation of a building is constructed, a structure will not stand if the floors and sills are poorly assembled.

2.0.0 ◆ METHODS OF FRAMING HOUSES

Various areas of the country have had different methods of constructing a wood frame dwelling. This variation in methods can be attributed to economic conditions, availability of material, or climate in different parts of the country. We will discuss the various types of framing and show the approved method of construction today. Structures that are framed and constructed entirely

Audiovisual

Show Transparency 1, Course Objectives.

Show Transparency 2, Performance Profile Tasks.

Classroom

Identify the four basic framing methods.

Teaching Tip

Acquire photographs or drawings from architectural firms or building material manufacturers that show specific types of building construction in various stages of completion. Pass them around for the trainees to examine.

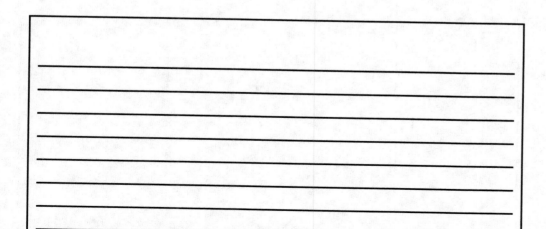

Describe platform and brace framing and identify the framing members. Point out the settling (shrinkage) problem with platform framing.

Show Transparency 3 (Figure 1).

of wood above the foundation fall into several classifications:

- Platform frame (also known as *western* and *box frame*)
- Braced frame
- Balloon frame
- Post-and-beam frame

2.1.0 Platform Frame

The platform frame, sometimes called the *western frame,* is used in most modern residential and light commercial construction. In this type of construction, each floor of a structure is built as an individual unit (*Figure 1*). The subfloor is laid in place prior to the exterior walls being put in place. The

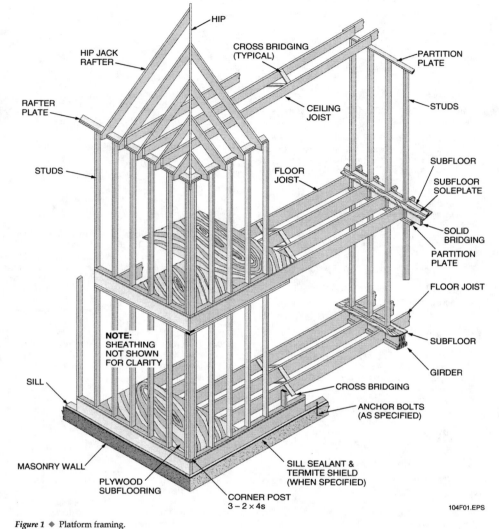

Figure 1 ◆ Platform framing.

Instructor's Notes:

Platform Framing

Explain that terms shown in bold (blue) are defined in the Glossary at the back of this module.

soleplate and top plate are nailed to the studs. Window openings are constructed and put in place, and studs are notched in order to let in a 1 × 4 diagonal brace.

In western or platform frame construction, the walls are built lying on the subfloor. Once a section or a part of a wall is complete, it is then lifted into place and nailed to the floor system and another plate is added to the top plate. This will tie in the second floor system. This method of construction allows workers to work safely on the floor while constructing the wall systems.

In the platform frame, there is a relatively large amount of settling caused by the shrinkage of a large number of horizontal, loadbearing frame members. Settling can result in various problems such as cracked plaster, cracked wallboard joints, uneven ceilings and floors, ill-fitting doors and windows, and nail pops (nails that begin to protrude through the wallboard).

2.2.0 Braced Frame

In early times, the brace frame method of construction was frequently used because most bearing lumber was in the vertical position. Very little shrinkage occurred in this type of construction.

In this method of construction, the framework does not rely on sheathing for rigidity. The framework, which in part had posts for corners and beams or girders, was mortised and tenoned together and held in place by pins and dowels. Diagonal braces were added to give the framework rigidity, thus making it stable. Therefore, the sheathing or planking could run up the sides of

the structure vertically. One common type of structure built in this fashion was the barn.

2.3.0 Balloon Frame

Balloon frame construction is rarely used today because of lumber and labor costs, but a substantial number of structures built with this type of frame are still in use. In the balloon frame method of construction, the studs are continuous from the sill to the rafter plate (*Figure 2*). The wall studs and first floor joists rest on a solid sill (usually a 4 × 6), with the floor joists being nailed to the sides of the studs. The second floor joists rest on a horizontal 1 × 4 or 1 × 6 board called a *ribbon* that is installed in notches (let-ins) to the faces of the studs. Braces (usually 1 × 4) are notched in on a diagonal into the outside face of the studs. This method of construction makes the frame self-supporting; therefore, it does not need to rely on the sheathing for rigidity.

In the balloon frame, firestops must be installed in the walls in several locations. A firestop is an approved material used in the space between frame members to prevent the spread of fire for a limited period of time. Typically, 2 × 4 wood blocks are installed between the studs for this purpose.

One advantage of balloon framing is that the shrinkage of the wood framing members is low, thus helping to reduce settling. This is because wood shrinks across its width, but practically no shrinkage occurs lengthwise. This provides for high vertical stability, making the balloon frame adaptable for two-story structures.

Briefly describe older style balloon framing and identify the framing members.

Show Transparency 4 (Figure 2).

Balloon Framing

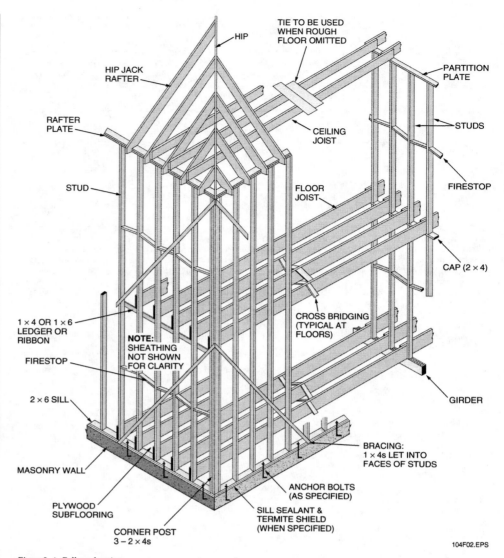

Figure 2 ◆ Balloon framing.

2.4.0 Post-and-Beam Frame

The post-and-beam method of framing floors and roofs has been used in heavy timber buildings for many years. It uses large, widely-spaced timbers for joists, posts, and rafters. Matched planks are often used for floors and roof sheathing (*Figure 3*).

Post-and-beam homes are normally designed for this method of framing. In other words, you cannot start a house under standard framing pro-

4.4

CARPENTRY LEVEL ONE—TRAINEE MODULE 27104

Instructor's Notes:

_____

_____

_____

_____

_____

_____

_____

_____

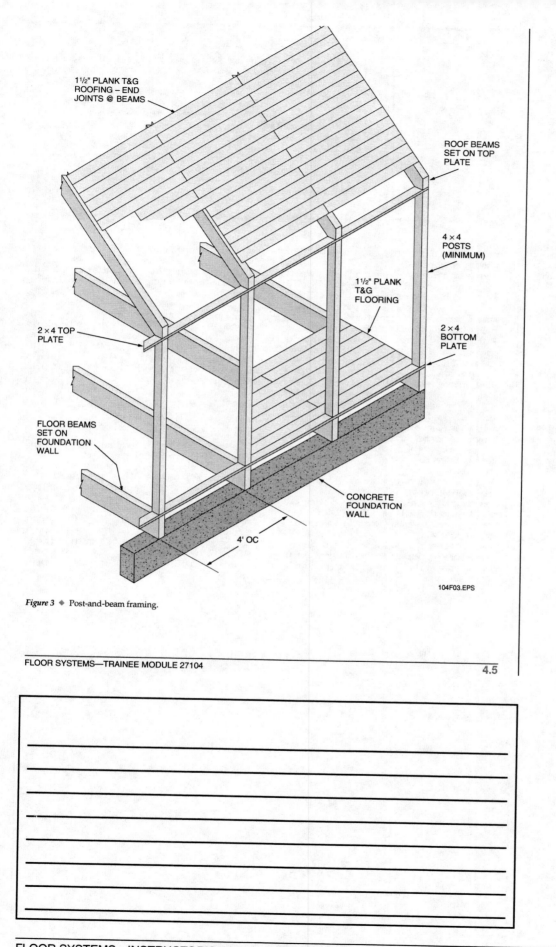

1½" PLANK T&G ROOFING – END JOINTS @ BEAMS

ROOF BEAMS SET ON TOP PLATE

4 × 4 POSTS (MINIMUM)

1½" PLANK T&G FLOORING

2 × 4 TOP PLATE

2 × 4 BOTTOM PLATE

FLOOR BEAMS SET ON FOUNDATION WALL

CONCRETE FOUNDATION WALL

4' OC

104F03.EPS

Figure 3 ◆ Post-and-beam framing.

Using illustrations or photographs, have the trainees identify the framing members for the various types of framing.

Briefly review the contents of a typical working drawing set.

Have each trainee complete to your satisfaction Performance Profile Task 1 on framing system identification. Fill out Performance Profile Sheets for each trainee.

Show Transparency 6 (Figure 4).

cedures and then alter the construction to post-and-beam framing.

In post-and-beam framing, plank subfloors or roofs are usually of 2" nominal thickness, supported on beams spaced up to 8' apart. The ends of the beams are supported on posts or piers. Wall spaces between posts are provided with supplementary framing as required for attachment of exterior and interior finishes. This additional framing also serves to provide lateral bracing for the building. Consider this versus the conventional framing that utilizes joists, rafters, and studs placed from 16" to 24" on center. Post-and-beam framing requires fewer but larger framing members spaced further apart. The most efficient use of 2" planks occurs when the lumber is continuous over more than one span. When standard lengths of lumber such as 12', 14', or 16' are used, beam spacings of 6', 7', or 8' are indicated. This factor has a direct bearing on the overall dimensions of the building.

If local building codes allow end joints in the planks to fall between supports, planks of random lengths may be used and beam spacing can be adjusted to fit the house dimensions. Windows and doors are normally located between posts in the exterior walls, eliminating the need for headers over the openings. The wide spacing between posts permits ample room for large glass areas. Consideration should be given to providing an adequate amount of solid wall siding. The siding must also provide ample and adequate lateral bracing.

A combination of conventional framing with post-and-beam framing is sometimes used where the two adjoin each other. On a side-by-side basis, no particular problems will be encountered.

Where a post-and-beam floor or roof is supported on a stud wall, a post should be placed under the end of the beam to carry the conventional load. A conventional roof can be used with post-and-beam construction by installing a header between the posts to carry the load from the rafters to the posts.

3.0.0 ◆ BUILDING WORKING DRAWINGS AND SPECIFICATIONS

Construction blueprints (working drawings) and related written specifications contain all the information and dimensions needed to build or remodel a structure. The interpretation of blueprints is critically important in the construction of floor systems, walls, and roof systems. It is recommended, therefore, that the material on blueprints studied earlier in the Core Curriculum module, *Introduction to Blueprints*, be reviewed. *Figure 4* shows the contents and sequence of a typical set of working drawings. The drawings that apply when building a floor system are briefly reviewed here.

Instructor's Notes:

_____
_____
_____
_____
_____
_____
_____
_____

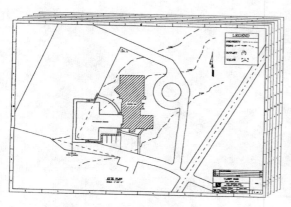

Figure 4 ◆ Typical format of a working drawing set.

TITLE SHEET(S)
ARCHITECTURAL DRAWINGS
• SITE (PLOT) PLAN
• FOUNDATION PLAN
• FLOOR PLANS
• INTERIOR/EXTERIOR ELEVATIONS
• SECTIONS
• DETAILS
• SCHEDULES

STRUCTURAL DRAWINGS
PLUMBING PLANS
MECHANICAL PLANS
ELECTRICAL PLANS

104F04.EPS

3.1.0 Architectural Drawings

The architectural drawings contain most of the detailed information needed by carpenters to build a floor system. The specific categories of architectural drawings commonly used include the following:

• Foundation plan
• Floor plan
• Section and detail drawings

3.1.1 Foundation Plan

The foundation plan is a view of the entire substructure below the first floor or frame of the building. It gives the location and dimensions of footings, grade beams, foundation walls, stem walls, piers, equipment footings, and foundations. Generally, in a detail view, it also shows the location of the anchor bolts or straps in foundation walls or concrete slabs.

3.1.2 Floor Plan

The floor plan is a cutaway view (top view) of the building, showing the length and breadth of the building and the layout of the rooms on that floor. It shows the following kinds of information:

• Outside walls, including the location and dimensions of all exterior openings
• Types of construction materials
• Location of interior walls and partitions
• Location and swing of doors
• Stairways
• Location of windows
• Location of cabinets, electrical and mechanical equipment, and fixtures
• Location of cutting plane line

3.1.3 Section and Detail Drawings

Section drawings are cutaway vertical views through an object or wall that show its interior makeup. They are used to show the details of construction and information about walls, stairs, and other parts of construction that may not show clearly on the plan. A section view is limited to the specific portion of the building construction that the architect wishes to clarify. It may be drawn on the same sheet as an elevation or plan view or it may appear on a separate sheet. *Figure 5* shows a typical example of a section drawing. Detail views are views that are normally drawn to a larger scale. They are often used to show aspects of a design that are too small to be shown in sufficient detail on a plan or elevation drawing. Like section drawings, detail drawings may be drawn on the same sheet as an elevation or plan drawing or may appear on a separate sheet in the set of plans.

Identify the categories of architectural drawings.

Emphasize the importance of reviewing all section and detail views for construction information.

Show Transparency 7 (Figure 5).

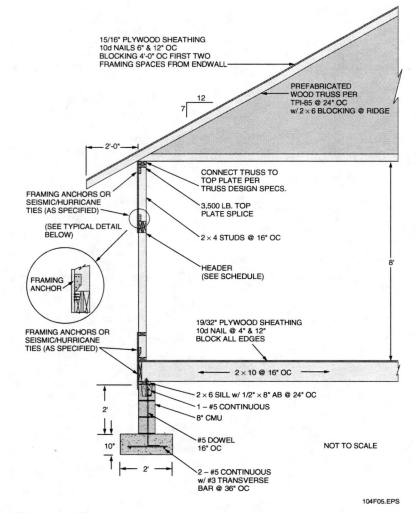

15/16" PLYWOOD SHEATHING
10d NAILS 6" & 12" OC
BLOCKING 4'-0" OC FIRST TWO
FRAMING SPACES FROM ENDWALL

PREFABRICATED
WOOD TRUSS PER
TPI-85 @ 24" OC
w/ 2 × 6 BLOCKING @ RIDGE

12
7

2'-0"

CONNECT TRUSS TO
TOP PLATE PER
TRUSS DESIGN SPECS.

FRAMING ANCHORS OR
SEISMIC/HURRICANE
TIES (AS SPECIFIED)

(SEE TYPICAL DETAIL
BELOW)

3,500 LB. TOP
PLATE SPLICE

2 × 4 STUDS @ 16" OC

FRAMING
ANCHOR

HEADER
(SEE SCHEDULE)

8'

FRAMING ANCHORS OR
SEISMIC/HURRICANE
TIES (AS SPECIFIED)

19/32" PLYWOOD SHEATHING
10d NAIL @ 4" & 12"
BLOCK ALL EDGES

2 × 10 @ 16" OC

2 × 6 SILL w/ 1/2" × 8" AB @ 24" OC

1 – #5 CONTINUOUS

2'

8" CMU

#5 DOWEL
16" OC

NOT TO SCALE

10"

2'

2 – #5 CONTINUOUS
w/ #3 TRANSVERSE
BAR @ 36" OC

104F05.EPS

Figure 5 ◆ Typical section drawing.

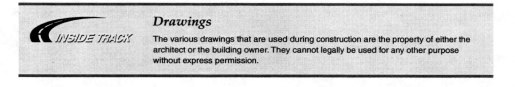

Drawings

The various drawings that are used during construction are the property of either the architect or the building owner. They cannot legally be used for any other purpose without express permission.

Instructor's Notes:

_____

_____

_____

_____

_____

_____

_____

_____

_____

Floor Plan

A typical floor plan provides a top view that details the layout of the rooms for each floor in the building.

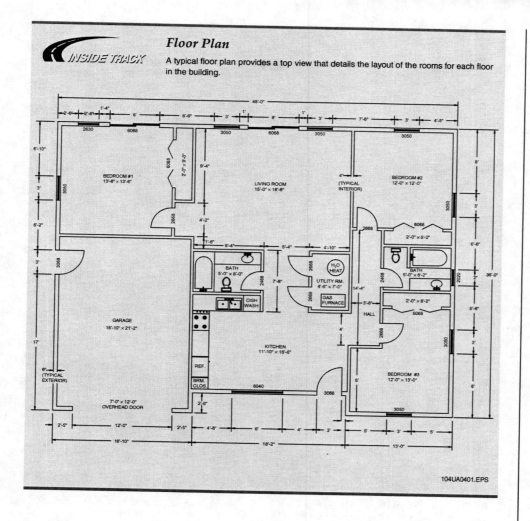

104UA0401.EPS

3.1.4 Structural Drawings

Structural drawings are created by a structural engineer and accompany the architect's plans. They are usually drawn for large structures such as an office building or factory. They show requirements for structural elements of the building including columns, floor and roof systems, stairs, canopies, bearing walls, etc. They include such details as:

- Height of finished floors and walls
- Height and bearing of bar joists or steel joists
- Location of bearing steel materials
- Height of steel beams, concrete planks, concrete Ts, and poured-in-place concrete
- Bearing plate locations
- Location, size, and spacing of anchor bolts
- Stairways

Explain the purpose of structural drawings and show an example.

FLOOR SYSTEMS—TRAINEE MODULE 27104

4.9

Dimension vs. Scale

The scale of a drawing refers to the amount or percentage that a document has been reduced in relation to reality (full scale). Specific dimensions on documents, however, should always take precedence over the scaled graphic representation (drawing) found on the plans. The drawing itself is meant to give a general idea of the overall layout. The dimensions found on the drawing are what you should use when planning and building.

Classroom

Explain the purpose of plumbing, mechanical, and electrical plans. Show an example of each.

Demonstrate how to read a typical drawing set. Point out the information required for floor systems.

3.2.0 Plumbing, Mechanical, and Electrical Plans

Plumbing plans show the size and location of water and gas systems if they are not included in the mechanical section. Mechanical plans show temperature control and ventilation equipment including ducts, louvers, and registers. Electrical plans show all electrical equipment, lighting, outlets, etc.

It is important to note that while carpenters usually work with architectural and structural drawings, there are useful notes and views on drawings found in other sections, especially the mechanical section. For example, typical items not found on architectural drawings but that may be found on mechanical drawings are exposed heating, ventilating, and air conditioning (HVAC) ductwork; heating convectors; and fire sprinkler piping. Any one or all of these items may require the carpenter to build special framing, so make a habit of reviewing all drawings. Also, make sure to coordinate any such work with the appropriate other trades to make sure that the proper framing is done to accommodate ductwork, piping, wiring, etc.

3.3.0 Reading Blueprints

The following general procedure is suggested as a method of reading any set of blueprints for understanding:

Step 1   Read the title block. The title block tells you what the drawing is about. It contains critical information about the drawing such as the scale, date of last revision, drawing number, and architect or engineer. If you have to remove a sheet from a set of drawings, be sure to fold the sheet with the title block facing up.

Step 2   Find the north arrow. Always orient yourself to the structure. Knowing where north is enables you to more accurately describe the locations of walls and other parts of the building.

Step 3   Always be aware that blueprints work together as a team. The reason the architect or engineer draws plans, elevations, and sections is that it requires more than one type of view to communicate the whole project. Learn how to use more than one drawing, when necessary, to find the information you need.

Step 4   Check the list of blueprints in the set. Note the sequence of the various types of plans. Some blueprints have an index on the front cover. Notice that the prints are broken into several categories:
  • Architectural
  • Structural
  • Mechanical
  • Electrical
  • Plumbing

Step 5   Study the site plan (plot plan) to observe the location of the building. Notice that the geographic location of the building may be indicated on the site plan.

Step 6   Check the foundation and floor plans for the orientation of the building. Observe the location and features of entries, corridors, offsets, and any special features.

Step 7   Study the features that extend for more than one floor, such as plumbing and vents, stairways, elevator shafts, heating and cooling ductwork, and piping.

Step 8   Check the floor and wall construction and other details relating to exterior and interior walls.

Step 9   Check the foundation plan for size and types of footings, reinforcing steel, and loadbearing substructures.

Step 10   Study the mechanical plans for the details of heating, cooling, and plumbing.

Step 11   Observe the electrical entrance and distribution panels, and the installation of the lighting and power supplies for special equipment.

Instructor's Notes:

_____

_____

_____

_____

_____

_____

_____

_____

Step 12 Check the notes on the various pages and compare the specifications against the construction details. Look for any variations.

Step 13 Thumb through the sheets of drawings until you are familiar with all the plans and structural details.

Step 14 Recognize applicable symbols and their relative locations in the plans. Note any special construction details or variations that will affect your job.

When you are building a floor system, the building plans (or specifications) should provide all the information you need to know about the floor system. Important information you should look for regarding the floor system includes:

- Type of wood or other materials used for sills, posts, girders, beams, joists, subfloors, etc.
- Size, location, and spacing of support posts or columns
- Direction of both joists and girders
- Manner in which joists connect to girders
- Location of any loadbearing interior walls that run parallel to joists
- Location of any toilet drains
- Rough opening sizes and locations of all floor openings for stairs, etc.
- Any cantilevering requirements
- Changes in floor levels
- Any special metal fasteners needed in earthquake areas
- Types of blocking or bridging
- Clearances from the ground to girder(s) and joists for floors installed over crawlspaces

3.4.0 Specifications

Written specifications are equally as important as the drawings in a set of plans. They furnish what the drawings cannot, in that they give detailed and accurate written descriptions of work to be done. They include quality and quantity of materials, methods of construction, standards of construction, and manner of conducting the work. Specifications will be studied in detail later in your training. The basic information found in a typical specification includes:

- Contract
- Synopsis of the work
- General requirements
- Owner's name and address
- Architect's name
- Location of structure
- Completion date
- Guarantees

- Insurance requirements
- Methods of construction
- Types and quality of building materials
- Sizes

4.0.0 ◆ THE FLOOR SYSTEM

Floor systems provide a base for the remainder of the structure to rest on. They transfer the weight of people, furniture, materials, etc., from the subfloor, to the floor framing, to the foundation wall, to the footing, then finally to the earth. Floor systems are constructed over basements or crawlspaces. Single-story structures built on slabs do not have floor systems; however, multi-level structures may have both a slab and a floor system. *Figure 6* shows a typical platform floor system and identifies the various parts.

4.1.0 Sills

Sills, also called *sill plates,* are the lowest members of a structure's frame. They rest horizontally on the foundation and support the floor joists. The foundation is the supporting portion of a structure below the first floor construction, including the footings. Sills serve as the attachment point to the concrete or block foundation for all of the other wood framing members. The sills provide a means of leveling the top of the foundation wall and also prevent the other wood framing lumber from making contact with the concrete or masonry, which can cause the lumber to rot.

Today, sills are normally made using a single layer of 2 × 6 lumber (*Figure 7*). Local codes normally require that pressure-treated lumber and/or foundation-grade redwood lumber be used for constructing the sill whenever the sill plate comes in direct contact with any type of concrete. However, where codes allow, untreated softwood can be used.

Sills are attached to the foundation wall using either anchor bolts (*Figure 7*) or straps (*Figure 8*) embedded in the foundation. The exposed portion of the strap-type anchor is nailed to the sill. Some types must be bent over the top of the sill, while others are nailed to the sides. The size, type, and spacing between the anchor bolts or straps used must be in compliance with local building codes. Their location and other related data are normally shown on the building blueprints.

In structures where the underfloor areas are used as part of the HVAC system, for storage, or as a basement, a glass-wool insulating material, called a *sill sealer* (*Figure 7*), should be installed to account for irregularities between the foundation wall and the sill. It seals against drafts, dirt, and

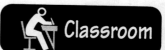

Classroom

Explain the purpose of specifications and show examples. Point out that specifications provide detailed information not given in the drawings.

Have the trainees practice using drawing sets and specifications to determine requirements for instructor-selected floor systems.

Homework

Assign reading of Sections 4.0.0–4.5.3 for the next class session.

Classroom

Explain the purpose of floor systems and identify each part of a typical platform floor system.

Performance Profile Test

Have each trainee complete to your satisfaction Performance Profile Task 2 on building drawings and specifications. Fill out Performance Profile Sheets for each trainee.

Audiovisual

Show Transparencies 8, 9, and 10 (Figures 6, 7, and 8).

Classroom

Describe sills and sill anchoring. Emphasize the need for foundation checks and the requirements for sealers, shields, and pressure-treated wood.

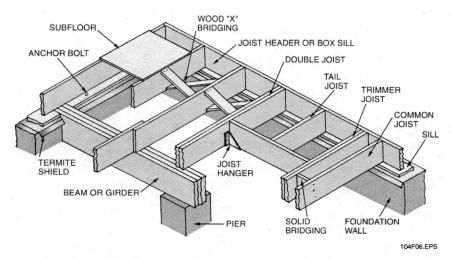

Figure 6 ◆ Typical platform frame floor system.

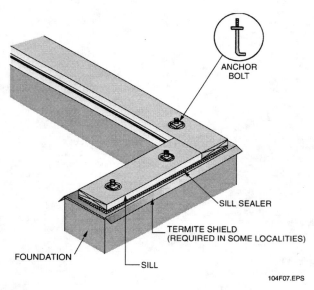

Figure 7 ◆ Typical sill installation.

Instructor's Notes:

_____

_____

_____

_____

_____

_____

_____

_____

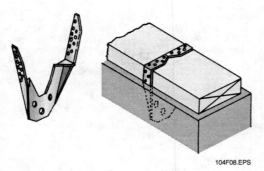

104F08.EPS

Figure 8 ◆ Typical sill anchor strap.

Sill Installation

Installing the sill plate is the first step in framing.

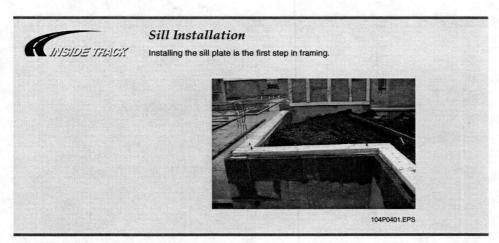

104P0401.EPS

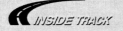

Foundation Checks

Always verify the foundation measurements and ensure that the foundation is square before you start framing a floor system. If the foundation is off by even a tiny amount, it can have a major impact on the framing.

Termite Shields

In areas where there is a high risk of termite infestation, a sheet metal termite shield should be installed below the sill. In some areas of the country, this is a code requirement. Termites live underground and come to the surface to feed on wood. They can enter through cracks in the masonry, tunnel through the hollow cells of concrete block, or build earthen tubes on the side of masonry to reach the wood.

insects. Sill sealer material is made in 6" wide, 50' rolls. Uncompressed, its thickness is 1". However, it can compress to as little as ½" when the weight of the structure is upon it. The sill sealer should be installed between the sill and the foundation wall, or between the sill and a termite shield (if used).

4.2.0 Beams/Girders and Supports

The distance between two outside walls is frequently too great to be spanned by a single joist. When two or more joists are needed to cover the span, support for the inboard joist ends must be provided by one or more beams, commonly called *girders*. Girders carry a very large proportion of the weight of a building. They must be well designed, rigid, and properly supported at the foundation walls and on the supporting posts or columns. They must also be installed so that they will properly support the joists. Girders may be made of solid timbers, built-up lumber, engineered lumber, or steel beams (*Figure 9*). Each type has advantages and disadvantages. Note that in some instances, precast reinforced concrete girders may also be used. A general procedure for determining how to size a girder is given later in this module.

4.2.1   Solid Lumber Girders

Solid timber girder stock used for beams is available in various sizes, with 4 × 6, 4 × 8, and 6 × 6

being typical sizes. If straight, large timbers are available, their use can save time by not having to make built-up girders. However, solid pieces of large timber stock are often badly bowed and can create a rise in the floor unless the crowns are pulled down. The crowns are the high points of the crooked edges of the framing members.

4.2.2   Built-Up Lumber Girders

Built-up girders are usually made using nominal 2" stock (2 × 8s, 2 × 10s) nailed together so that they act as one piece. Built-up girders have the advantage of not warping as easily as solid wooden girders and are less likely to have decayed wood in the center. The disadvantage is that a built-up girder is not capable of carrying the same load as an equivalent size solid timber girder. When constructing a built-up girder, the individual boards must be nailed together according to code requirements. Also, it is necessary to stagger the joints at least 4' in either direction. Construction of built-up girders is covered in more detail later in this module.

4.2.3   Engineered Lumber Girders

Laminated veneer lumber (LVL) and glue-laminated lumber (glulam) are engineered lumber products that are used for girders and other framing members. Their advantage is that they are stronger than the same size structural lumber;

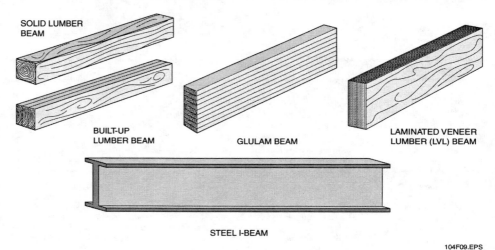

Figure 9 ◆ Types of girders.

Instructor's Notes:

_____

_____

_____

_____

_____

_____

_____

_____

I-Beams

The first plywood I-beam was created in 1969. In 1977, the first I-beam was created using laminated veneer lumber (LVL). This new construction offered superior strength and stability. In 1990, oriented strand board (OSB) web material, constructed of interlocking fibers, began to be used in I-beams, as shown here. OSB is less expensive than plywood and is not as prone to warping or cracking. Engineered wood products were once only available through a handful of companies that pioneered the industry. Today, engineered lumber and lumber systems are offered by a wide variety of companies.

104P0402.EPS

therefore, the same size piece can bear more weight. Or, looked at another way, a smaller-dimensioned piece of engineered lumber can bear equal weight. For a given length, the greater strength of engineered lumber products allows them to span a greater distance. Another advantage is that they are very straight with no crowns or warps.

LVL girders are made from laminated wood veneer like plywood. The veneers are laid up in a staggered pattern with the veneers overlapping to increase strength. Unlike plywood, the grain of each layer runs in the same direction as the other layers. The veneers are bonded with an exterior-grade adhesive, then pressed together and heated under pressure.

Glulam girders are made from lengths of solid, kiln-dried lumber glued together. They are commonly used where the beams are to remain exposed. Glulam girders are available in three appearance grades: industrial, architectural, and premium. For floor systems like those described in this module, industrial grade would normally be used because appearance is not a priority. Architectural grade is used where beams are exposed and appearance is important. Premium grade is used where the highest-quality appearance is needed. Glulam beams are available in

various widths and depths and in lengths up to 40' long.

4.2.4 Steel I-Beam Girders

Metal beams can span the greatest distances and are often used when there are few or no piers or interior supports in a basement. Also, they can span greater distances with smaller beam sizes, thereby creating greater headroom in a basement or crawlspace. For example, a 6" high steel beam may support the same load as an 8" or 10" high wooden beam. Two types of steel beams are available: standard flange (S-beam) and wide flange (W-beam). The wide beam is generally used in residential construction. Being metal, I-beams are more expensive than wood and are harder to work with. They are normally used only when the design or building code calls for it.

4.2.5 Beam/Girder Supports

Girders and beams must be properly supported at the foundation walls, and at the proper intervals in between, either by supporting posts, columns, or piers (*Figure 10*). Solid or built-up wood posts installed on pier blocks are commonly used to support floor girders, especially for floors built

Describe the various methods of supporting beams and girders. Point out the methods of tying the supports to the beams and girders.

Show Transparencies 12 and 13 (Figures 10 and 11).

over a crawlspace. Usually, 4 × 4 or 4 × 6 posts are used. However, all posts must be as wide as the girder. Where girder stock is jointed over a post, a 4 × 6 is normally required. To secure the wood posts to their footings, pieces of ½" reinforcing rod or iron bolts are often embedded in the support footings before the concrete sets. These project into holes bored in the bottoms of the posts. The use of galvanized steel post anchors is another widely used method of fastening the bottoms of wooden posts to their footings (*Figure 11*). The

tops of the posts are normally fastened to the girder using galvanized steel post caps. In addition to securing the post to the girder, these caps also provide for an even bearing surface.

Four-inch round steel columns filled and reinforced with concrete (*Figure 10*), called *lally columns*, are commonly used as support columns in floors built over basements. Some types of lally columns must be cut to the required height, while others have a built-in jackscrew that allows the column to be adjusted to the proper height. Metal

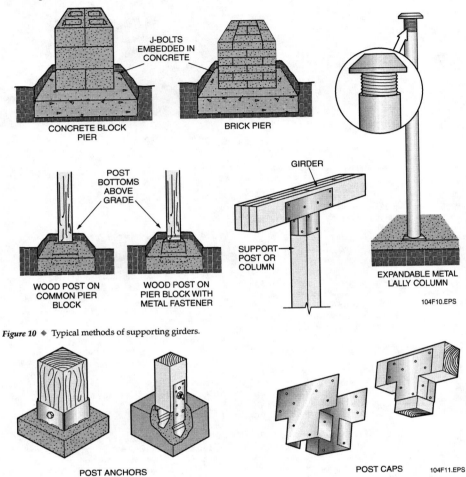

Figure 10 ◆ Typical methods of supporting girders.

Figure 11 ◆ Typical post anchors and caps.

CARPENTRY LEVEL ONE—TRAINEE MODULE 27104

Instructor's Notes:

_____

_____

_____

_____

_____

_____

_____

_____

CARPENTRY LEVEL ONE—INSTRUCTOR'S GUIDE MODULE 27104

plates are installed at the top and bottom of the columns to distribute the load over a wider area. The plates normally have predrilled holes so that they may be fastened to the girder.

Support piers made of brick or concrete block (*Figure 10*) are more difficult to work with because their level cannot be adjusted. The height of the related footings must be accurate so that when using 4" thick bricks or 8" tall blocks, their tops come out at the correct height to support the girder.

The spacing or interval required between the girder posts or columns is determined by local building codes based on the stress factor of the girder beam (i.e., how much weight is put on the girder beam). It is important to point out here that the farther apart the support posts or columns are spaced, the heavier the girder must be in order to carry the joists over the span between them. An example of a girder and supporting columns used in a 24' × 48' building is shown in *Figure 12*. In this example, column B supports one-half of the girder load existing between the building wall A and column C. Column C supports one-half of the girder load between columns B and D. Likewise, column D will share equally the girder loads with column C and the wall E.

As shown in *Figure 13*, support of girder(s) at the foundation walls can be done by constructing posts made from solid wood or piers made of concrete block or brick. Another widely used method is to construct girder (beam) pockets into the concrete or concrete block foundation walls (*Figure 14*). Provide steel reinforcement as required by the job specifications.

The specifications for girder pockets vary with the size of the girder being used. A rule of thumb

Explain beam support spacing and size requirements. Point out that wooden beams should not rest directly on concrete and that at least 4" at the ends of each beam should rest on wall supports.

Show Transparencies 14 through 16 (Figures 12 through 14).

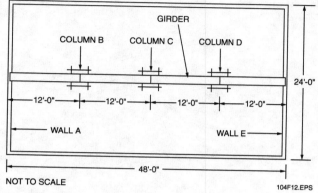

Figure 12 ◆ Example of column spacing.

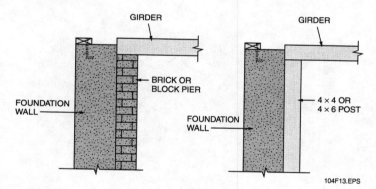

Figure 13 ◆ Post or pier support of a girder at the foundation wall.

FLOOR SYSTEMS—TRAINEE MODULE 27104

4.17

Explain joist sizing, placement, and support methods. Point out joist loading and deflection requirements. Emphasize that joists are doubled under partitions, around openings, and under points of extra load.

Show Transparencies 17 through 19 (Figures 15 through 17).

is that the pocket should be at least one inch wider than the beam and the beam must have at least 4" of bearing on the wall. Wooden girders placed in the pocket should not be allowed to come in direct contact with the concrete or masonry foundation. This is because the chemicals in the concrete or masonry can deteriorate the wood. The end of the wooden girder should sit on a steel plate that is at least ¼" thick. Some carpenters also use metal flashing to line the girder pocket to help protect the wood from the concrete. In some applications, any one of several types of galvanized steel girder hangers can be used to secure the girder to the foundation. *Figure 14* shows one common type.

It should be pointed out here that it is normal to use temporary supports, such as jacks and/or 2 × 4 studs nailed together with braces, to support the girder(s) while the floor is being constructed. After the floor is assembled, but before the subflooring is installed, the permanent support posts or columns are put into place.

4.3.0 Floor Joists

Floor joists are a series of parallel, horizontal framing members that make up the body of the floor frame (*Figure 15*). They rest on and transfer the building load to the sills and girders. The flooring or subflooring is attached to them. The span determines the length of the joist that must be used. Safe spans for joists under average loads can be found using the latest tables available from wood product manufacturers or sources such as the National Forest Products Association and the Southern Forest Products Association. For floors, this is usually figured on a basis of 50 lbs. per sq. ft. (10 lbs. dead load and 40 lbs. live load). Dead load is the weight of permanent, stationary construction and equipment included in a building. Live load is the total of all moving and variable loads that may be placed upon a building.

Joists must not only be strong enough to carry the load that rests on them; they must also be stiff

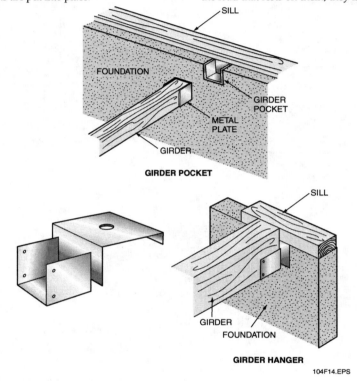

GIRDER POCKET

GIRDER HANGER

104F14.EPS

Figure 14 ◆ Girder pocket and girder hanger girder supports.

4.18

CARPENTRY LEVEL ONE—TRAINEE MODULE 27104

Instructor's Notes:

_____

_____

_____

_____

_____

_____

_____

_____

4.18

CARPENTRY LEVEL ONE—INSTRUCTOR'S GUIDE MODULE 27104

enough to prevent undue bending (deflection) or vibration. Too much deflection in joists is undesirable because it can make a floor noticeably springy. Building codes typically specify that the deflection downward at the center of a joist must not exceed 1/360th of the span with normal live load. For example, for a joist with a 15' span, this would equal a maximum of ½" of downward deflection (15' span × 12 = 180" ÷ 360 = .5"). A general procedure for sizing joists is given later in this module.

Joists are normally placed 16" on center (OC) and are always placed crown up. However, in some applications joists can be set as close as 12" OC or as far apart as 24" OC. All these distances are used because they accommodate 4' × 8' subfloor panels and provide a nailing surface where two panels meet. Joists can be supported by the top of the girder or be framed to the side. *Figure 16*

shows three methods for joist framing at the girder. Check your local code for applicability. Note that if joists are lapped over the girder, the minimum amount of lap is 4" and the maximum amount of lap is 12". *Figure 17* shows some examples of the many different types of joist hangers that can be used to fasten joists to girders as well as other support framing members. Joist hangers are used where the bottom of the girder must be flush with the bottoms of the joists. At the sill end of the joist, the joist should rest on at least 1½" of wood. In platform construction, the ends of all the joists are fastened to a header joist, also called a *band joist* or *rim joist*, to form the box sill.

Joists must be doubled where extra loads need to be supported. When a partition runs parallel to the joists, a double joist is placed underneath. Joists must also be doubled around all openings in

Point out that joist hangers require special high-strength nails called *joist nails* or *stub nails*.

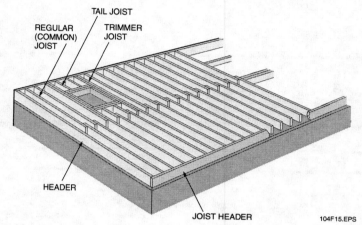

Figure 15 ◆ Floor joists.

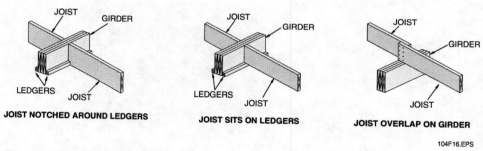

JOIST NOTCHED AROUND LEDGERS   **JOIST SITS ON LEDGERS**   **JOIST OVERLAP ON GIRDER**

104F16.EPS

Figure 16 ◆ Methods of joist framing at a girder.

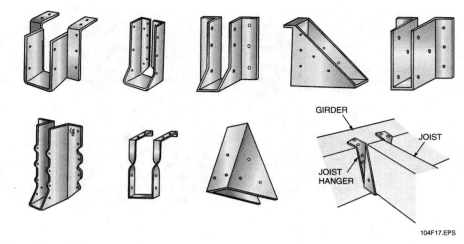

Figure 17 ◆ Typical types of joist hangers.

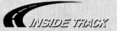
Maintaining a Flush Top Surface

If you are securing a joist to a girder with a ledger, you must first make sure that you maintain a flush top surface for the subflooring. Not all joists are the same size. To account for these small discrepancies, toenail the joist to the top of the girder prior to installing the ledger. Once you have established a smooth, flat surface, you can install the ledger under the joist.

Special Nails

Joist hangers require special nails to secure them to joists and girders. These nails are 1½" long and stronger than common nails. They are often referred to as *joist* or *stub* nails. The picture below shows various types of hangers as well as joist nails.

104P0403.EPS

Instructor's Notes:

_____

_____

_____

_____

_____

_____

_____

_____

the floor frame for stairways, chimneys, etc., to reinforce the rough opening in the floor. These additional joists used at such openings are called trimmer joists. They support the headers that carry short joists called tail joists. Double joists should spread where necessary to accommodate plumbing.

In residential construction, floors traditionally have been built using wooden joists. However, the use of prefabricated engineered wood products such as wood I-beams and various types of trusses is also becoming common.

4.3.1 Notching and Drilling of Wooden Joists

When it is necessary to notch or drill through a floor joist, most building codes will stipulate how deep a notch can be made. For example, the Standard Building Code specifies that notches on the ends of joists shall not exceed one-fourth the depth. Therefore, in a 2 × 10 floor joist, the notch could not exceed 2½" (see *Figure 18*).

This code also states that notches for pipes in the top or bottom shall not exceed one-sixth the depth, and shall not be located in the middle third of the span. Therefore, when using a 2 × 10 floor joist, a notch cannot be deeper than 1⅝". This notch can be made either in the top or bottom of the joist, but it cannot be made in the middle third of the span. This means that if the span is 12', the middle span from 4' to 8' could not be notched.

This code further requires that holes bored for pipe or cable shall not be within 2" of the top or bottom of the joist, nor shall the diameter of any such hole exceed one-third the depth of the joist. This means that if a hole needs to be drilled, it may not exceed 3" in diameter if a 2 × 10 floor joist is used. Always check the local codes.

Some wood I-beams are manufactured with perforated knockouts in their web, approximately 12" apart. Never notch or drill through the beam flange or cut other openings in the web without checking the manufacturer's specification sheet.

Also, do not drill or notch other types of engineered lumber (e.g., LVL, PSL, and glulam) without first checking the specification sheets.

4.3.2 Wood I-Beams

Wood I-beams, sometimes called *solid-web trusses*, are made in various depths and with lengths up to 80'. These manufactured joists are not prone to shrinking or warping. They consist of an oriented strand board (OSB) or plywood web (*Figure 19*) bonded into grooves cut in the wood flanges on the top and bottom. This arrangement provides a joist that has a strength-to-weight ratio much greater than that of ordinary lumber. Because of their increased strength, wood I-beams can be used in greater spans than a comparable length of dimension lumber. They can be cut, hung, and nailed like ordinary lumber. Special joist hangers and strapping, similar to those used

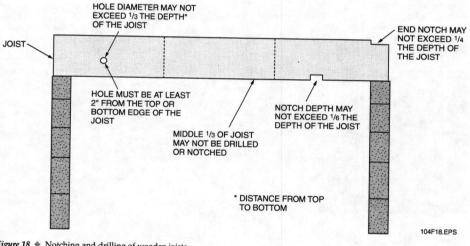

Figure 18 ◆ Notching and drilling of wooden joists.

Explain the restrictions for notching or drilling joists. Emphasize that codes must be checked before drilling or notching is attempted.

Show Transparency 20 (Figure 18).

Describe the construction of wood I-beams. Emphasize that they should not be notched and the manufacturer's instructions must be followed for installation and drilling.

Show Transparencies 21 and 22 (Figures 19 and 20).

Wood I-Beams

I-beams have specific guidelines and instructions to follow for cutting, blocking, and installation. It is important to always follow the manufacturer's instructions when installing these materials. Otherwise, you may create a very dangerous situation by compromising the structural integrity of the beam.

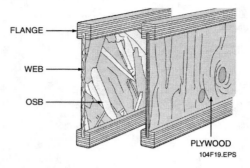

FLANGE

WEB

OSB

PLYWOOD

104F19.EPS

Figure 19 ◆ I-beams.

with wood joists, are used to fasten wood I-beam joists to girders and other framing members. Wood I-beam joists are typically manufactured with 1½" diameter, prestamped knockout holes in the web about 12" OC that can be used to accommodate electrical wiring. Other holes or openings can be cut into the web, but these can only be of a size and at the locations specified by the I-beam manufacturer. Under no circumstances should the flanges of I-beam joists be cut or notched. *Figure 20* shows a typical floor system constructed with wood I-beams.

104F20.EPS

Figure 20 ◆ Typical floor system constructed with engineered I-beams (second floor shown).

4.3.3 Trusses

Trusses are manufactured joist assemblies made of wood or a combination of steel and wood (*Figure 21*). Solid light-gauge steel and open-web steel trusses are also made, but these are used mainly in commercial construction. Like the wood I-beams, trusses are stronger than comparable lengths of dimension lumber, allowing them to be used over longer spans. Longer spans allow more freedom in building design because interior loadbearing walls and extra footings can often be eliminated. Trusses generally erect faster and easier with no need for trimming or cutting in the field. They also provide the additional advantage of permitting ducting, plumbing, and electrical wires to be run easily between the open webs.

Floor trusses consist of three components: chords, webs, and connector plates. The wood chords (outer members) are held rigidly apart by either wood or metal webs. The connector plates are toothed metal plates that fasten the truss web and chord components together at the intersecting points. The type of truss used most frequently in residential floor systems is the parallel-chord 4 × 2 truss. This name is derived from the chords being made of 2 × 4 lumber with the wide surfaces facing each other. Webs connect the chords. Diagonal webs positioned at 45° to the chords mainly resist the shearing stresses in the truss. Vertical webs, which are placed at right angles to

4.22

Instructor's Notes:

_____

_____

_____

_____

_____

_____

_____

_____

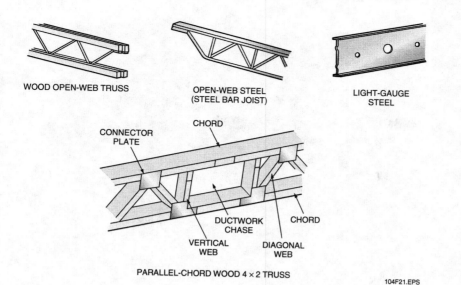

WOOD OPEN-WEB TRUSS

OPEN-WEB STEEL
(STEEL BAR JOIST)

LIGHT-GAUGE
STEEL

CONNECTOR PLATE

CHORD

CHORD

DUCTWORK CHASE

VERTICAL WEB

DIAGONAL WEB

PARALLEL-CHORD WOOD 4 × 2 TRUSS

104F21.EPS

Figure 21 ◆ Typical trusses.

the chords, are used at critical load transfer points where additional strength is required. Wood is used most frequently for webs, but galvanized steel webs are also used. Trusses made with metal webs provide greater clear spans for any given truss depth than wood-web trusses. The openings in the webs are larger too, which allows more room for HVAC ducting.

Note that there are several different kinds of parallel-chord trusses. What makes each one different is the arrangement of its webs. Typically, parallel-chord floor trusses with wood webs are available in depths ranging from 12" to 24" in 1" increments. The most common depths are 14" and 16". Some metal-web trusses are available with the same actual depth dimensions as 2 × 8, 2 × 10, and 2 × 12 solid wood joists, making them interchangeable with an ordinary joist-floor system. *Figure 22* shows a typical floor system constructed with trusses.

4.4.0 Bridging

Bridging is used to stiffen the floor frame to prevent unequal deflection of the joists and to enable an overloaded joist to receive some support from the joists on either side. Most building codes require that bridging be installed in rows between the floor joists at intervals of not more than 8'. For

example, floor joists with spans of 8' to 16' need one row of bridging in the center of the span.

Three types of bridging are commonly used (*Figure 23*): wood cross-bridging, metal cross-bridging, and solid bridging. Wood and metal cross-bridging are composed of pieces of wood or metal set diagonally between the joists to form an X. Wood cross-bridging is typically 1 × 4 lumber placed in double rows that cross each other in the joist space.

Metal cross-bridging is installed in a similar manner. Metal cross-bridging comes in a variety of styles and different lengths for use with a particular joist size and spacing. It is usually made of 18-gauge steel and is ¾" wide. When using cross-bridging, you may nail the top, but do not nail the bottom until the subfloor is installed. Solid bridging, also called *blocking*, consists of solid pieces of lumber, usually the same size as the floor joists, installed between the joists. It is installed in an offset fashion to enable end nailing.

4.5.0 Subflooring

Subflooring consists of panels or boards laid directly on and fastened to floor joists (*Figure 24*) in order to provide a base for underlayment and/or the finish floor material. Underlayment is a material, such as particleboard or plywood, laid

FLOOR SYSTEMS—TRAINEE MODULE 27104

4.23

Describe the various types of bridging. Point out that subflooring should be installed before solid bridging (blocking) is installed or the bottoms of cross bridging are secured.

Show Transparency 25 (Figure 23).

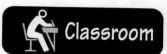

Explain the difference between underlayment and subfloors. Cover the types of materials used.

Show Transparency 26 (Figure 24).

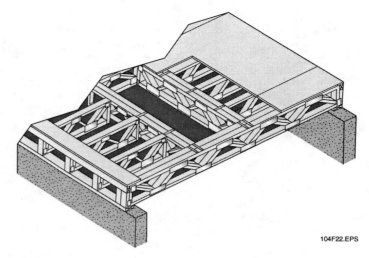

104F22.EPS

Figure 22 ◆ Typical floor system constructed with trusses.

WOOD CROSS BRIDGING   SOLID WOOD BRIDGING   STEEL CROSS BRIDGING

104F23.EPS

Figure 23 ◆ Types of bridging.

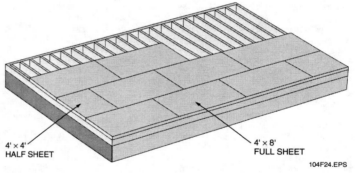

4' × 4'
HALF SHEET

4' × 8'
FULL SHEET

104F24.EPS

Figure 24 ◆ Subflooring installation.

Instructor's Notes:

_____

_____

_____

_____

_____

_____

_____

_____

on top of the subfloor to provide a smoother surface for some finished flooring. This surface is normally applied after the structure is built but before the finished floor is laid. The subfloor adds rigidity to the structure and provides a surface upon which wall and other framing can be laid out and constructed. Subfloors also act as a barrier to cold and dampness, thus keeping the building warmer and drier in winter. Either plywood, OSB or other manufactured board panels, or common wooden boards can be used as subflooring.

4.5.1 Plywood Subfloors

Butt-joint or tongue-and-groove plywood is widely used for residential subflooring. Used in 4' × 8' panels, typically ⅝" to ¾" thick when the joists are placed 16" OC, it goes on quickly and provides great rigidity to the floor frame. APA-rated sheathing plywood panels (Table 1) are generally used for subflooring in two-layer floor systems. APA-rated Sturd-I-Floor® tongue-and-groove plywood panels are commonly used in combined subfloor-underlayment (single-layer) floor systems where direct application of carpet, tile, etc., to the floor is intended.

Traditionally, plywood panels have been fastened to the floor joists using nails. Today, it is becoming more common to use a glued floor system in which the subfloor panels are both glued and nailed to the joists. This method helps stiffen the floors. It also helps eliminate squeaks and nail popping. Procedures for installing plywood subfloors, including gluing, are described later in this module.

4.5.2 Manufactured Board Panel Subfloors

Manufactured panels made of materials such as composite board, waferboard, OSB, and structural particleboard can also be used for subflooring. Detailed information on the construction and composition of these manufactured wood products is contained in an earlier module. Panels made of these materials have been rated by the American Plywood Association and meet all standards for subflooring. The method for installing these kinds of panels is basically the same as that used for plywood.

4.5.3 Board Subfloors

There are some instances when 1 × 6 or 1 × 8 boards are used as subflooring. Boards can be laid either diagonally or perpendicular to the floor joists. However, it is more common to lay them diagonally across the floor frame at a 45° angle. This provides for more rigidity of the floor and also assists in the bracing of the floor joists. Also, if laid perpendicular to the joist in a subfloor where oak flooring is to be laid over it, the oak flooring (instead of the subflooring) would have to be laid diagonally to the floor joist. This is necessary to prevent the shrinkage of the subfloor from affecting the joints in the finished oak floor, which would cause the oak flooring to pull apart. Board subflooring is nailed at each joist. Typically, two nails are used in each 1 × 6 board and three nails for wider boards. Note that a subfloor made of boards normally is not as rigid as one made of plywood or other manufactured panels.

INSIDE TRACK

OSB Subfloors

Today, many builders prefer to use OSB for subfloors. It offers acceptable structural strength at a reduced price.

104P0404.EPS

Table 1 Guide to APA Performance-Rated Plywood Panels

Grade Destination	Description and Common Uses	Typical Trademarks
APA-Rated Sheathing Exposure 1 or 2	Specially designed for subflooring and wall and roof sheathing, but can also be used for a broad range of other construction and industrial applications. Can be manufactured as conventional veneered plywood, as a composite, or as a nonveneered panel. For special engineered applications, including high-load requirements and certain industrial uses, veneered panels conforming to PS 1 may be required. Specify Exposure 1 when long construction delays are anticipated. Common thicknesses: $\frac{5}{16}$, $\frac{3}{8}$, $\frac{7}{16}$, $\frac{15}{32}$, $\frac{1}{2}$, $\frac{19}{32}$, $\frac{23}{32}$, $\frac{3}{4}$.	_____APA_____ Rated Sheathing **32/16** $\frac{15}{32}$" Sized for Spacing Exposure 1 ___**000**___ NER-108
APA Structural 1 Rated Sheathing Exposure 1	Unsanded all-veneer plywood grades for use where strength properties are of maximum importance; structural diaphragms, box beams, gusset plates, stressed-skin panels, containers, pallet bins. Made only with exterior glue (Exposure 1). Common thicknesses: $\frac{5}{16}$, $\frac{3}{8}$, $\frac{15}{32}$, $\frac{19}{32}$, $\frac{5}{8}$, $\frac{23}{32}$, $\frac{3}{4}$.	_____APA_____ Rated Sheathing Structural 1 **48/24** $\frac{23}{32}$" Sized for Spacing Exterior ___**000**___ PS 1-83 NER-108
APA-Rated Sturd-I-Floor Exposure 1 or 2	For combination subfloor-underlayment. Provides smooth surface for application of carpet and possesses high concentrated and impact load resistance. Can be manufactured as conventional veneered plywood, as a composite, or as a nonveneered panel. Available square edge or tongue-and-groove. Specify Exposure 1 when long construction delays are anticipated. Common thicknesses: $\frac{19}{32}$, $\frac{5}{8}$, $\frac{23}{32}$, $\frac{3}{4}$.	_____APA_____ Rated Sturd-I-Floor **20 OC** $\frac{19}{32}$" Sized for Spacing Exposure 1 ___**000**___ NER-108
APA-Rated Sturd-I-Floor 48 OC (2-4-1) Exposure 1	For combination subfloor-underlayment on 32" and 48" spans and for heavy timber roof construction. Provides smooth surface for application of resilient floor coverings and possesses high concentrated and impact load resistance. Manufactured only as conventional veneered plywood and only with exterior glue (Exposure 1). Available square edge or tongue-and-groove. Thickness: $1\frac{1}{8}$.	_____APA_____ Rated Sturd-I-Floor **48 OC** 2-4-1 $1\frac{1}{8}$" Sized for Spacing Exposure 1 Tag ___**000**___ Underlayment PS1-83 NER-108

American Plywood Association

Instructor's Notes:

_____

_____

_____

_____

_____

_____

_____

_____

5.0.0 ◆ LAYING OUT AND CONSTRUCTING A PLATFORM FLOOR ASSEMBLY

After the foundation is completed and the concrete or mortar has properly set up, assembly of the floor system can begin. Framing of the floors and sills is usually done before the foundation is backfilled. This is because the floor frame helps the foundation withstand the pressure placed on it by the soil. This section gives an overview of the procedures and methods used for laying out and constructing a basic platform floor assembly. When building any floor system, always coordinate your work with that of the other trades to ensure the framing is properly done to accommodate ductwork, piping, wiring, etc.

The construction of a platform floor assembly is normally done in the sequence shown below:

Step 1 Check the foundation wall for squareness.

Step 2 Lay out and install the sill plates.

Step 3 Build and/or install the girders and supports.

Step 4 Lay out the sills and girders for the floor joists.

Step 5 Lay out the joist locations for partitions and floor openings.

Step 6 Cut and attach the joist headers to the sill.

Step 7 Install the joists.

Step 8 Frame the openings in the floor.

Step 9 Install the bridging.

Step 10 Install the subflooring.

5.1.0 Checking the Foundation for Squareness

Before installing the sill plates, the foundation wall must be checked to make sure that it meets the dimensions specified on the blueprints and that the foundation is square. However, keep in mind that parallel and plumb take precedence over being square. This is easily done by making measurements of the foundation with a 100' steel measuring tape. First, the lengths of each of the foundation walls must be measured and recorded (*Figure 25*). The measurements must be as exact as possible. Following this, the foundation is measured diagonally from one outside corner to the opposite outside corner. A second diagonal measurement is then made between the outsides of the remaining two corners. If the measured lengths of the opposite walls are equal and the diagonals are equal, the foundation is square. For buildings where the foundation is other than a simple rectangle, a good practice is to divide the area into two or more individual square or rectangular areas and measure each area as described above.

5.2.0 Installing the Sill

For floors where the sill is installed flush with the outside of the foundation walls, installation of the sill begins by snapping chalklines on the top of the foundation walls in line with the inside edge of the sill (*Figure 26*). If the sill must be set in to accommodate the thickness of wall sheathing, brick veneer, etc., the chalklines may be snapped for the outside edge of the sill plates or the inside edge (if the foundation size allows). At each corner, the true location of the outside corner of the

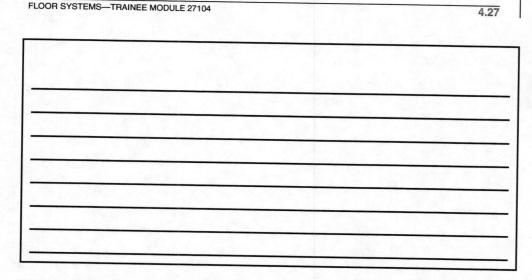

- MEASURE THE FOUR WALLS
- MEASURE THE DIAGONALS

IF OPPOSITE WALLS ARE EQUAL AND THE DIAGONALS ARE EQUAL, THE FOUNDATION IS SQUARE

104F25.EPS

Figure 25 ◆ Checking the foundation for squareness.

4.27

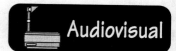

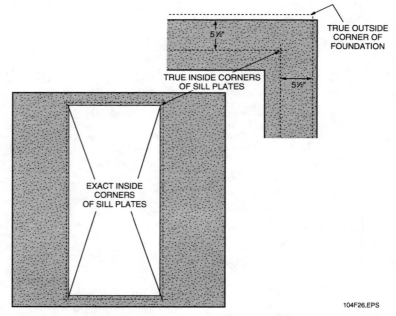

Figure 26 ◆ Inside edges of sill plates marked on the top of the foundation wall.

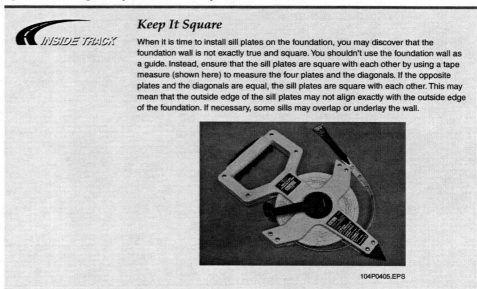

Keep It Square

When it is time to install sill plates on the foundation, you may discover that the foundation wall is not exactly true and square. You shouldn't use the foundation wall as a guide. Instead, ensure that the sill plates are square with each other by using a tape measure (shown here) to measure the four plates and the diagonals. If the opposite plates and the diagonals are equal, the sill plates are square with each other. This may mean that the outside edge of the sill plates may not align exactly with the outside edge of the foundation. If necessary, some sills may overlap or underlay the wall.

104P0405.EPS

Instructor's Notes:

_____

_____

_____

_____

_____

_____

_____

_____

sill is used as a reference point to mark the corresponding inside corner of the sill on the foundation wall. To do this, the exact width of the sill stock being used must be determined. For example, if using 2 × 6 sills, 5½" is the sill width measurement. After the exact inside corners are located and marked on the sill, chalklines are snapped between these points. This gives an outline on the top of the foundation wall of the exact inside edges of the sill plates. At this point, a good practice is to double-check the dimensions and squareness of these lines to make sure that they are accurate.

After the location of the sill is marked on the foundation, the sill pieces can be measured and cut. Take into consideration that there must be an anchor bolt within 12" of the end of any plate. Also, sill plates cannot butt together over any opening in the foundation wall. When selecting the lumber, choose boards that are as straight as possible for making the sills. Badly bowed pieces should not be used.

Holes must be drilled in the sill plates so that they can be installed over the anchor bolts embedded in the foundation wall. To lay out the location of these holes, hold the sill sections in place on top of the foundation wall against the anchor bolts (*Figure 27*). At each anchor bolt, use a combination square to scribe lines on the sill corresponding to both sides of the bolt. On the foundation, measure the distance between the center of each anchor bolt and the chalkline, then transfer this distance to the corresponding bolt location on the sill by measuring from the inside edge. After the sill hole layout is done, the holes in the sill are drilled. They should be drilled about ⅛" to ¼" larger than the diameter of the anchor bolt in order to allow for some adjustment of the sill plates, if necessary. Also, make sure all holes are drilled straight.

Before installing the sill plates, the termite shield (if used) and sill sealer are installed on the foundation. Following this, the sill sections are placed in position over the anchor bolts, making sure that the inside edges of the sill plate sections are aligned with the chalkline on top of the foundation wall and that the inside corners are aligned with their marks. The sill plates are then loosely fastened to the foundation with the anchor bolt nuts and washers and the sill checked to make sure it is level. An 8' level can be used for this task. However, using a transit or builder's level and checking the level every 3' or 4' along the sill is more accurate. It cannot be emphasized enough how important it is that the sill be level. If it is not level, it will throw off the building's floors and walls. Low spots can be shimmed with plywood wedges or filled with grout or mortar. If the sill is too high, the high areas of the concrete foundation will need to be ground or chipped away. After the sill has been made level, the anchor bolt nuts can be fully tightened. Be careful not to overtighten the nuts, especially if the concrete is not thoroughly dry and hard, because this can crack the wall.

5.3.0 Installing a Beam/Girder

In preparation for installing a girder, it is necessary to use the job specifications to determine the details related to its installation. For the purpose of an explanation, assume you are working with a structure that has a foundation that is 24' wide and 48' long (*Figure 28*). The foundation is poured concrete that is 12" thick with 6"-deep and 7"-wide

Show Transparencies 28 and 29 (Figures 26 and 27).

Demonstrate sill plate squaring and leveling. Emphasize the importance of this activity.

Explain the method for placement of a built-up girder.

Show Transparency 30 (Figure 28).

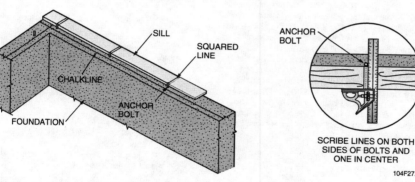

Figure 27 ◆ Square lines across the sill to locate the anchor bolt hole.

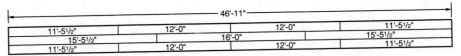

BUILT-UP GIRDER

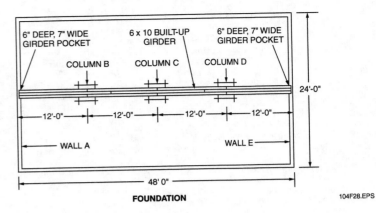

FOUNDATION

104F28.EPS

Figure 28 ◆ Example girder and support column data.

girder (beam) pockets centered in the short walls. The girder is to be a built-up beam containing three thicknesses of 2 × 10. The columns used to support the girder when the floor system is complete are three 4" lally columns, concrete-filled, with ¼" steel plates top and bottom. The distance between these columns is 12'-0" OC.

To lay out the distances for each of the support columns, first use a steel tape to measure from one end of the foundation to the other in the precise location where the girder will sit. Using a plumb bob, hold the line at the 24'-0" mark. This locates the center of the middle support column. Then, measure 12'-0" to the left of center and 12'-0" to the right of center to locate the center of the other two support columns. The distance from the center of the two end columns into their girder pockets in the foundation walls is 11'-5½". This allows for ½" of space between the back of each girder pocket and the end of the girder. Given the dimensions above, the finished built-up girder for our example needs to measure 46'-11" in length. When constructing this girder, remember that the joints should fall directly over the support columns and the girder crown must face upward.

When framing floors, use the fasteners as indicated below:

- *16d nails*—Used to attach the header to joists, to install solid bridging, and to construct beams and girders.
- *Special nails*—Furnished to attach the joist hangers.
- *8d nails*—Used to install the wood cross-bridging and subfloor.
- *Pneumatic, ring shank, and screw nails, or etched galvanized staples*—Used to apply the subfloor and underlayment.
- *Construction adhesive*—Used to apply the subfloor and underlayment.

Instructor's Notes:

As shown in *Figure 28*, the 6 × 10 built-up girder can be constructed using eight 12' long 2 × 10s and three 16' long 2 × 10s. Four of the 12' long 2 × 10s are cut to 11'-5½" and two of the 16' long 2 × 10s are cut to 15'-5½". The 16' pieces are used in order to provide for an overlap of at least 4'-0" at the joints. To make the girder, nail the 2 × 10s together in the pattern shown using 16d nails spaced about 24" apart. Note that the nailing schedule will vary at different locations, so make sure to consult local codes for the proper schedule. Drive the nails at an angle for better holding power. Be sure to butt the joints together so they form a tight fit. Continue this process until the 46'-11" girder is finished. Once completed, the girder is put in place, supported by temporary posts or A-frames, and made level. Note that the temporary supports are removed and replaced by the permanent lally columns after the floor system joists are all installed.

5.4.0 Laying Out Sills and Girders for Floor Joists

Joists should be laid out so that the edges of standard size subfloor panels fall over the centers of the joists. There are different ways to lay out a sill plate to accomplish this. One method for laying out floor joists 16" OC is described here. Begin by using a steel tape to measure out from the end of the sill exactly 15¼" (*Figure 29*). At this point, use a speed square to square a line on the sill. To make sure of accurate spacing, drive a nail into the sill at the 15¼" line, then hook the steel tape to the nail and stretch it the length of the sill. At every point on the tape marked as a multiple of 16" (most tapes highlight these numbers), make a mark on the sill. It is important that the spacing be laid out accurately; otherwise, the subfloor panels may not fall in the center of some joists.

After the sill is marked, use your speed square to square a line at each mark. Next, mark a narrow X next to each line on the sill to show the actual position where each joist is to be placed. Note that the lines marked on the sill mark the edge of the joists, not the center. Be sure to mark the X on the proper side. If the layout has been started from the left side of the sill, as shown in *Figure 29*, the X should be placed to the right of the line. If the layout has been started from the right side, the X should be placed to the left of the line. After the

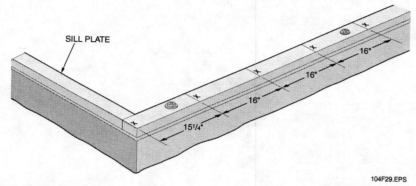

SILL PLATE

15¼"   16"   16"   16"   16"

104F29.EPS

Figure 29 ◆ Marking the sill for joist locations.

Measurement Tip

The reason 15¼", not 16", is used in *Figure 29* as the first measurement is so that the first panel of flooring will come to the outside edge of the first joist, not the center of it. All flooring panels except the first and last need to fall on the center of the joists to provide a nailing surface for the adjoining piece. The first and last panels do not have adjoining panels on one side. By reducing the first measurement ¾" (half the thickness of the joist), you shift the first piece of floor paneling from the center of the first joist to the outside edge of it.

FLOOR SYSTEMS—TRAINEE MODULE 27104

4.31

Explain the alternate method of marking joist locations using the box sill.

Demonstrate the layout marking of extra joists for partitions and openings that are parallel with the joists.

Show Transparency 32 (Figure 30).

Demonstrate box sill construction. Point out that some codes require metal straps or fasteners to secure the box sill to the sill plate.

locations for all common 16" OC joists have been laid out, the locations for any double joists, trimmer joists, etc., should be marked on the sill and identified with a T (or other letter) instead of an X.

After the first sill has been laid out, the process is repeated on the girder and the opposite sill. If the joists are in line, Xs should be marked on the same side of the mark on both the girder and the sill plate on the opposite wall. If the joists are lapped at the girder, an X should be placed on both sides of the mark on the girder and on the opposite side of the mark on the sill plate on the opposite wall.

The location of the floor joists can be laid out directly on the sills as described above. However, in platform construction, some carpenters prefer to lay them out on the header joists rather than the sill. If done on the header, the procedure is basically the same. Also, instead of making a series of individual measurements, some carpenters make a layout rod marked with the proper measurements and use it to lay out the sills and girders.

5.5.0 Laying Out Joist Locations for the Partition and Floor Openings

After the locations of all the common 16" OC floor joists are laid out, it is necessary to determine the locations of additional joists needed to accommodate loadbearing partitions, floor openings, etc., as shown on the blueprints. Typically, these include:

- Double joists needed under loadbearing interior walls that run parallel with the joists (*Figure 30*). Depending on the structure, the double joists may need to be separated by 2 × 4 (or larger) blocks placed every 4' to allow for plumbing and electrical wires to pass into the wall from below. Loadbearing walls that run perpendicular to the joist system normally do not need additional joists added.
- Double joists needed for floor openings for stairs, chimneys, etc.

The sill plates and girder should be marked where the joists are doubled on each side of a large floor opening. Also, the sill and girder should be marked for the locations of the shorter tail joists at the ends of floor openings. They can be identified by marking their locations with a T instead of an X.

5.6.0 Cutting and Installing Joist Headers

After the sills and girder are laid out for the joist locations, the box sill can be built (*Figure 31*). The

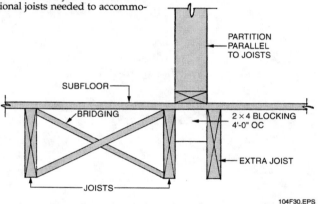

Figure 30 ◆ Double joists at a partition that runs parallel to the joists.

INSIDE TRACK

Double Joists

In addition to supporting parallel loadbearing interior walls, joists are doubled under extremely heavy objects. Whirlpools, bathtubs, and oversize refrigerators put additional load factors on the floor and require a suitable joist system to support the weight and prevent the objects from damaging the structural integrity of the building.

Instructor's Notes:
_____
_____
_____
_____
_____
_____
_____
_____

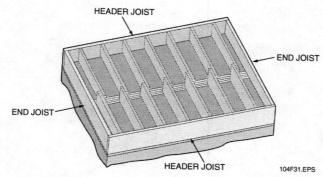

Figure 31 ◆ Box sill installed on foundation.

box sill encloses the joist system and is made of the same size stock (typically 2 × 10s). It consists of the two header joists that run perpendicular to the joists and the first and last joists (end joists) in the floor system. The headers are placed flush with the outside edges of the sill. Good straight stock should be used so that the headers do not rise above the sill plates. They need to sit flat on the sill so that they do not push up the wall. After the joist headers are cut, they are toenailed to the sill and face-nailed where the first and last joists meet the header joist. Any splices that need to be made in the header joist can meet either in the center of a joist or be joined by a scab. Note that some carpenters do not use header joists; they use blocks placed between the ends of the joists instead. Also, in areas subject to earthquakes, hurricanes, or tornadoes, codes may require that metal fasteners be used to further attach the header and end joists to the sill.

5.7.0 Installing Floor Joists

With the box sill in place, floor joists are placed at every spot marked on the sill. This includes the extra joists needed at the locations of partition walls and floor openings. When installing each joist, it is important to locate the crown and always point the crown up. With the joist in position at the header joist, hold the end tightly against the header and along the layout line so the sides are plumb, then end nail it to the header and toenail it to the sill (*Figure 32*). This procedure is repeated until all the joists are attached to their associated header. To facilitate framing of openings in the floor, a good practice is to leave the full-length joists out where the floor openings occur.

Following this, the joists are fastened at the girder. If they join end-to-end without overlapping at the girder, they should be joined with a scarf or metal fastener. Where the joists overlap at the girder, they should be face-nailed together and toenailed to the girder.

5.8.0 Framing Opening(s) in the Floor

Floor openings are framed by a combination of headers and trimmer joists (*Figure 33*). Headers run perpendicular to the direction of the joists and are doubled. Full-length trimmer joists and short tail joists run parallel to the common joists. The blueprints show the location in the floor frame and the size of the rough opening (RO). This represents the dimensions from the inside edge of one trimmer joist to the inside edge of the other trimmer joist and from the inside edge of one header to the inside edge of the other header. The method used to frame openings can vary depending on the particular situation. Trimmers and headers at openings must be nailed together in a certain sequence so that there is never a need to nail through a double piece of stock.

A typical procedure for framing an opening like the one shown in *Figure 33* is given here:

Step 1 First install full-length trimmer joists A and C, then cut four header pieces with a length corresponding to the distance between the trimmer joists A and C.

Step 2 Nail two of these header pieces (headers No. 1 and No. 2) between trimmer joists A and C at the required distances.

Demonstrate floor joist placement. Point out that joist crowns should always be positioned pointing up and must always be fastened together at the girder. Also mention that any joists that are to be located near or across floor openings should not be placed until later in the process.

Show Transparency 34 (Figure 32).

Demonstrate how to frame a floor opening. Emphasize that the correct procedure must be followed to avoid nailing through double pieces of stock.

Show Transparency 35 (Figure 33).

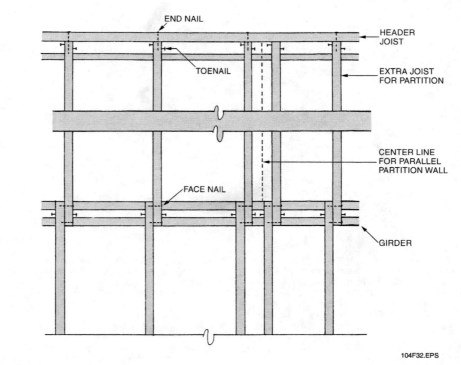

Figure 32 ◆ Installing joists.

104F32.EPS

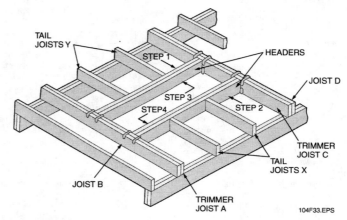

Figure 33 ◆ Floor opening construction.

104F33.EPS

4.34

Instructor's Notes:

_____

_____

_____

_____

_____

_____

_____

_____

Step 3 Following this, cut short tail joists X and Y and nail them to headers No. 1 and No. 2, as shown. Check the code to see if hangers are required.

Step 4 After headers No. 1 and No. 2 and tail joists X and Y are securely nailed, headers No. 3 and No. 4 can be installed and nailed to headers No. 1 and No. 2. Then, joists B and D can be placed next to and nailed to trimmer joists A and C, respectively.

5.9.0 Installing Bridging

Three types of bridging can be installed: wood cross-bridging, metal cross-bridging, and solid bridging. Wood cross-bridging is typically made of 1 × 4 pieces of wood installed diagonally between the joists to form an X. Normally, the bridging is installed every 8' or in the middle of the joist span. For example, joists with a 12' span would have a row of bridging installed at 6'.

A framing square can be used to lay out the bridging. First, determine the actual distance between the floor joists and the actual depth of the joist. For example, 2 × 10 joists 16" OC measure 14½" between them. The actual depth of the joist is 9¼". To lay out the bridging, position the framing square on a piece of bridging stock, as shown in *Figure 34*. This will give the proper length and angle to cut the bridging. Make sure to use the same side of the framing square in both places. Once the required length and angle of the bridging are determined, many carpenters build a jig to cut the numerous pieces of bridging.

Metal cross-bridging comes in a variety of styles and different lengths for use with a particular joist size and spacing. Its installation should be done in accordance with the manufacturer's instructions. Layout of the joist system for the location of installing metal cross-bridging is done

in the same way as described for wood cross-bridging.

Solid bridging consists of solid pieces of lumber joist stock installed between the joists. Layout of the joist system for the location of solid bridging is done in the same way as described for wood cross-bridging. The solid bridging is installed between pairs of joists, first on one side of the chalkline and then in the next pair of joists on the other. This staggered method of installation enables end nailing. Note that because of variations in lumber thickness and joist spacing, the length of the individual pieces of solid bridging placed between each joist pair may have to be adjusted.

5.10.0 Installing Subflooring

Installation of the subfloor begins by measuring 4' in from one side of the frame and snapping a chalkline across the tops of the floor joists from one end to the other. When installing 4' × 8' plywood, OSB, or similar floor panels, the long (8') dimension of the panels must be run across (perpendicular) to the joists (*Figure 35*). Also, the panels must be laid so that the joints are staggered in each successive course (row). *There should never be an intersection of four corners.* This is done by starting the first course with a full panel and continuing to lay full panels to the opposite end. Following this, the next course is started using half a panel (4' × 4'), then continuing to lay full panels to the opposite end. This procedure is repeated until the surface of the floor is covered. The ends of the panels that overhang the end of the building are then cut off flush with the floor frame. When butt-joint plywood panels are used, at least ⅛" of space should be left between each head joint and side joint for expansion. If installing a single-layer floor, blocking is also

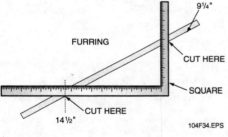

Figure 34 ◆ Framing square used to lay out bridging.

Hand out Job Sheet 27104-1. Under your supervision, have the trainees perform the tasks on the Job Sheet. Note the proficiency of each trainee on his or her Job Sheet and Skill Test Record.

Demonstrate how to construct various types of bridging.

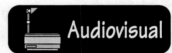

Show Transparency 36 (Figure 34).

Hand out Job Sheet 27104-2. Under your supervision, have the trainees perform the tasks on the Job Sheet. Note the proficiency of each trainee on his or her Job Sheet and Skill Test Record.

Describe subflooring layout and installation. Point out that panels must be positioned to a chalkline, staggered, run lengthwise across joists, and spaced for expansion.

Show Transparency 37 (Figure 35).

Bridging Installation

To install the bridging, a straight chalkline must be snapped across the top of the joists at the center of their span. Then, one end of a piece of bridging is nailed flush with the top of the joist on one side of the line. Nail only the top end. The bottom ends are not nailed until the subfloor is installed and its weight is applied. Following this, nail another piece of bridging to the other joist in the same space. Make sure it is flush with the top and on the opposite side of the line. Install bridging between the remaining pairs of joists until finished. When installing bridging between joists, make sure that the two pieces do not touch because this can cause the floor to squeak.

104P0406.EPS

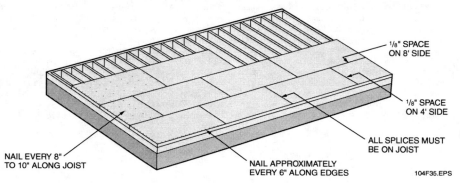

1/8" SPACE ON 8' SIDE

1/8" SPACE ON 4' SIDE

ALL SPLICES MUST BE ON JOIST

NAIL EVERY 8" TO 10" ALONG JOIST

NAIL APPROXIMATELY EVERY 6" ALONG EDGES

104F35.EPS

Figure 35 ◆ Installing butt-joint floor panels.

Subflooring

When you install the floor panels that comprise the subfloor, you should stagger the layout of each course (row). From the top view, the floor should have a traditional brick wall–style layout in which the vertical edges of the panels meet an adjoining row at the midpoint. There should never be an intersection of four corners.

Instructor's Notes:

_____

_____

_____

_____

_____

_____

_____

_____

_____

Floor Joist Transitions

If the floor joists are overlapped at a girder, the layout for the paneling will differ from one side of the floor to the other by 1½". If the lap occurs at a natural break for the panels, the next course can be slid back 1½" so that it continues to butt on a joist. However, if the panels cover the joist lap, then 2 × 4 scabs must be nailed to the joists to provide a nailing surface for the panels at this transition.

required under the joints of the butt-edged panels. Specifications for nailing panels to floor members vary with local codes. Typically, when nailing ⅝"-thick panels to joists on 16" centers, the nailing would be done every 6" along the edges and 8" to 10" along intermediate members using ring-shank or screw nails. To avoid fatigue, pneumatic nailers are commonly used to nail the subfloors. Some carpenters use power screw guns with extended handles to screw the subfloor to the joists.

Traditionally, plywood panels have been fastened to the floor joists using nails. Today, it is more common to use a glued floor system in which the subfloor panels are both glued and nailed (or screwed) to the joists, as previously discussed. Before each of the panels is placed, a ¼" bead of subflooring adhesive is applied to the

joists using a caulking gun. Two beads are applied on joists where panel ends butt together. Following this, the panel should immediately be nailed to the joists before the adhesive sets. Be sure all nails hit the floor joists.

Building a subfloor with tongue-and-groove panels is done in basically the same way as described for butt-joint panels, with the following exceptions. Begin the first course of paneling with the tongue (*Figure 36*) of the panels facing the outside of the house, not the inside. This leaves the groove as the leading edge. The next course of panels has to be interlocked into the previous course by driving the new sheets with a 2 × 4 block and a sledgehammer. The grooved edge of the panels can take this abuse; the tongued edge cannot.

Describe the methods of fastening subflooring to joists. Point out that fastening specifications may vary with local codes.

Describe T&G panel installation. Point out that the starting panel tongue must face outside the building and that expansion spacing is required.

Show Transparency 38 (Figure 36).

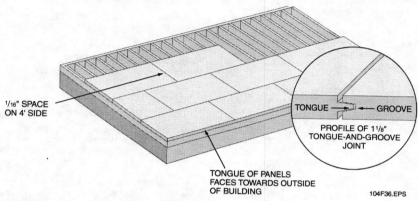

1/16" SPACE ON 4' SIDE

TONGUE — GROOVE

PROFILE OF 1⅛" TONGUE-AND-GROOVE JOINT

TONGUE OF PANELS FACES TOWARDS OUTSIDE OF BUILDING

104F36.EPS

Figure 36 ◆ Installing tongue-and-groove floor panels.

Hand out Job Sheets 27104-3 and 27104-4. Under your supervision, have the trainees perform the tasks on the Job Sheets. Note the proficiency of each trainee on his or her Job Sheets and Skill Test Record.

Avoiding Panel Pinch

When nailing the individual panels, leave the last 6" of width along the groove unnailed. Nail this portion once the next course is driven in place or else the groove will be pinched, making threading of the tongue very difficult. If you are also gluing the subfloor, apply a ⅛" bead of subflooring adhesive along the groove of the panel in addition to applying it to all the joist surfaces.

Have each trainee complete to your satisfaction Performance Profile Task 4 on floor assembly and Task 5 on bridging. Fill out Performance Profile Sheets for each trainee.

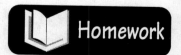

Assign reading of Section 6.0.0 for the next class session.

Describe joist installation for projections and cantilever floors. Point out that these joists usually extend inward a distance that is twice the overhang unless local codes specify otherwise.

Hand out Job Sheet 27104-5. Under your supervision, have the trainees perform the tasks on the Job Sheet. Note the proficiency of each trainee on his or her Job Sheet and Skill Test Record.

Explain the methods of estimating flooring materials. Point out that if boards are used for the subfloor, a waste allowance must be added.

Have each trainee complete to your satisfaction Performance Profile Task 6 on cantilever floor joist installation. Fill out Performance Profile Sheets for each trainee.

Assign reading of Sections 7.0.0–7.5.0 for the next class session.

6.0.0 ◆ INSTALLING JOISTS FOR PROJECTIONS AND CANTILEVERED FLOORS

Porches, decks, and other projections from a building present some special floor-framing situations. Projections overhang the foundation wall and are suspended without any vertical support posts. When constructing a projection, it makes a difference whether the joists run parallel or perpendicular to the common joists in the floor system. If they are parallel, longer joists are simply run out past the foundation wall (*Figure 37*). If they must be run perpendicular to the common joists, the projection must be framed with cantilevered joists. This means that you have to double up a common joist, then tie the cantilevered joists into this double joist. Regardless of whether you are using parallel or cantilevered joists, there should always be at least a ⅓ to ⅔ relationship between the length of the total joist and the distance it can project past the foundation wall. Check local codes for exact requirements. Stated another way, the joist should extend inward a distance equal to at least twice the overhang.

7.0.0 ◆ ESTIMATING THE QUANTITY OF FLOOR MATERIALS

Because of the importance of the floor in carrying the weight of the structure, it is important to correctly determine the materials required. You must be able to recognize special needs such as floor openings, cantilevers, and partition supports that affect material requirements. Once the needs are determined, you must be able to estimate the quantities of materials needed in order to construct the floor without delays or added expense from too much material. The process begins by checking the building specifications for the kinds and dimensions of materials to be used. It also requires that the blueprints be checked or scaled to determine the dimensions of the various components needed. These include:

- Sill sealer, termite shield, and sill
- Girder or beam
- Joists
- Joist headers

- Bridging
- Subflooring

For the purpose of an example in the sections below, we will use the floor system shown in *Figure 38* to determine the quantity of floor and sill framing materials needed.

7.1.0 Sill, Sill Sealer, and Termite Shield

To determine the amount of sill, sill sealer, and/or termite shield materials required, simply measure the perimeter of the foundation. For our example in *Figure 38*, the amount of material needed is 192 lineal feet [2 × (32' + 64') = 192'].

7.2.0 Beams/Girders

The quantity of girder material needed is determined by the type of girder and its length. For our example, if using a solid girder, the length of material needed would be 64 lineal feet. If a built-up beam made of three 2 × 12s is used as shown, the length of 2 × 12 material needed is 192 lineal feet (3 × 64' = 192').

7.3.0 Joists and Joist Headers

To determine the number of floor joists in a frame, divide the length of the building by the joist spacing and add one joist for the end and one joist for each partition that runs parallel to the joists. For our example, there are no partitions; therefore, the number of 2 × 8 joists is 49 [(64' × 12") ÷ 16" OC = 48 + 1 = 49]. Because there are two rows of joists (one on each side of the girder), the total number of joists needed is 98 (2 × 49 = 98). Each of these joists would be about 18' long. The amount of 2 × 8 material needed for the header joists is 128 lineal feet (2 × 64' = 128').

7.4.0 Bridging

Codes require one row of bridging in spans over 8' and less than 16' in length. Two rows of bridging are required in spans over 16'.

Instructor's Notes:

_____

_____

_____

_____

_____

_____

_____

_____

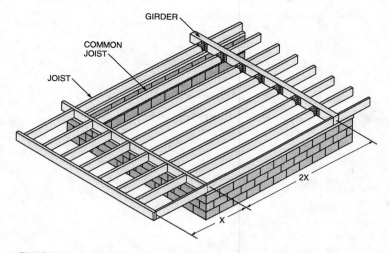

EXAMPLE OF PROJECTION JOIST RUNNING IN SAME DIRECTION AS COMMON JOISTS

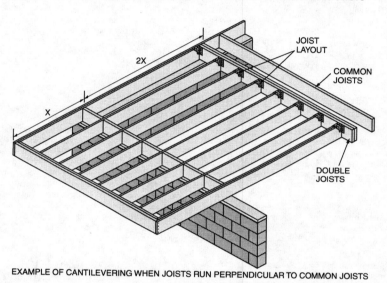

EXAMPLE OF CANTILEVERING WHEN JOISTS RUN PERPENDICULAR TO COMMON JOISTS

104F37.EPS

Figure 37 ◆ Cantilevered joists.

FLOOR SYSTEMS—INSTRUCTOR'S GUIDE MODULE 27104

4.39

Show Transparency 40
(Figure 38).

**Hand out Worksheet
27104-3. Have the trainees
complete the tasks on the
Worksheet. Record the
proficiency of each trainee.**

**Have each trainee complete
to your satisfaction
Performance Profile Task 9
on estimation. Fill out
Performance Profile Sheets
for each trainee.**

**Assign reading of Sections
8.0.0–8.2.0 for the next
class session.**

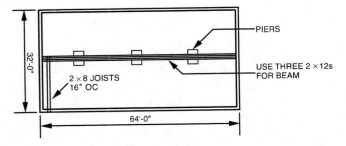

DETERMINE MATERIALS NEEDED:

- SILL
- SILL SEALER
- TERMITE SHIELD
- GIRDER
- JOISTS
- JOIST HEADERS
- BRIDGING
- SUBFLOOR

104F38.EPS

Figure 38 ◆ Determining floor system materials.

To find the amount of wood cross-bridging needed, determine the number of rows of bridging needed and the length of each row of bridging to find the total lineal footage for the bridging rows. For our example, this is 128 lineal feet (2 rows × 64' = 128'). Next, multiply the total lineal footage of bridging rows by the appropriate factor given in *Table 2* to get the total lineal feet of bridging needed. For our example, we are using 2 × 8 joists; therefore, the total amount of bridging needed is 256 lineal feet (2 × 128' = 256').

To find the amount of solid bridging needed, determine the number of rows of bridging needed and the length of each row of bridging. Then, multiply the number of rows by the length of each row to determine the total lineal feet needed. For our example, we need 128 lineal feet (2 × 64' = 128').

To determine the amount of steel bridging needed, multiply the number of rows of bridging needed by the length of each row of bridging to determine the total lineal footage of bridging rows. For our example, this is 128 lineal feet (2 × 64' = 128'). Then, multiply the total lineal footage

of bridging rows by .75 (¾) to find the number of spaces between joists that are 16" OC. For our example, there are 96 spaces (.75 × 128 = 96). Then, multiply the number of spaces by 2 to determine the total number of steel bridging pieces needed. For our example, we need 192 pieces (2 × 96 = 192).

7.5.0 Flooring

To determine the number of 4' × 8' plywood/OSB sheets needed to cover a floor, divide the total floor area by 32 (the area in square feet of one panel). For our example, the total floor area is 2,048 square feet (64 × 32 = 2,048). Therefore, we need 64 panels (2,048 ÷ 32 = 64). For any fractional sheets, round up to the next whole sheet.

If using lumber boards for flooring, calculate the total floor area to be covered. To this amount, add a quantity of material to allow for waste. When using 1 × 6 lumber, a rule of thumb is to add ⅛ to the total area for waste; for 1 × 8 lumber, add ⅙ for waste.

8.0.0 ◆ GUIDELINES FOR DETERMINING PROPER GIRDER AND JOIST SIZES

Normally, the sizes of the girder(s) and joists used in a building are specified by the architect or structural engineer who designs the building. However, a carpenter should be familiar with the procedures used to determine girder and joist sizes.

Table 2 Wood Cross Bridging Multiplication Factor

Joist Size	Spacing (Inches OC)	Lineal Feet of Material (per Foot of Bridging Row)
2 × 6, 2 × 8, 2 × 10	16	2
2 × 12	16	2.25
2 × 14	16	2.5

4.40

Instructor's Notes:

_____

_____

_____

_____

_____

_____

_____

8.1.0 Sizing Girders

For the purpose of explanation, the following discussion on how to size a girder is keyed to the example first-floor plan shown in *Figure 39*. Assume that a built-up girder is to be used and the ceilings are drywall (not plaster).

Step 1 Determine the distance (span) between girder supports. For our example floor, the span is 8'-0".

Step 2 Find the girder load width. The girder must be able to carry the weight of the floor on each side to the midpoint of the joist that rests upon it. Therefore, the girder load is half the length of the joist span on each side of the girder multiplied by 2. For our example floor, the girder load width is 12'-0" (6'-0" on each side of the girder).

Step 3 Find the total floor load per square foot carried by the joists and bearing partitions. This is the sum of the loads per square foot as shown in *Figure 40*, with the exception of the roof load. Roof loads are not included because these are carried on the outside walls unless braces or partitions are placed under the rafters. With a drywall ceiling and no partitions, the total load per square foot for our example floor is 50 pounds per square foot (40 pounds live load + 10 pounds dead load).

Step 4 Find the total load on the girder. This is the product of the girder span times the girder width times the total floor load. For our example floor, the total load on the girder is 4,800 pounds ($8 \times 12 \times 50 = 4,800$).

Step 5 Select the proper size of girder according to local codes. *Table 3* is typical. It indicates safe loads on standard-size girders for spans from 6' to 10'. For our example floor, a 6×8 built-up girder is needed to carry a 4,800-pound load at an 8' span. Note that shortening the span is the most economical way to increase the load that a girder will carry.

8.2.0 Sizing Joists

For the purpose of explanation, the following discussion on how to size joists is keyed to the example first-floor plan shown in *Figure 41*. Assume that the plan falls into the 40-pound live load category and will be built with joists on 16" centers and drywall ceilings.

Step 1 Determine the length of the joist span. For our example, the span is 16'.

Step 2 Determine if there is a dead load on the ceiling. For our example, the ceiling is drywall; therefore, there is a dead load of 10 pounds per square foot.

Step 3 Select the proper size of joists according to local codes or by using the latest tables available from wood product manufacturers or sources such as the National Forest Products Association and the Southern Forest Products Association. *Table 4* indicates maximum safe spans for various sizes of wood joists under ordinary load conditions. For floors, this is usually figured on a basis of 50 pounds per square foot (10 pounds dead load and 40 pounds live load). For our example, *Table 4* shows that for a 40-pound live load, a 2×10 joist 16" OC will carry the load up to a span of 19'-2". Note that a 2×8 joist 16" OC will carry the required load up to 15'-3", which is not long enough for the required span.

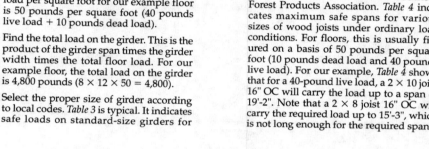

- DETERMINE LENGTH OF SPAN AND GIRDER WIDTH
- FIND TOTAL FLOOR LOAD PER SQUARE FOOT CARRIED BY JOIST AND BEARING PARTITIONS
- CALCULATE TOTAL LOAD ON GIRDER
- SELECT PROPER SIZE OF GIRDER IN ACCORDANCE WITH LOCAL CODES

104F39.EPS

Figure 39 ◆ Sizing girders.

Demonstrate girder sizing using the example provided in the Trainee Module or another foundation plan.

Show Transparencies 41 and 42 (Figures 39 and 40).

Demonstrate joist sizing using the example provided in the Trainee Module or another foundation plan.

Show Transparency 43 (Figure 41).

Have the trainees select the proper girder/beam and joist size from the tables in the Trainee Module for various floor plans, floor loads, and span data.

Have each trainee complete to your satisfaction Performance Profile Tests 10–11 on proper girder/beam and joist sizing.

GIRDER LOADS (POUNDS PER SQUARE FOOT)

	LIVE LOAD*	DEAD LOAD
ROOF	20	10
ATTIC FLOOR	20	20 (FLOORED)
		10 (NOT FLOORED)
SECOND FLOOR	40	20
PARTITIONS		20
FIRST FLOOR	40	20 (CEILING PLASTERED)
		10 (CEILING NOT PLASTERED)
PARTITIONS		20

*USUAL LOCAL REQUIREMENTS

EXAMPLE:

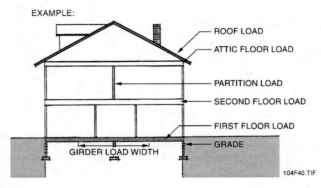

Figure 40 ◆ Floor loads.

Table 3 Safe Girder Loads

Nominal Girder Size	Safe Load in Pounds for Spans Shown				
	6 Ft.	7 Ft.	8 Ft.	9 Ft.	10 Ft.
6 × 8 solid	8,306	7,118	6,220	5,539	4,583
6 × 8 built-up	7,359	6,306	5,511	4,908	4,062
6 × 10 solid	11,357	10,804	9,980	8,887	7,997
6 × 10 built-up	10,068	9,576	8,844	7,878	7,086
8 × 8 solid	11,326	9,706	8,482	7,553	6,250
8 × 8 built-up	9,812	8,408	7,348	6,554	5,416
8 × 10 solid	15,487	14,732	13,608	12,116	10,902
8 × 10 built-up	13,424	12,968	11,792	10,504	9,448

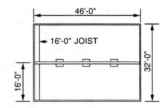

- FIND LENGTH OF SPAN
- DETERMINE LIVE LOAD PER SQUARE FOOT
- DETERMINE IF THERE IS A DEAD LOAD ON CEILING
- SELECT PROPER SIZE JOISTS IN ACCORDANCE WITH LOCAL CODES

104F41.EPS

Figure 41 ◆ Sizing joists.

Instructor's Notes:

_____

_____

_____

_____

_____

_____

_____

_____

Table 4 Safe Joist Spans

Nominal Joist Size	Spacing	30# Live Load	40# Live Load	50# Live Load	60# Live Load
2 × 6	12"	14'-10"	13'-2"	12'-0"	11'-1"
	16"	12'-11"	11'-6"	10'-5"	9'-8"
	24"	10'-8"	9'-6"	8'-7"	7'-10"
2 × 8	12"	19'-7"	17'-5"	15'-10"	14'-8"
	16"	17'-1"	15'-3"	13'-10"	12'-9"
	24"	14'-2"	13'-6"	11'-4"	10'-6"
2 × 10	12"	24'-6"	21'-10"	19'-11"	18'-5"
	16"	21'-6"	19'-2"	17'-5"	16'-1"
	24"	17'-10"	15'-10"	14'-4"	13'-3"
2 × 12	12"	29'-4"	26'-3"	24'-0"	22'-2"
	16"	25'-10"	23'-0"	21'-0"	19'-5"
	24"	21'-5"	19'-1"	17'-4"	16'-9"
3 × 8	12"	24'-3"	21'-8"	19'-10"	18'-4"
	16"	21'-4"	19'-1"	17'-4"	16'-0"
	24"	17'-9"	15'-9"	14'-4"	13'-3"
3 × 10	12"	30'-2"	27'-1"	34'-10"	23'-0"
	16"	26'-8"	23'-10"	21'-9"	20'-2"
	24"	22'-3"	19'-10"	18'-1"	16'-8"

Summary

A great majority of a carpenter's time is devoted to building floor systems. It is important that a carpenter not only be knowledgeable about both traditional and modern floor framing techniques, but, more importantly, be able to construct modern flooring systems.

The construction of a platform floor assembly involves the tasks listed below and is normally done in the sequence listed:

- Check the foundation for squareness.
- Lay out and install the sill plates.
- Build and/or install the girders and supports.
- Lay out the sills and girders for the floor joists.
- Lay out the joist locations for partitions and floor openings.
- Cut and attach the joist headers to the sill.
- Install the joists.
- Frame the openings in the floor.
- Install the bridging.
- Install the subflooring.

Have the trainees complete the Review Questions and go over the answers prior to administering the Module Examination.

Review Questions

1. In braced frame construction, the corner posts and beams are _____.
   a. nailed
   b. mortised and tenoned
   c. pinned
   d. glued

2. In balloon framing, the second floor joists sit on a _____.
   a. plate nailed to the top plate of the wall assembly below
   b. 1 × 4 or 1 × 6 ribbon let in to the wall studs
   c. sill attached to the top of the wall assembly below
   d. 2 × 4 let in to the wall studs

3. Shrinkage in wood framing members occurs mainly _____.
   a. lengthwise
   b. both lengthwise and on the wide width
   c. on the wide width
   d. in the middle

4. The most common framing method used in modern residential and light commercial construction is _____ framing.
   a. brace
   b. balloon
   c. platform
   d. post-and-beam

5. The method of construction that experiences a relatively large amount of settling as a result of shrinkage is _____ framing.
   a. platform
   b. brace
   c. balloon
   d. post-and-beam

6. The method of construction that features widely spaced, heavy framing members is _____ framing.
   a. brace
   b. balloon
   c. platform
   d. post-and-beam

7. In a set of working drawings, the details about the floor used in a building most likely will be defined in the _____.
   a. architectural drawings
   b. structural drawings
   c. mechanical plans
   d. site plans

Refer to the following illustration to answer Questions 8 through 12.

8. The letter *H* on the diagram is pointing to the _____.
   a. sill
   b. termite shield
   c. bearing plate
   d. sill sealer

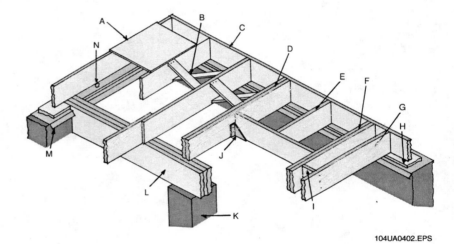

104UA0402.EPS

Instructor's Notes:

_____

_____

_____

_____

_____

_____

_____

_____

9. The letter *C* on the diagram is pointing to the _____.
   a. tail joist
   b. trimmer joist
   c. joist header
   d. common joist

10. The letter *F* on the diagram is pointing to the _____.
    a. tail joist
    b. trimmer joist
    c. joist header
    d. common joist

11. The letter *G* on the diagram is pointing to the _____.
    a. tail joist
    b. trimmer joist
    c. joist header
    d. common joist

12. The letter *L* on the diagram is pointing to the _____.
    a. joist header
    b. beam or girder
    c. column
    d. triple joist

13. For a given size, the type of girder with the least strength is the _____ girder.
    a. built-up
    b. solid lumber
    c. LVL
    d. glulam

14. For a given total load, the size of the girder _____ if the span between its support columns is increased.
    a. can be decreased
    b. must be increased
    c. can remain the same
    d. must be decreased

15. The weight of all moving and variable loads that may be placed on a building is referred to as _____ weight.
    a. dead
    b. live
    c. variable
    d. permanent

16. Notches for pipes made in the top or bottom flanges of wood I-beams shall not exceed _____ of the depth.
    a. one-fourth
    b. one-half
    c. one-sixth
    d. None of the above; notches should never be made in the flanges of wood I-beams.

17. Which of the following panels is *not* used for subflooring?
    a. Plywood
    b. Oriented strand board (OSB)
    c. Tongue-and-groove plywood
    d. Gypsum drywall

18. The first task that should be done when constructing a floor is to _____.
    a. lay out and install sill plates
    b. build and install girders and supports
    c. check the foundation for squareness
    d. lay out the sill plates and girders for joist locations

19. When nailing floor systems, nails should not be spaced closer to one another than _____ their length nor closer to the edge of a framing member than _____ their length.
    a. one-half; one-half
    b. one-half; one-quarter
    c. one-quarter; one-quarter
    d. one-quarter; one-half

20. When laying out the sill for joist locations 16" OC, the first measurement on the sill should be at _____.
    a. 14¼"
    b. 15⅜"
    c. 16"
    d. 15¼"

21. If building a floor to the plan shown in *Figure 28,* how much sill material is needed?
    a. 144'
    b. 128'
    c. 96'
    d. 1,152'

22. For the same structure, how many feet of 2 × 10s are needed for a built-up girder?
    a. 48'
    b. 96'
    c. 144'
    d. 192'

Administer the Module Examination. Be sure to record the results on Craft Training Report Form 200 and submit the form to the Training Program Sponsor.

Performance Profile Test

Ensure that all Performance Profile Tests have been completed and Performance Profile Sheets for each trainee are filled out. Be sure to record the results of the testing on Craft Training Report Form 200 and submit the results to the Training Program Sponsor.

23. For the same structure, what is the total number of 2 × 10 joists needed if the joists are spaced 16" OC?
  a. 36
  b. 37
  c. 72
  d. 74

24. For the same structure, how many feet of wood cross-bridging are needed?
  a. 96'
  b. 192'
  c. 48'
  d. 216'

25. For the same structure, how many 4' × 8' panels of subflooring material are needed?
  a. 1,152
  b. 72
  c. 36
  d. 48

4.46   CARPENTRY LEVEL ONE—TRAINEE MODULE 27104

Instructor's Notes:

_____

_____

_____

_____

_____

_____

_____

_____

Trade Terms Introduced in This Module

Crown: The high point of the crooked edge of a framing member.

Dead load: The weight of permanent, stationary construction and equipment included in a building.

Firestop: An approved material used to fill air passages in a frame to retard the spread of fire.

Foundation: The supporting portion of a structure, including the footings.

Header joist: A framing member used in platform framing into which the common joists are fitted, forming the box sill. Header joists are also used to support the free ends of joists when framing openings in a floor.

Joist hanger: A metal stirrup secured to the face of a structural member, such as a girder, to support and align the ends of joists flush with the member.

Let-in: Any type of notch in a stud, joist, etc., which holds another piece. The item that is supported by the notch is said to be *let in*.

Live load: The total of all moving and variable loads that may be placed upon a building.

Pier: A column of masonry used to support other structural members, typically girders or beams.

Rafter plate: The top or bottom horizontal member at the top of a wall.

Scab: A length of lumber applied over a joint to strengthen it.

Scarf: To join the ends of stock together with a sloping lap joint so there appears to be a single piece.

Sheathing: Boards or sheet material fastened to floors, roofs, or walls.

Soleplate: The bottom horizontal member of a wall frame.

Span: The distance between structural supports such as walls, columns, piers, beams, or girders.

Tail joist: Short joists that run from an opening to a bearing.

Top plate: See *rafter plate*.

Trimmer joist: A full-length joist that reinforces a rough opening in the floor.

Truss: An engineered assembly made of wood, or wood and metal members, that is used to support floors and roofs.

Underlayment: A material, such as particleboard or plywood, laid on top of the subfloor to provide a smoother surface for the finished flooring.

Answers to Review Questions

Answer	Section
1. b	2.2.0
2. b	2.3.0
3. c	2.3.0
4. c	2.1.0
5. a	2.1.0
6. d	2.4.0
7. a	3.1.0
8. a	4.0.0, Figure 6
9. c	4.0.0, Figure 6
10. b	4.0.0, Figure 6
11. d	4.0.0, Figure 6
12. b	4.0.0, Figure 6
13. a	4.2.2
14. b	4.2.5
15. b	4.3.0
16. d	4.3.1
17. d	4.5.0
18. c	5.0.0
19. b	5.3.0
20. d	5.4.0
21. a	7.1.0
22. c	7.2.0
23. d	7.3.0
24. b	7.4.0
25. c	7.5.0

Instructor's Notes:

_____

_____

_____

_____

_____

_____

_____

_____

Additional Resources

This module is intended to present thorough resources for task training. The following reference works are suggested for further study. These are optional materials for continued education rather than for task training.

Carpentry. Homewood, IL: American Technical Publishers.
Carpentry. New York, NY: Delmar Publishers.
Modern Carpentry. Tinley Park, IL: Goodheart-Willcox Company, Inc.

The NCCER makes every effort to keep these textbooks up-to-date and free of technical errors. We appreciate your help in this process. If you have an idea for improving this textbook, or if you find an error, a typographical mistake, or an inaccuracy in the NCCER's Craft Training textbooks, please write us, using this form or a photocopy. Be sure to include the exact module number, page number, a detailed description, and the correction, if applicable. Your input will be brought to the attention of the Technical Review Committee. Thank you for your assistance.

Instructors – If you found that additional materials were necessary in order to teach this module effectively, please let us know so that we may include them in the Equipment/Materials list in the Instructor's Guide.

Write: Curriculum Revision and Development Department
National Center for Construction Education and Research
P.O. Box 141104, Gainesville, FL 32614-1104

Fax: 352-334-0932

E-mail: curriculum@nccer.org

Craft _____ Module Name _____

Copyright Date _____ Module Number _____ Page Number(s) _____

Description _____

_____

_____

_____

(Optional) Correction _____

_____

_____

(Optional) Your Name and Address _____

_____

_____

Wall and Ceiling Framing

27105-01

MODULE OVERVIEW

This module introduces the carpentry trainee to the materials and general procedures used in wall and ceiling framing.

PREREQUISITES

Please refer to the Course Map in the Trainee Module. Prior to training with this module, it is suggested that the trainee shall have successfully completed the following modules:

Core Curriculum; Carpentry Level One, Modules 27101 through 27104

LEARNING OBJECTIVES

Upon completion of this module, the trainee will be able to:

1. Identify the components of a wall and ceiling layout.
2. Describe the procedure for laying out a wood frame wall, including plates, corner posts, door and window openings, partition Ts, bracing, and firestops.
3. Describe the correct procedure for assembling and erecting an exterior wall.
4. Describe the common materials and methods used for installing sheathing on walls.
5. Lay out, assemble, erect, and brace exterior walls for a frame building.
6. Describe wall framing techniques used in masonry construction.
7. Explain the use of metal studs in wall framing.
8. Describe the correct procedure for laying out a ceiling.
9. Cut and install ceiling joists on a wood frame building.
10. Estimate the materials required to frame walls and ceilings.

PERFORMANCE OBJECTIVES

Under the supervision of the instructor, the trainee should be able to:

1. Lay out wall and partition locations on a floor using a blueprint as a guide.
2. Cut studs, trimmers, cripples, and headers to dimension.
3. Assemble corners, partition Ts, and headers.
4. Construct and erect wall sections, including sheathing.
5. Lay out and install ceiling joists.

NCCER STANDARDIZED CRAFT TRAINING PROGRAM

The National Center for Construction Education and Research (NCCER) provides a standardized national program of accredited craft training. Key features of the program include instructor certification, competency-based training, and performance testing. The program provides trainees, instructors, and companies with a standard form of recognition through a National Craft Training Registry. The program is described in full in the *Guidelines for Accreditation*, published by the NCCER. For more information on standardized craft training, contact the NCCER at P.O. Box 141104, Gainesville, FL 32614-1104, 352-334-0911, visit our Web site at www.NCCER.org, or e-mail info@NCCER.org.

HOW TO USE THIS ANNOTATED INSTRUCTOR'S GUIDE

Each page presents two sections of information. The larger section displays each page exactly as it appears in the Trainee Module. The narrow column ties suggested trainee and instructor actions to each page and provides icons to call your attention to material, safety, audiovisual, or testing requirements. The bottom of each page includes space for your notes.

 **Teaching Tip**

If you see the Teaching Tip icon, that means there is a teaching tip associated with this section. Also refer to any suggested teaching tips at the end of the module.

SAFETY CONSIDERATIONS

Ensure that the trainees are equipped with appropriate personal protective equipment.

PREPARATION

Before teaching this module, you should review the Module Outline, the Learning and Performance Objectives, and the Materials and Equipment List. Be sure to allow ample time to prepare your own training or lesson plan and gather all required equipment and materials.

MATERIALS AND EQUIPMENT LIST

Materials:
Transparencies
Markers/chalk
8d common nails
16d box nails
Floor plan
2 × 4 or 2 × 6 framing lumber for studs and joists
2 × 12 header material
1/4" CD plywood for header spacers
1/2" CD plywood
Stock for blocking
Metal brace material
Sheathing material
Joist lumber
Copies of Job Sheets 1 through 5*
Module Examinations*
Performance Profile Sheets*
Videotape (optional) *Wall Framing*
Videotape (optional) *Roof Framing, Part One*

Equipment:
Overhead projector and screen
Whiteboard/chalkboard
Appropriate personal protective equipment
Chalkline
25' tape
Steel tape
Framing hammer
Framing square or speed square
Circular saw
Extension cord
4' level
6' stepladder
Videocassette recorder (VCR)/TV set (optional)

*Packaged with this Annotated Instructor's Guide.

ADDITIONAL RESOURCES

This module is intended to present thorough resources for task training. The following reference works are suggested for both instructors and motivated trainees interested in further study. These are optional materials for continued education rather than for task training.

Carpentry, Leonard Koel. Homewood, IL: American Technical Publishers, 1997.

Carpentry, Gasper J. Lewis. Albany, NY: Delmar Publishers, 2000.

Modern Carpentry, Willis H. Wagner and Howard Bud Smith. Tinley Park, IL: The Goodheart-Willcox Company, Inc., 2000.

Wall Framing, videotape. Gainesville, FL: The National Center for Construction Education and Research.

Roof Framing, Part One, videotape. Gainesville, FL: The National Center for Construction Education and Research.

TEACHING TIME FOR THIS MODULE

An outline for use in developing your lesson plan is presented below. Note that each Roman numeral in the outline equates to one session of instruction. Each session has a suggested time period of 2½ hours. This includes 10 minutes at the beginning of each session for administrative tasks and one 10-minute break during the session. Approximately 20 hours are suggested to cover *Wall and Ceiling Framing*. You will need to adjust the time required for hands-on activity and testing based on your class size and resources.

Topic   **Planned Time**

Session I. Introduction; Components of a Wall; Laying Out a Wall; Measuring and Cutting Studs; Assembling and Erecting Walls

   A. Introduction

   B. Components of a Wall   _____

      1. Corners   _____

      2. Partition Intersections   _____

      3. Headers   _____

         a. Built-Up Headers   _____

         b. Other Types of Headers   _____

   C. Laying Out a Wall   _____

      1. Laying Out Wall Openings   _____

   D. Measuring and Cutting Studs   _____

   E. Assembling the Wall   _____

      1. Firestops   _____

   F. Erecting the Wall   _____

      1. Plumbing and Aligning Walls   _____

         a. Bracing   _____

         b. Sheathing   _____

         c. Panelized Walls   _____

Session II. Laying Out a Wall; Laboratory

   A. Laying Out a Wall

   B. Laboratory   _____

   Hand out Job Sheet 27105-1. Under your supervision, have the trainees perform the tasks on the Job Sheet. Note the proficiency of each trainee on his or her Job Sheet and Skill Test Record.

Session III. Measuring and Cutting Studs; Laboratory

   A. Measuring and Cutting Studs

   B. Laboratory   _____

   Hand out Job Sheet 27105-2. Under your supervision, have the trainees perform the tasks on the Job Sheet. Note the proficiency of each trainee on his or her Job Sheet and Skill Test Record.

Session IV. Assembling Walls; Laboratory

   A. Assembling Walls

   B. Laboratory   _____

   Hand out Job Sheet 27105-3. Under your supervision, have the trainees perform the tasks on the Job Sheet. Note the proficiency of each trainee on his or her Job Sheet and Skill Test Record.

Session V. Erecting Walls; Laboratory

A. Erecting Walls _____

B. Laboratory _____

Hand out Job Sheet 27105-4. Under your supervision, have the trainees perform the tasks on the Job Sheet. Note the proficiency of each trainee on his or her Job Sheet and Skill Test Record.

Session VI. Ceiling Layout and Framing

A. Ceiling Layout and Framing _____

1. Cutting and Installing Ceiling Joists _____

Session VII. Laboratory

A. Laboratory _____

Hand out Job Sheet 27105-5. Under your supervision, have the trainees perform the tasks on the Job Sheet. Note the proficiency of each trainee on his or her Job Sheet and Skill Test Record.

Session VIII. Estimating Materials; Wall Framing in Masonry; Metal Studs in Framing; Module Examination and Performance Testing

A. Estimating Materials _____

B. Wall Framing in Masonry _____

1. Framing Door and Window Openings in Masonry _____

C. Metal Studs in Framing _____

1. Fabrication _____

D. Summary _____

1. Summarize module.

2. Answer questions.

E. Module Examination _____

1. Trainees must score 70% or higher to receive recognition from the NCCER.

2. Record the testing results on Craft Training Report Form 200 and submit the results to the Training Program Sponsor.

F. Performance Testing _____

1. Trainees must perform each task to the satisfaction of the instructor to receive recognition from the NCCER.

2. Record the testing results on Craft Training Report Form 200 and submit the results to the Training Program Sponsor.

Wall and Ceiling Framing

Instructor's Notes:

_____

_____

_____

_____

_____

_____

_____

_____

Course Map

This course map shows all of the modules in the first level of the Carpentry curriculum. The suggested training order begins at the bottom and proceeds up. Skill levels increase as a trainee advances on the course map. The training order may be adjusted by the local Training Program Sponsor.

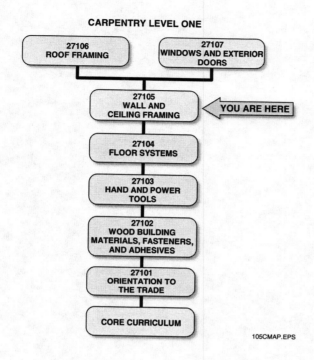

CARPENTRY LEVEL ONE

27106
ROOF FRAMING

27107
WINDOWS AND EXTERIOR DOORS

27105
WALL AND CEILING FRAMING ◄ YOU ARE HERE

27104
FLOOR SYSTEMS

27103
HAND AND POWER TOOLS

27102
WOOD BUILDING MATERIALS, FASTENERS, AND ADHESIVES

27101
ORIENTATION TO THE TRADE

CORE CURRICULUM

105CMAP.EPS

Assign reading of Module 27105.

MODULE 27105 CONTENTS

1.0.0 INTRODUCTION 5.1

2.0.0 COMPONENTS OF A WALL 5.1

    2.1.0 Corners 5.2

    2.2.0 Partition Intersections 5.4

    2.3.0 Headers 5.4

    2.3.1 Built-Up Headers 5.5

    2.3.2 Other Types of Headers 5.7

3.0.0 LAYING OUT A WALL 5.7

    3.1.0 Laying Out Wall Openings 5.8

4.0.0 MEASURING AND CUTTING STUDS 5.11

5.0.0 ASSEMBLING THE WALL 5.14

    5.1.0 Firestops 5.15

6.0.0 ERECTING THE WALL 5.18

    6.1.0 Plumbing and Aligning Walls 5.18

    6.1.1 Bracing 5.20

    6.1.2 Sheathing 5.20

    6.1.3 Panelized Walls 5.23

7.0.0 CEILING LAYOUT AND FRAMING 5.23

    7.1.0 Cutting and Installing Ceiling Joists 5.25

8.0.0 ESTIMATING MATERIALS 5.27

9.0.0 WALL FRAMING IN MASONRY 5.30

    9.1.0 Framing Door and Window Openings in Masonry 5.32

10.0.0 METAL STUDS IN FRAMING 5.32

    10.1.0 Fabrication 5.32

SUMMARY 5.32

REVIEW QUESTIONS 5.34

PROFILE IN SUCCESS 5.35

GLOSSARY 5.36

ANSWERS TO REVIEW QUESTIONS 5.37

ADDITIONAL RESOURCES 5.38

Figures

Figure 1   Wall and partition framing members 5.2

Figure 2   Corner construction typical of western platform framing 5.3

Figure 3   Alternative method of corner construction 5.3

Figure 4   Constructing nailing surfaces for partitions 5.4
Figure 5   Two more ways to construct nailing surfaces 5.5
Figure 6   Partition layout 5.5
Figure 7   Two types of built-up headers 5.6
Figure 8   Other types of headers 5.7
Figure 9   Soleplate and top plate positioned and tacked in place 5.8
Figure 10   Marking stud locations 5.8
Figure 11   Sample floor plan 5.9
Figure 12   Door, window, and header schedules 5.10
Figure 13   Laying out a wall opening 5.12
Figure 14   Window and door framing 5.12
Figure 15   Example of soleplate and top plate marked for layout 5.13
Figure 16   Calculating the length of a regular stud 5.13, 5.14
Figure 17   Firestops 5.15
Figure 18   Wall lifting jack 5.18
Figure 19   Plumbing and aligning a wall 5.19
Figure 20   Double top plate layout 5.19
Figure 21   Use of metal bracing 5.20
Figure 22   Marking joist and rafter locations on the double top plate 5.24
Figure 23   Splicing ceiling joists 5.25
Figure 24   Cutting joist ends to match roof pitch 5.26
Figure 25   Reinforcing ceiling joists 5.26
Figure 26   Sample floor plan 5.27
Figure 27   Partition backing using a 1 × 6 . . . . .5.31
Figure 28   Partition backing using 2 × 4 blocks 5.31
Figure 29   Placement of furring strips at corners of masonry walls 5.32
Figure 30   Metal framing 5.33

Tables

Table 1   Maximum Span for Exterior Headers 5.7
Table 2   Ceiling Joist Spacing and Allowable Spans 5.24

Instructor's Notes:

_____

_____

_____

_____

_____

_____

_____

_____

Wall and Ceiling Framing

Ensure you have everything required to teach the course. Check the Materials and Equipment List at the front of this Instructor's Guide.

Objectives

Upon completion of this module, the trainee will be able to:

1. Identify the components of a wall and ceiling layout.
2. Describe the procedure for laying out a wood frame wall, including plates, corner posts, door and window openings, partition Ts, bracing, and firestops.
3. Describe the correct procedure for assembling and erecting an exterior wall.
4. Describe the common materials and methods used for installing sheathing on walls.
5. Lay out, assemble, erect, and brace exterior walls for a frame building.
6. Describe wall framing techniques used in masonry construction.
7. Explain the use of metal studs in wall framing.
8. Describe the correct procedure for laying out a ceiling.
9. Cut and install ceiling joists on a wood frame building.
10. Estimate the materials required to frame walls and ceilings.

Prerequisites

Successful completion of the following Task Modules is recommended before beginning study of this Task Module: Core Curriculum; Carpentry Level One, Modules 27101 through 27104.

Required Trainee Materials

1. Trainee Task Module
2. Appropriate personal protective equipment

1.0.0 ◆ INTRODUCTION

In an earlier module, you learned about laying out and erecting a wood frame floor. In this module, you will learn about laying out and erecting walls with openings for windows and doors. Also covered in this module are instructions for laying out and installing ceiling joists.

Keep in mind when framing walls and ceilings that precise layout of framing members is extremely important. Finish material such as sheathing, drywall, paneling, etc., is sold in 4 × 8 sheets. If the studs, rafters, and joists are not straight and evenly spaced for their entire length, you will not be able to fasten the sheet material to them. A tiny error on one end becomes a large error as you progress toward the other end. Spacings of 16" and 24" on center are used because they will divide evenly into 48".

2.0.0 ◆ COMPONENTS OF A WALL

Figure 1 identifies the structural members of a wood frame wall. Each of the members shown on the illustration is then described. You will need to know these terms as you proceed through this module.

- *Blocking (spacer)*—A wood block used as a filler piece and support between framing members.
- *Cripple stud*—In wall framing, a short framing stud that fills the space between a header and a top plate or between the sill and the soleplate.
- *Double top plate*—A plate made of two members to provide better stiffening of a wall. It is also used for connecting splices, corners, and partitions that are at right angles (perpendicular) to the wall.
- *Header*—A horizontal structural member that supports the load over an opening such as a door or window.

Show Transparency 1, Course Objectives.

Show Transparency 2, Performance Profile Tasks.

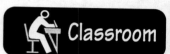

Identify the basic components of a wall.

Show Transparency 3 (Figure 1).

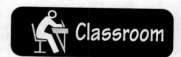

Point out that terms shown in bold (blue) are defined in the Glossary at the back of the module.

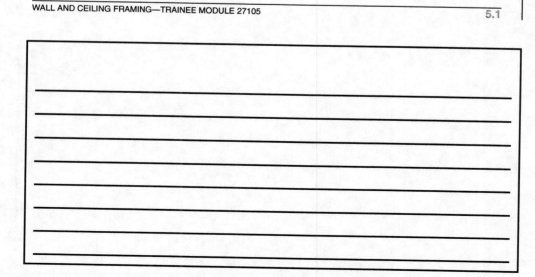

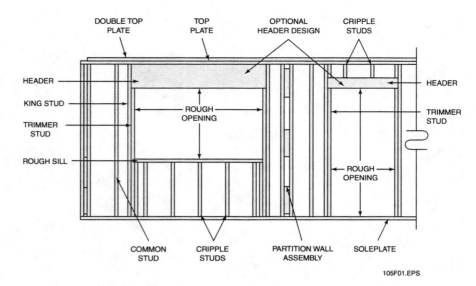

Figure 1 ◆ Wall and partition framing members.

- *King stud*—The full-length stud next to the trimmer stud in a wall opening.
- *Partition*—A wall that subdivides space within a building. A *bearing* partition or wall is one that supports the floors and roof directly above in addition to its own weight.
- *Rough opening*—An opening in the framing formed by framing members, usually for a window or door.
- *Rough sill*—The lower framing member attached to the top of the lower cripple studs to form the base of a rough opening for a window.
- *Soleplate*—The lowest horizontal member of a wall or partition to which the studs are nailed. It rests on the rough floor.
- *Stud*—The main vertical framing member in a wall or partition.
- *Top plate*—The upper horizontal framing member of a wall used to carry the roof trusses or rafters.
- *Trimmer stud*—The vertical framing member that forms the sides of rough openings for doors and windows. It provides stiffening for the frame and supports the weight of the header.

This section contains an overview of the layout and assembly requirements and procedures for walls.

2.1.0 Corners

When framing a wall, you must have solid corners that can take the weight of the structure. In addition to contributing to the strength of the structure, corners must provide a good nailing surface for sheathing and interior finish materials. Carpenters generally select the straightest, least defective studs for corner framing.

Figure 2 shows the method typically used in western platform framing. In one wall assembly, there are two common studs with blocking between them. This provides a nailing surface for the first stud in the adjoining wall. Notice the use of a double top plate at the top of the wall to provide greater strength.

Figure 3 shows a different way to construct a corner. It has several advantages:

- It doesn't require blocking, which saves time and materials.
- It results in fewer voids in the insulation.
- It promotes better coordination among trades. For example, an electrician running wiring through the corner shown in *Figure 2* would need to bore holes through two or three studs and, possibly, a piece of blocking. However, an electrician wiring through the corner shown in *Figure 3* would need to bore through only two studs.

Instructor's Notes:

_____

_____

_____

_____

_____

_____

_____

_____

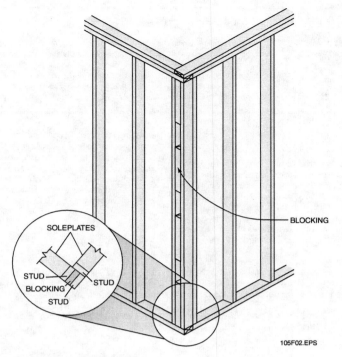

SOLEPLATES

STUD

BLOCKING

STUD

STUD

BLOCKING

105F02.EPS

Figure 2 ◆ Corner construction typical of western platform framing.

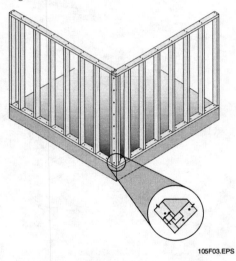

105F03.EPS

Figure 3 ◆ Alternative method of corner construction.

Explain the different methods of construction for partition intersections. Point out the differences in layout plan dimensions.

Show Transparencies 6 through 8 (Figures 4 through 6).

Lay out a partition T location.

Discuss various types of header construction. Emphasize the importance of selecting the correct header type and size for the support of overhead loads.

INSIDE TRACK

Framing Methods

Western platform framing is a method of construction in which a first floor deck is built on top of the foundation walls. Then, the first floor walls are erected on top of the platform. Upper floor platforms are built on top of the first floor walls, and upper floor walls are erected on top of the upper floor platforms. In balloon framing, which is a method seldom used today, the studs extend from the sill plate to the rafter plate. Balloon framing requires the use of much longer studs.

2.2.0 Partition Intersections

Interior partitions must be securely fastened to outside walls. For that to happen, there must be a solid nailing surface where the partition intersects the exterior frame. *Figure 4* shows a common way to construct a nailing surface for the partition intersections or Ts. The nailing surface can be a full stud nailed perpendicular between two other full studs, or it can be short pieces of 2 × 4 lumber, known as *blocking*, nailed between the two other full studs.

Figure 5 shows two other ways to prepare a nailing surface. Compare *Figures 4* and *5*. Notice that in *Figure 4* the spacing between studs differs, but in *Figure 5* the spacing between studs remains the same.

To lay out a partition location, measure from the end of the wall to the centerline of the partition opening, then mark the locations for the partition studs on either side of the centerline (*Figure 6*).

Although plans normally show stud-to-stud dimensions, they sometimes show finish-to-finish dimensions. Center-to-center dimensions are typically used in metal framing, which will be discussed later in this module.

2.3.0 Headers

When wall framing is interrupted by an opening such as a window or door, a method is needed to distribute the weight of the structure around the opening. This is done by the use of a header. The header is placed so that it rests on the trimmer studs, which transfer the weight to the soleplate or subfloor, and then to the foundation.

The width of a header should be equal to the rough opening, plus the width of the trimmer

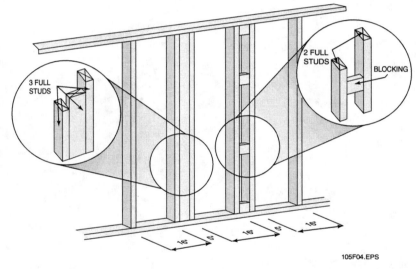

Figure 4 ◆ Constructing nailing surfaces for partitions.

CARPENTRY LEVEL ONE—TRAINEE MODULE 27105

Instructor's Notes:

_____

_____

_____

_____

_____

_____

_____

_____

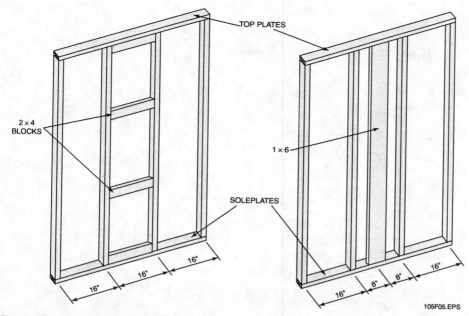

Figure 5 ◈ Two more ways to construct nailing surfaces.

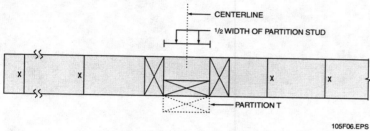

Figure 6 ◈ Partition T layout.

studs. For example, if the rough opening for a 3' wide window is 38", the width of the header would be 41" (1½" trimmer stud plus 38" rough opening plus the other 1½" trimmer stud, or 41" total).

2.3.1 Built-Up Headers

Headers are usually made of built-up lumber (although solid wood beams are sometimes used as headers). Built-up headers are usually made from 2 × lumber separated by ½" plywood spacers (*Figure 7*). A full header is used for large openings and fills the area from the rough opening to the bottom of the top plate. A small header with cripple studs is suitable for average-size windows and doors and is usually made from 2 × 4 or 2 × 6 lumber.

Table 1 gives the maximum span typically used for various load conditions.

Show Transparencies 9 and 10 (Figures 7 and 8).

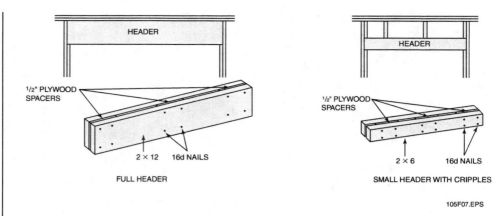

FULL HEADER

SMALL HEADER WITH CRIPPLES

105F07.EPS

Figure 7 ◆ Two types of built-up headers.

Headers

Headers can be constructed in many ways, some of which are shown here.

HEADER WITH CRIPPLES 105P0501.EPS

SOLID HEADER 105P0502.EPS

GARAGE DOOR HEADER 105P0503.EPS
USING ENGINEERED LUMBER

Instructor's Notes:

_____

_____

_____

_____

_____

_____

_____

_____

Table 1 Maximum Span for Exterior Headers

Header Material Two Members on Edge	Load		
	Single Story	Two Stories	Three Stories
2 × 4	3'–6"	2'–6"	2'
2 × 6	6'	5'	4'
2 × 8	8'	7'	6'
2 × 10	10'	8'	7'
2 × 12	12'	9'	8'

2.3.2 Other Types of Headers

Figure 8 shows some other types of headers that are used in wall framing. Carpenters often use truss headers when the load is very heavy or the span is extra wide. The architect's plans usually show the design of the trusses.

Other types of headers used for heavy loads are wood or steel I-beams, box beams, and engineered wood products such as laminated veneer lumber (LVL), parallel strand lumber (PSL), and laminated lumber (glulam).

3.0.0 ◆ LAYING OUT A WALL

This section covers the basic procedures for laying out wood frame walls with correctly sized window and door openings and partition Ts. Later in this module, you will be introduced to methods for framing with metal studs, and framing window and door openings in masonry walls.

Wall framing is generally done with 2 × 4 studs spaced 16" on center. In a one-story building, 2 × 4 spacing can be 24" on center. If 24" spacing is used in a two-story building, the lower floor must be framed with 2 × 6 lumber. The following provides an overview of the procedure for laying out a wall.

Step 1 Mark the locations of the soleplates by measuring in the width of the soleplate (e.g., 3½") from the outside edge of the sill on each corner. Snap a chalkline to mark the soleplate location, then repeat this for each wall.

Step 2 The top plate and soleplate are laid out together. Start by placing the soleplate as indicated by the chalkline and tacking it in place (*Figure 9*). Lay the top plate against the soleplate so that the location of framing members can be transferred from the soleplate to the top plate. Also tack the top plate. Tacking prevents the plates from moving, which would make the critical layout lines inaccurate.

Step 3 Lay out the common stud positions. To begin, measure and square a line 15¼" from one end. Subtracting this ¾" ensures that sheathing and other panels will fall at the center of the studs. Drive a nail at that point and use a continuous tape to measure and mark the stud locations every 16" (*Figure 10*). Align your framing square at each mark. Scribe a line along each side of the framing square tongue across both the soleplate and top plate. These lines will show the outside edges of each stud, centered on 16" intervals.

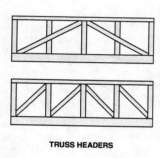

TRUSS HEADERS

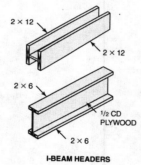

2 × 12
2 × 12
2 × 6
½ CD PLYWOOD
2 × 6

I-BEAM HEADERS

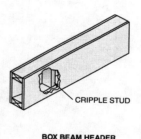

CRIPPLE STUD

BOX BEAM HEADER

105F08.EPS

Figure 8 ◆ Other types of headers.

WALL AND CEILING FRAMING—TRAINEE MODULE 27105

5.7

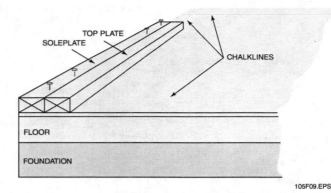

105F09.EPS

Figure 9 ◆ Soleplate and top plate positioned and tacked in place.

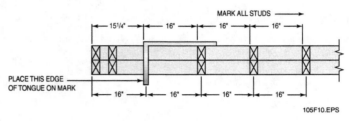

105F10.EPS

Figure 10 ◆ Marking stud locations.

3.1.0 Laying Out Wall Openings

The floor plan drawings for a building (*Figure 11*) show the locations of windows and doors. Notice that each window and door on the floor plan is identified by size. In this case, the widths are shown in feet and inches, but the complete dimension information is coded. Look at the window in bedroom #1 coded 2630. This means the window is 2'–6" wide (the width is always given first) and 3'–0" high. Similar codes give widths and heights for the doors.

The window and door schedules (*Figure 12*) provided in the architect's drawings will list the dimensions of windows and doors, along with types, manufacturers, and other information.

Placement of windows is important, and is normally dealt with on the architectural drawings. A good rule of thumb is to avoid placing horizontal framework at eye level. This means you have to consider whether people will generally be sitting or standing when they look out the window. Unless they are architectural specialty windows, the tops of all windows should be at the same height. The bottom height will vary depending on the use. For example, the bottom of a window over the kitchen sink should be higher than that of a living room or dining room window. The standard height for a residential window is 6'–8" from the floor to the bottom of the window top (head) jamb.

CARPENTRY LEVEL ONE—TRAINEE MODULE 27105

Explain how to interpret floor plans and schedules. Point out that floor plans normally use centerline dimensions from wall centers to wall openings.

Audiovisual

Show Transparencies 13 and 14 (Figures 11 and 12).

Instructor's Notes:

5.8

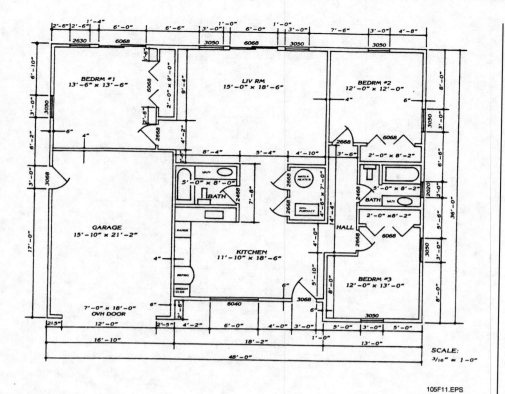

Figure 11 ◆ Sample floor plan.

105F11.EPS

Rough Opening Dimensions

The residential plan shown in *Figure 11* shows dimensions to the sides of rough openings—that is, from the corner of the building to the near side of the first rough opening, then from the far side of the first rough opening to the near side of the second rough opening. However, it is more common for plans to show centerline dimensions—that is, from the corner of the building to the center of the first rough opening, then from the center of the first rough opening to the center of the second rough opening.

DOOR SCHEDULE

NO.	SIZE	DOOR MAT'L	TYPE	H.W.#	FRAME MAT'L	TYPE	HEAD	JAMB	SILL	REMARKS
104A	PR. 3'-0" x 7-8⁵/₁₆" x 1-³/₄"	W.D.	IV	11	W.D.	—	8/16	4/16	3/16	
104B	PR. 3'-0" x 8'-0" x 1-³/₄"	GLASS	VI	10	—	—	3/17	15/16, 16/16	4/7	W/ FULL GLASS SIDELITE
105A	3'-0" x 7'-2" x 1-³/₄"	W.D.	I	1	H.M.	II	5/15	6/17, 7/7	—	
105B			I	1		I	SIM. 3/16, 10	SIM. 4/16, 10	—	
106			I	1		I	1/15	2/15	—	
107			I	1		I	1/15	2/15	—	
108	PR. 3'-0" x 7'-2" x 1-³/₄"		II	4		II	8/15	9/15	—	"C" LABEL
109	3'-0" x 7'-2" x 1-³/₄"		I	7		I	3/15	4/15	—	
110			I	7		I	3/15	4/15	—	
112			I	13		I	SIM. 3/16, 10	SIM. 4/16, 10	—	
113A			I	13		III	5/15	6/7	—	"C" LABEL
113B			I	13		I	SIM. 1/16, 10	SIM. 2/16, 10	—	
114	2'-6" x 7'-2" x 1-³/₄"		I	13		I	SIM. 1/16, 10	SIM. 2/16, 10	—	
115A	PR. 3'-0" x 7'-0" x 1-³/₄"	ALUM.	V	14	ALUM.	—	28/15	33/15	38/15	
115B	PR. 3'-0" x 7'-2" x 1-³/₄"	W.D.	III	5	H.W.	I	10/15	11/16, 12/16	—	"C" LABEL NOTE 1
121D				17			4/15	28/10	—	

WINDOW SCHEDULE

SYMBOL	WIDTH	HEIGHT	MAT'L	TYPE	SCREEN & DOOR	QUANTITY	REMARKS	MANUFACTURER	CATALOG NUMBER
A	3'-8"	3'-0"	ALUM.	DOUBLE HUNG	YES	2	4 LIGHTS, 4 HIGH	LBJ WINDOW CO.	141 PW
B	3'-8"	5'-0"	ALUM.	DOUBLE HUNG	YES	1	4 LIGHTS, 4 HIGH	LBJ WINDOW CO.	145 PW
C	3'-0"	5'-0"	ALUM.	STATIONARY	STORM ONLY	2	SINGLE LIGHTS	H & J GLASS CO.	59 PY
D	2'-0"	3'-0"	ALUM.	DOUBLE HUNG	YES	1	4 LIGHTS, 4 HIGH	LBJ WINDOW CO.	142 PW
E	2'-0"	6'-0"	ALUM.	STATIONARY	STORM ONLY	2	20 LIGHTS	H & J GLASS CO.	37 TS
F	3'-6"	5'-0"	ALUM.	DOUBLE HUNG	YES	1	16 LIGHTS, 4 HIGH	LBJ WINDOW CO.	141 PW

HEADER SCHEDULE

HEADER SIZE	EXTERIOR 26' + UNDER	EXTERIOR 26' TO 32'	INTERIOR 26' + UNDER	INTERIOR 26' TO 32'
(2) 2 x 4	3'-6"	3'-0"	USE (2) 2 x 6	
(2) 2 x 6	6'-6"	6'-0"	4'-0"	3'-0"
(2) 2 x 8	8'-6"	8'-0"	5'-6"	5'-0"
(2) 2 x 10	11'-0"	10'-0"	7'-0"	6'-6"
(2) 2 x 12	13'-6"	12'-0"	8'-6"	8'-0"

105F12.EPS

Figure 12 ◆ Door, window, and header schedules.

Instructor's Notes:

_____

_____

_____

_____

_____

_____

_____

_____

The window and door schedules will sometimes provide the rough opening dimensions for windows and doors. Another good source of information is the manufacturer's catalog. It will provide rough and finish opening dimensions, as well as the unobstructed glass dimensions.

When roughing-in a window, the rough opening width equals the width of the window plus the thickness of the jamb material; this is usually 1½" (¾" on each side), plus the shim clearance (½" on each side). Therefore, the rough opening for a 3' window would be 38½". The height of the rough opening is figured in the same way. Be sure to check the manufacturer's instructions for the dimensions of the windows you are using.

To lay out a wall opening, proceed as follows:

Step 1 Measure from the corner to the start of the opening, then add half the width of the window or door to determine the centerline (*Figure 13*). Mark the locations of the full studs and trimmer studs by measuring in each direction from the centerline. Mark the cripples 16" on center starting with one trimmer.

Each window opening requires a common stud (king stud) on each side, plus a header, cripple studs, and a sill (*Figure 14*). If the window is more than 4' wide, local codes may require a double sill. Door openings also require trimmer studs, king studs, and a header. Cripple studs will be needed unless the door is double-wide. In that case, a full header may be called for.

Step 2 Mark the location of each regular stud and king stud (X), trimmer (T), and cripple (C), as shown in *Figure 15*. (This is a suggested marking method. The only important thing is to mark the locations with codes that you and other members of your crew will recognize.)

4.0.0 ◆ MEASURING AND CUTTING STUDS

It is extremely important to precisely measure the first one of each type of stud that will be used (common, trimmer, and cripple) as a template for the others.

Regular and king studs — *Figure 16* shows the methods for determining the exact length of a regular stud for installation on a slab or wood floor.

To determine the stud length when the installation is directly on a concrete slab, simply subtract the thickness of the soleplate (1½") and double top plate (3") from the desired ceiling height and add the thickness of the ceiling material. In the case shown in *Figure 16(A)*, the length of the stud is based on the ceiling height, which is 96", plus the ½" thickness of the ceiling material, less the combined plate thicknesses of 4½", or 92". (This assumes that the flooring material has no appreciable thickness.)

In the example shown in *Figure 16(B)*, the thickness of the underlayment must also be considered. Therefore, the length of the stud should be 92⅝"; i.e., ceiling height plus the combined thicknesses of the ceiling material and underlayment (½" + ⅝"), less the combined thicknesses of the plates (4½"). (Again, we are assuming a flooring material of no appreciable thickness.)

Figure 16(C) shows an interior, non-bearing wall that does not require the use of a double top plate. Therefore, the calculated stud length is 1½" longer (94⅛").

Trimmers—The length of a window or door trimmer stud is determined by subtracting the thickness of the soleplate from the height of the header. If the installation is on a wood floor, the thickness of the underlayment must also be subtracted.

Cripples—To determine the length of a cripple stud above a door or window, combine the height of the trimmer and the thickness of the header, then subtract that total from the length of a regular stud. To determine the length of a cripple stud below a window, determine the height of the rough opening from the floor, then subtract the combined thicknesses of the rough sill and soleplate.

INSIDE TRACK

Check Stud Lengths

Precut studs in various lengths are available from many lumberyards. These precut studs are ideal for walls on built-up wood floors for the 96" finished ceiling, a very common finished ceiling height. For example, studs precut to 92⅝" are used with exterior or loadbearing interior walls with double top plates.

Sometimes lumberyards deliver the wrong size of precut studs. Unless you look closely, a precut stud doesn't look much different from a standard 8' (96") stud. Always check the precut studs to make sure they are the right length. Taking a few seconds to measure before you start might save hours of rebuilding later on.

Demonstration

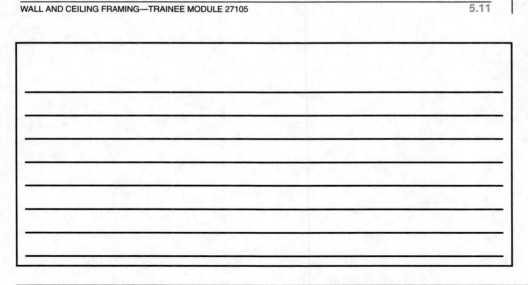

Demonstrate wall opening layout procedures.

Audiovisual

Show Transparencies 15 through 17 (Figures 13 through 15).

Demonstration

Demonstrate how to measure and cut various types of studs for a wall with a window opening. Point out that it is better to measure and cut cripples after the walls are assembled.

Audiovisual

Show Transparencies 18 through 20 (Figures 16A through 16C).

Teaching Tip

For wall and ceiling framing demonstrations or laboratory exercises, you may wish to use half-size (half-scale) layout dimensions and lumber (for example, 1 × 2s = 2 × 4s, 1 × 4s = 2 × 8s, etc.).

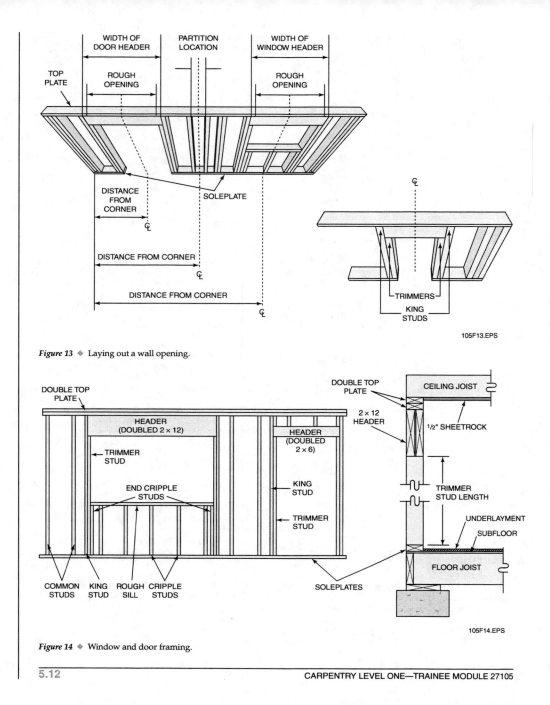

Figure 13 ◆ Laying out a wall opening.

Figure 14 ◆ Window and door framing.

CARPENTRY LEVEL ONE—TRAINEE MODULE 27105

Instructor's Notes:

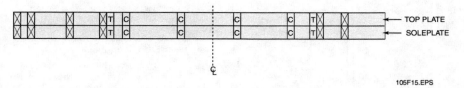

Figure 15 ◈ Example of soleplate and top plate marked for layout.

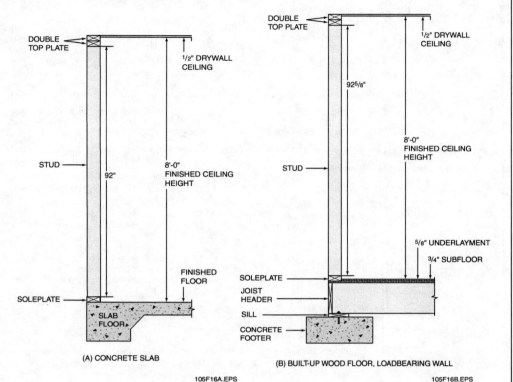

(A) CONCRETE SLAB

105F16A.EPS

(B) BUILT-UP WOOD FLOOR, LOADBEARING WALL

105F16B.EPS

Figure 16 (1 of 2) ◈ Calculating the length of a regular stud.

Measure Twice, Cut Once

Before cutting all the studs, double-check your measurements, or have someone else do it. You do not want to find out there was an error in calculation after you have cut 200 pieces of 2 × 4. When you are satisfied with the measurements, cut the required amounts for each type of stud and stack them neatly near the wall locations. Be sure to allow ample room to assemble the walls.

Emphasize the importance of rechecking dimensions of the first (template) stud of each type. Point out that dimensions of precut studs must also be checked.

Demonstrate wall assembly. Point out that wall studs with slight crowns should all face in the same direction and bowed studs should not be used.

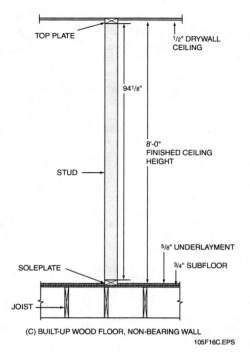

TOP PLATE

½" DRYWALL CEILING

94⅛"

STUD

8'-0" FINISHED CEILING HEIGHT

⅝" UNDERLAYMENT

¾" SUBFLOOR

SOLEPLATE

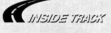

JOIST

(C) BUILT-UP WOOD FLOOR, NON-BEARING WALL

105F16C.EPS

Figure 16 (2 of 2) ◆ Calculating the length of a regular stud.

5.0.0 ◆ ASSEMBLING THE WALL

The preferred procedure for assembling the wall is to lay out and assemble the wall on the floor with the inside of the wall facing down.

Step 1 Start by laying the soleplate near the edge of the floor. Then, place the top plate about a regular stud length away from the soleplate. Be sure to use treated lumber if the soleplate is in contact with a masonry floor.

Step 2 Assemble the corners and partition Ts using the straightest pieces to ensure that the corners are plumb. Also, save some of the straightest studs for placement in the wall where countertops or fixtures will hit the centers of studs (such as in kitchens, bathrooms, and laundry rooms).

Step 3 Lay a regular stud at each X mark with the crown up. If a stud is bowed, replace it and use it to make cripples.

Step 4 Assemble the window and door headers and put them in place with the crowns up.

Step 5 Lay out and assemble the rough openings, making sure that each opening is the correct size and that it is square.

Step 6 Nail the framework together. For 2 × 4 framing, use two 16d nails through the plate into the end of each stud. For 2 × 6 framing, use three nails. The use of a nail gun is recommended for this purpose; however, *do not attempt this if you have not received proper training in the use of this tool.*

INSIDE TRACK

Cripple Studs

It's good to know how to calculate the proper lengths as the text describes; however, in practice, lumber dimensions and assembly tolerances can vary. Therefore, it's better to hold off until after the headers, trimmers, and rough sills are assembled, and then actually measure and cut the cripples to the lengths needed.

THINK ABOUT IT

Determining Stud Length

1. If the finish ceiling height of a building with built-up wood floors is supposed to be 9'-6", how long are the studs in exterior walls?
2. If the top of the rough sill of a window is supposed to be 32" above the floor, how long are the cripple studs below the sill?

Discuss the "Think About It."

Discuss the "Think About It."

CARPENTRY LEVEL ONE—TRAINEE MODULE 27105

Instructor's Notes:

_____

_____

_____

_____

_____

_____

_____

_____

CARPENTRY LEVEL ONE—INSTRUCTOR'S GUIDE MODULE 27105

Coping with Natural Defects

Because wood is a natural material, it will have variations that can be considered defects. Lumber is almost never perfectly straight. Even when a piece of lumber is sawn straight, it will most likely curve, twist, or split as it dries.

In the old days, wood cut at a sawmill was just stacked and left to air dry. Normal changes in daily air temperature, humidity, and other conditions would result in lumber with very interesting shapes. Nowadays, lumber is dried in a kiln (a large, low-temperature oven), which will reduce the amount of twisting and curving. But even modern lumber is still somewhat distorted.

Virtually all lumber is slightly (or, sometimes, greatly) curved along its narrow side (the 1½" dimension in 2 × 4s). This is called a *crown*. It's normal, and, unless it is extreme, the crown doesn't prevent the lumber from being used.

When assembling a floor, you should sight down each piece of lumber to determine which side has the crown. Mark the crowned side. When using the lumber in a floor, position all the lumber crown side up. Weight on the floor will cause the crown to flatten out.

In a wall, there's no force that will flatten the crown. But, for a more uniform nailing surface for the sheathing, you should position all crowns in the same direction. In the example to follow, the crowns are placed up, so when the wall is erected the crown side will be toward the exterior of the building.

A curve along the wide side of lumber (the 3½" dimension in 2 × 4s) is called a *bow*. If a noticeably bowed stud was used in a wall, sheathing could not be nailed to it in a straight line. Either the nail would miss the bowed stud, or you would have to take extra time to lay out a curved line to nail through. That's not good use of your time, so discard the bowed stud and use a straight one instead.

5.1.0 Firestops

In some areas, local building codes may require firestops. Firestops are short pieces of 2 × 4 blocking (or 2 × 6 pieces if the wall is framed with 2 × 6 lumber) that are nailed between studs. See *Figure 17*.

Without firestops, the space between the studs will act like a flue in a chimney. Any holes drilled through the soleplate and top plate create a draft, and air will rush through the space. In a fire, air, smoke, gases, and flames can race through the chimney-like space.

The installation of firestops has two purposes. First, it slows the flow of air, which feeds a fire through the cavity. Second, it can actually block flames (temporarily, at least) from traveling up through the cavity.

If the local code requires firestops, it may also require that holes through the soleplate and top plate (for plumbing or electrical runs) be plugged with a firestopping material to prevent airflow.

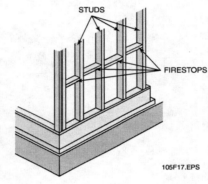

STUDS

FIRESTOPS

105F17.EPS

Figure 17 ◆ Firestops.

Explain that local codes may require firestops. Mention that firestop blocking can also be used to stabilize long studs in partitions of high ceiling rooms in new construction.

Show Transparency 21 (Figure 17).

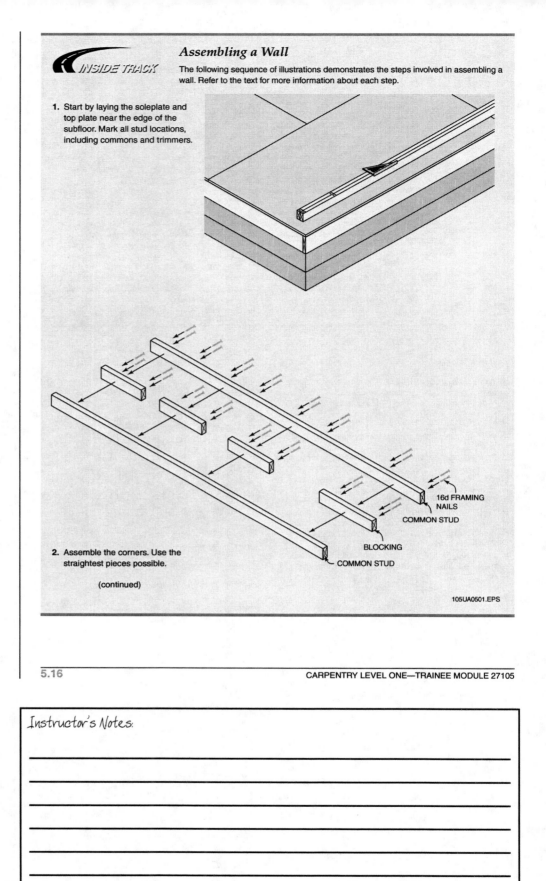

Assembling a Wall

The following sequence of illustrations demonstrates the steps involved in assembling a wall. Refer to the text for more information about each step.

INSIDE TRACK

1. Start by laying the soleplate and top plate near the edge of the subfloor. Mark all stud locations, including commons and trimmers.

2. Assemble the corners. Use the straightest pieces possible.

(continued)

16d FRAMING NAILS

COMMON STUD

BLOCKING

COMMON STUD

105UA0501.EPS

Instructor's Notes:

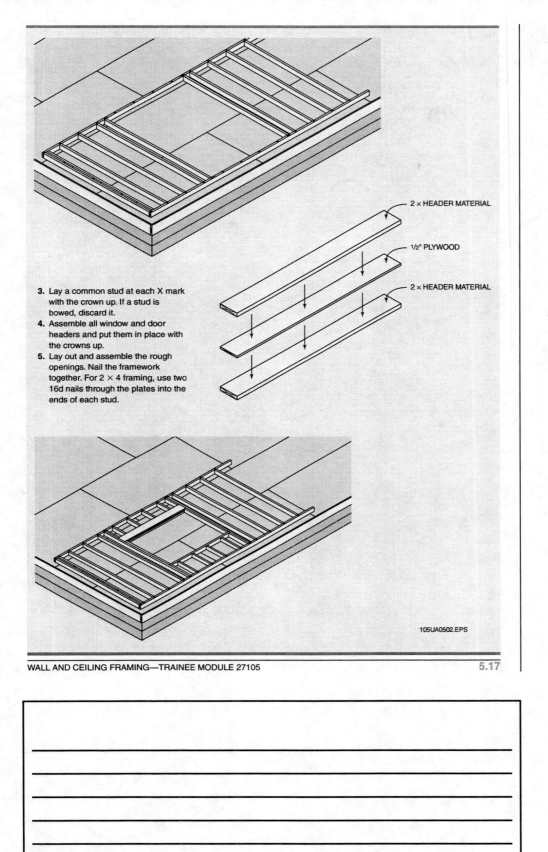

3. Lay a common stud at each X mark with the crown up. If a stud is bowed, discard it.
4. Assemble all window and door headers and put them in place with the crowns up.
5. Lay out and assemble the rough openings. Nail the framework together. For 2 × 4 framing, use two 16d nails through the plates into the ends of each stud.

2 × HEADER MATERIAL

1/2" PLYWOOD

2 × HEADER MATERIAL

105UA0502.EPS

Demonstrate the erection of a wall with the help of the trainees. Emphasize that banding or cleats should be used to prevent the wall from skidding when it is raised.

Show Transparencies 22 through 24 (Figures 18 through 20).

Demonstrate how to plumb/align walls using temporary bracing.

6.0.0 ◆ ERECTING THE WALL

There are four primary steps in erecting a wall:

Step 1 If the sheathing was installed with the wall laying down, or if the wall is very long, it will probably be too heavy to be lifted into place by the framing crew. In that case, use a crane or the special lifting jacks made for that purpose (*Figure 18*). Use cleats to prevent the wall from sliding.

Step 2 Raise the wall section and nail it in place using 16d nails on every other floor joist. On a concrete slab, use preset anchor bolts or powder-actuated pins. *Do not attempt this if you have not received proper training and certification in the use of these tools.*

Step 3 Plumb the corners and apply temporary exterior bracing. Then erect, plumb, and brace the remaining walls. The bracing helps keep the structure square and will prevent the walls from being blown over by the wind. Generally, the braces remain in place until the roof is complete.

Step 4 As the walls are erected, straighten the walls and nail temporary interior bracing in place.

6.1.0 Plumbing and Aligning Walls

Accurate plumbing of the corners is possible only after all the walls are up. Always use a straightedge along with a hand level (*Figure 19*). The straightedge can be a piece of 2 × 4 lumber. Blocks ¾" thick are nailed to each end of the 2 × 4. The blocks make it possible to accurately plumb the wall from the bottom plate to the top plate. (If you just placed the level directly against the wall, any bow or crown in the end stud would give a false reading.)

The plumbing of corners requires two people working together. One carpenter releases the nails at the bottom end of the corner brace so that the top of the wall can be moved in or out. At the same time, the second carpenter watches the level. The bottom end of the brace is renailed when the level shows a plumb wall.

Install the second plate of all the double top plates (*Figure 20*). In addition to adding strength in bearing walls, the second plate helps to straighten a bowed or curved wall. If bows are turned opposite each other, intersections of walls should be double plated after walls are erected.

Figure 18 ◆ Wall lifting jack.

105F18.EPS

Instructor's Notes:

_____

_____

_____

_____

_____

_____

_____

_____

Raising a Wall

When a wall is being erected, either by hand or with a lifting jack, the bottom of the wall can slide forward. If the wall slides off the floor platform, the wall or objects on the ground below it can be damaged and workers can be injured.

Use wood blocks or cleats securely nailed to the outside of the rim joist to catch the wall as it slides forward, preventing damage or injury.

Some carpenters use metal banding (used to bundle loads of wood from the mill) to achieve the same effect. One end of a short length of banding is nailed to the floor platform. The other end is bent up 90° and nailed to the bottom of the soleplate. When the wall is raised, the flexible band straightens out horizontally, much like a hinge, but the wall cannot slide forward.

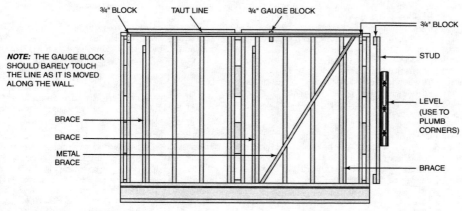

NOTE: THE GAUGE BLOCK SHOULD BARELY TOUCH THE LINE AS IT IS MOVED ALONG THE WALL.

Figure 19 ◆ Plumbing and aligning a wall.

105F19.EPS

Figure 20 ◆ Double top plate layout.

105F20.EPS

Overlap the corners and partition Ts. Drive two 16d nails at each end, then drive one 16d nail at each stud location.

After you have plumbed all the corners, line up the tops of the walls. This must be done before you nail the ceiling to the tops of the walls. To line up the walls, proceed as follows:

Step 1 Start at the top plate at one corner of the wall. Fasten a string at that corner. Stretch the string to the top plate at the corner at the opposite end of the wall, and fasten the string.

Step 2 Cut three small blocks from 1 × 2 lumber. Place one block under each end of the string so that the line is clear of the wall. Use the third block as a gauge to check the wall at 6' or 8' intervals.

Step 3 At each checkpoint, nail one end of a temporary brace near the top of a wall stud. Also attach a short 2 × 4 block to the subfloor. Adjust the wall (by moving the top of the wall in or out) so the string is barely touching the gauge block. When the wall is in the right position, nail the other end of the brace to the floor block.

Classroom

Emphasize that temporary bracing should be retained until the ceiling(s) and roof are in place and permanent bracing is installed.

Describe various methods of permanent bracing and demonstrate the installation of bracing acceptable to local codes.

Show Transparency 25 (Figure 21).

Describe the various types of sheathing and demonstrate the application of plywood sheathing.

Hand out Job Sheets 27105-1 through 27105-4. Under your supervision, have the trainees perform the tasks on the Job Sheets. Note the proficiency of each trainee.

Have each trainee complete to your satisfaction Performance Profile Tests 1 through 4. Fill out Performance Profile Sheets for each trainee.

> **Note**
> Do not remove the temporary braces until you have completed the framing—particularly the floor or roof diaphragm that sits on top of the walls—and sheathing for the entire building.

6.1.1 Bracing

Permanent bracing is important in the construction of exterior walls. Many local building codes will require bracing when certain types of sheathing are used. In some areas where high winds are a factor, lateral bracing is a requirement even when ½" plywood is used as the sheathing.

Several methods of bracing have been used since the early days of construction. One method is to cut a notch or let-in for a 1 × 4 or 1 × 6 at a 45° angle on each corner of the exterior walls. Another method is to cut 2 × 4 braces at a 45° angle for each corner. Still another type of bracing used (where permitted by the local code) is metal wall bracing (*Figure 21*). This product is made of galvanized steel.

Metal strap bracing is easier to use than let-in wood bracing. Instead of notching out the studs for a 1 × 4 or 2 × 4, you simply use a circular saw to make a diagonal groove in the studs, top plate, and soleplate for the rib of the bracing strap and nail the strap to the framing.

With the introduction of plywood, some areas of the country have done away with corner bracing. However, along with plywood came different types of sheathing that are by-products of the wood industry and do not have the strength to withstand wind pressures. When these are used,

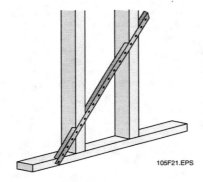

105F21.EPS

Figure 21 ◆ Use of metal bracing.

permanent bracing is needed. Building codes in some areas will allow a sheet of ½" plywood to be used on each corner of the structure in lieu of diagonal bracing, when the balance of the sheathing is fiberboard. In other areas of the country, the codes require the use of bracing except where ½" plywood is used throughout for sheathing. In still other areas, the use of bracing is required regardless of the type of sheathing used.

6.1.2 Sheathing

Sheathing is the material used to close in the walls. APA-rated material, such as plywood and non-veneer panels such as OSB and other reconstituted wood products, are generally used for sheathing.

Some carpenters prefer to apply the sheathing to a squared wall while the wall frame is still lying on the subfloor. Although this helps to ensure that the wall is square, it has two drawbacks:

• It may make the wall too heavy for the framing crew to lift.
• If the floor is not perfectly straight and level, it will be a lot more difficult to square and plumb the walls once they are erected.

When plywood is used, the panels will range from ⁵⁄₁₆" to ¾" thick. A minimum thickness of ⅜" is recommended when siding is to be applied. The higher end of the range is recommended when the sheathing acts as the exterior finish surface. The panels may be placed with the grain running horizontally or vertically. If they are placed horizontally, local building codes may require that blocking be used along the top edges.

Typical nailing requirements call for 6d nails for panels ½" thick or less and 8d nails for thicker panels. Nails are spaced 6" apart at the panel edges and 12" apart at intermediate studs.

Other materials that are sometimes used as sheathing are fiberboard (insulation board), exterior-rated gypsum wallboard, and rigid foam sheathing. A major disadvantage of these materials is that siding cannot be nailed to them. It must either be nailed to the studs or special fasteners must be used. If you are installing any of these materials, keep in mind that the nailing pattern is different from that of rated panels. In addition, they take roofing nails instead of common nails. Check the manufacturer's literature for more information.

When material other than rated panels is used as sheathing, rated panels can be installed vertically at the corners to eliminate the need for corner bracing in some applications. Check the plans and local codes.

CARPENTRY LEVEL ONE—TRAINEE MODULE 27105

Instructor's Notes:

_____
_____
_____
_____
_____
_____
_____
_____

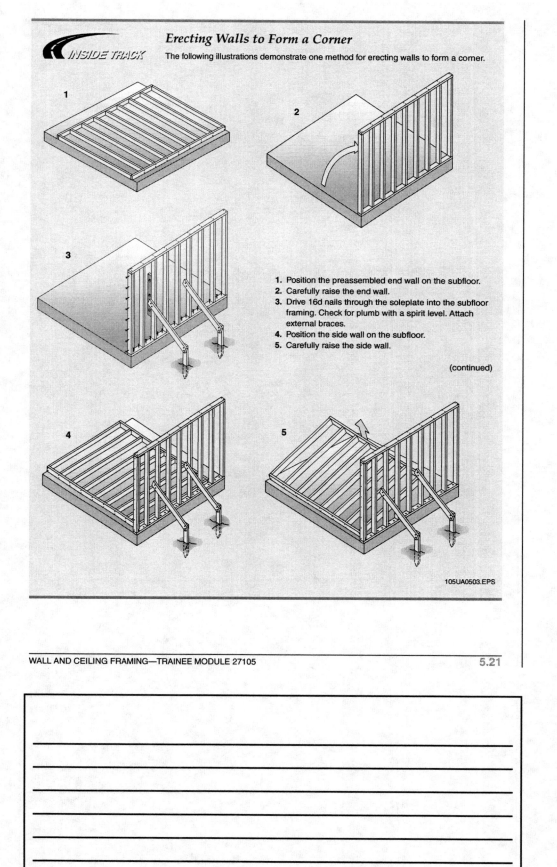

Erecting Walls to Form a Corner

INSIDE TRACK The following illustrations demonstrate one method for erecting walls to form a corner.

1. Position the preassembled end wall on the subfloor.
2. Carefully raise the end wall.
3. Drive 16d nails through the soleplate into the subfloor framing. Check for plumb with a spirit level. Attach external braces.
4. Position the side wall on the subfloor.
5. Carefully raise the side wall.

(continued)

105UA0503.EPS

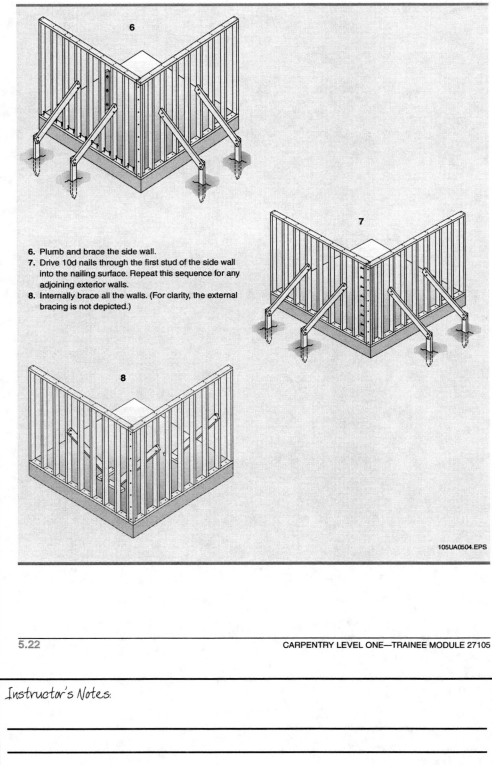

6. Plumb and brace the side wall.
7. Drive 10d nails through the first stud of the side wall into the nailing surface. Repeat this sequence for any adjoining exterior walls.
8. Internally brace all the walls. (For clarity, the external bracing is not depicted.)

105UA0504.EPS

Instructor's Notes:

_____
_____
_____
_____
_____
_____
_____
_____

6.1.3 Panelized Walls

Instead of building walls on the job site, they can be prefabricated in a shop and trucked to the job site. The walls, or *panels* as they are called, are either set with a small crane or by hand. The wall sections or panels will vary in length from 4' to 16'.

When working with a pre-engineered structure, the drying-in time is much quicker than with a field-built structure. A 1,200-square-foot residence, for example, can be dried-in within two working days. The siding would be applied at the factory and the walls erected the first day. The soffit and fascia would be installed on the morning of the second day, and the roof dried-in and ready to shingle by the morning of the third day. The residence would be ready for rough-in plumbing, electrical, heating, and cooling by the third day.

7.0.0 ◆ CEILING LAYOUT AND FRAMING

After you assemble and erect the walls, what is next? Traditionally, in a one-story building, carpenters would install ceiling joists on top of the structure, spanning the narrow dimension of the building from top plate to top plate. Then, if the carpenters intended to build a common gable roof, they would install rafters that extend from the ends of the ceiling joists to the peak, forming a triangle.

Nowadays, it is more likely that carpenters will install roof trusses instead of joists in a one-story building. The lowest member (or *bottom chord*) of the truss serves the same purpose as the ceiling joist. Alternatively, if the building is taller than one story, carpenters will install floor joists or floor trusses and build a platform and erect the walls for the second story. Again, the floor joist or truss, when viewed from below, serves the same purpose as the ceiling joist.

You will learn more about roof trusses and modern roof framing methods in later modules. For now, let's focus on the older, simpler way of framing with ceiling joists.

Ceiling joists have two important purposes:

- The joists are the top of the six-sided box structure of a building. They keep opposite walls from spreading apart.
- The joists provide a nailing surface for ceiling material, such as drywall, which is attached to the underside of joists.

As noted previously, joists extend all the way across the structure, from the outside edge of the double top plate on one wall to the outside edge

of the double plate of the opposite wall. Ordinarily, carpenters lay the ceiling joists across the narrow width of a building, usually at the same positions as the wall studs.

If the spacing of the ceiling joists is the same as that of the wall studs, lay out the joists directly above the studs. This makes it easier to run ductwork, piping, flues, and wiring above the ceiling.

Laying out ceiling joists for a gable roof is very similar to laying out floor joists and wall studs. Measure along the double top plate to a point 15¼" in from the end of the building. Mark it, square the line, then use a long steel tape to mark every 16" (or 24", depending on the architect's plans). To the right of each line, mark an X for the joist location. Then mark an R for each rafter on the left side of each mark (*Figure 22*). Repeat this procedure on the opposite wall and on any bearing partitions.

Note
If you are installing a hip roof, it is also necessary to mark the end double top plates for the locations of hip rafters. Start by marking the center of the end double top plate. Then measure and place marks for joists and rafters by measuring from each corner toward the center mark.

As you learned in the module on floor layout, the actual allowable span for joists depends on the species, size, and grade of lumber, as well as the spacing and the load to be carried. *Table 2* contains information on spacing and allowable spans for ceiling joists. The architect relies on this and other information in designing the structure.

If the joist exceeds the allowable span, two pieces of joist material must be spliced over a bearing wall or partition. *Figure 23* shows two ways to splice joists. In the first method, as shown in *Figure 23(A)*, two joists are overlapped above the center of the bearing partition. The overlap should be no less than 6".

Figure 23(B) shows another way to splice joists. Instead of overlapping, the two joists butt together directly over the center of the bearing partition. A shorter piece of joist material is nailed to both joists to make a strong joint. Other materials used to reinforce the joist splice include plywood, 1 × lumber, steel strapping, and special anchors. The architectural plans will specify the proper method(s) to be used.

Assign reading of Sections 7.0.0–7.1.0 for the next class session.

Describe the various methods of ceiling framing. Point out the purpose of ceiling joists.

Show Transparencies 26 and 27 (Figures 22 and 23).

If desired, portions of the video *Roof Framing, Part One* may be used to demonstrate the layout and installation of ceiling joists.

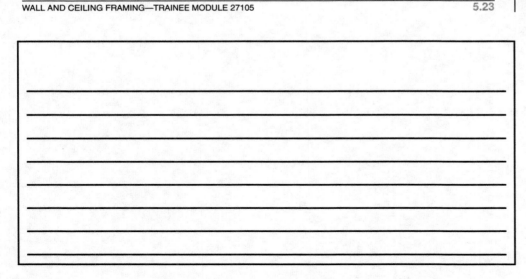

Discuss the layout considerations for ceiling joists.

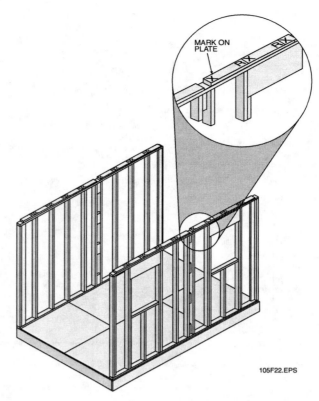

Figure 22 ◆ Marking joist and rafter locations on the double top plate.

105F22.EPS

Table 2 Ceiling joist spacing and allowable spans

| Species | Grade | Span (feet and inches) | | | | | |
| | | 2 × 4 | | 2 × 6 | | 2 × 8 | |
		16" OC	24" OC	16" OC	24" OC	16" OC	24" OC
Douglas Fir-Larch	2	11–6	10–0	18–1	15–7	23–10	20–7
	3	9–9	7–11	14–8	11–11	19–4	15–9
Douglas Fir South	2	10–6	9–2	16–6	14–5	21–9	19–0
	3	9–5	7–9	14–2	11–7	18–9	15–3
Hem-Fir2	10–9	9–5	16–11	13–11	22–4	18–4	
	3	8–7	7–0	13–1	10–8	17–2	14–0
Mountain Hemlock	2	9–11	8–8	15–7	13–8	20–7	18–0
	3	8–11	7–3	13–3	10–10	17–6	14–4
Mountain Hemlock-Hem-Fir	2	9–11	8–8	15–7	13–8	20–7	18–0
	3	8–7	7–0	13–1	10–8	17–2	14–0
Southern Pine	2	8–11	7–8	13–6	11–0	17–5	14–2
	3	7–1	5–9	10–5	8–6	13–3	10–10

Instructor's Notes:

_____

_____

_____

_____

_____

_____

_____

_____

Joist Installation

Two carpenters were working on the second floor of a house. There was no guardrail or cover over the floor opening for the stairway. While placing a joist in position, one of the carpenters fell through the stairway opening to the concrete basement below, receiving a fatal head injury.

The Bottom Line: Use fall protective equipment when working above floor level. Always install appropriate guardrails and cover floor openings.

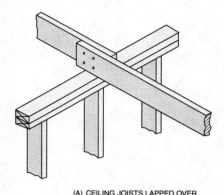

(A) CEILING JOISTS LAPPED OVER BEARING PARTITION

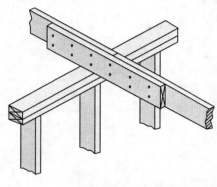

(B) CEILING JOISTS BUTTED OVER BEARING PARTITION

105F23.EPS

Figure 23 ◆ Splicing ceiling joists.

Overlapped Splices

An overlapped splice is superior to a butted splice because a butted splice reduces by one-half the critical joist-bearing surface, or the part of the joist that is supported by the bearing wall beneath it. Also, a butted splice requires more materials and labor than the simpler, stronger overlapped splice.

7.1.0 Cutting and Installing Ceiling Joists

The joists must be cut to the proper length, so the ends of the joists will be flush with the outside edge of the double top plate. As you learned in the previous section, you must allow enough extra length for any overlap of the joists above bearing partitions.

You must also cut the ends of the joists at an angle matching the rafter pitch, as shown in *Figure* 24. This is so the roof sheathing will lie flush on the roof framing. You will learn more about this in the module about roof framing.

After you have cut the joists to the right length and have positioned them in the right places, toenail them into the double top plate. The architectural plans might also call for the installation of metal anchors.

After you install the joists, you must nail a ribband or strongback across the joists to prevent twisting or bowing (*Figure 25*). The strongback is

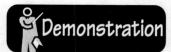

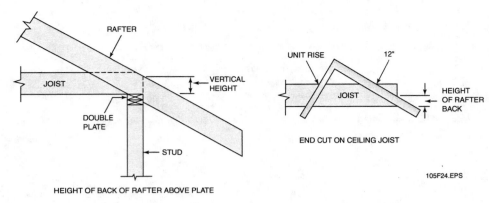

105F24.EPS

Figure 24 ◆ Cutting joist ends to match roof pitch.

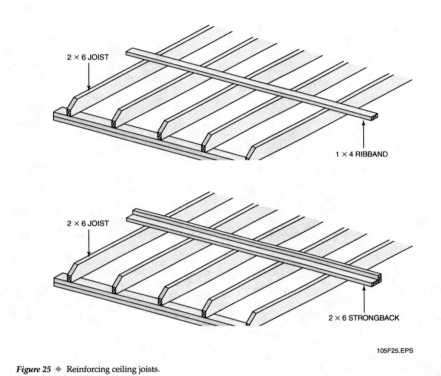

105F25.EPS

Figure 25 ◆ Reinforcing ceiling joists.

CARPENTRY LEVEL ONE—TRAINEE MODULE 27105

Instructor's Notes:

_____

_____

_____

_____

_____

_____

_____

_____

used for longer spans. In addition to holding the joists in line, the strongback provides support for the joists at the center of the span.

8.0.0 ◆ ESTIMATING MATERIALS

In this section, you will follow the basic steps to estimate the amount of lumber you will need to frame the walls and ceilings of a building. Here's some important information that you need:

- For this example, the structure is 24' × 30'. (See *Figure 26*.)
- The walls are 8' tall.
- You will construct the building with 2 × 4s with 16" on center spacing.

- You will construct the ceiling from 2 × 6s.
- The building has two 24" wide windows on each narrow side, and two 48" wide windows and one 36" wide door on each wide side.
- Inside, the building is divided into three rooms: one is 12' × 30', two are each 11'–1½" × 12', divided by a 4'–10" wide hallway.
- The long wall that separates the large room from the two smaller rooms and hallway is a loadbearing partition.
- Each of the smaller rooms has a 36" door leading to the hallway.
- There is a 36" opening, without a door, from the large room to the hallway.

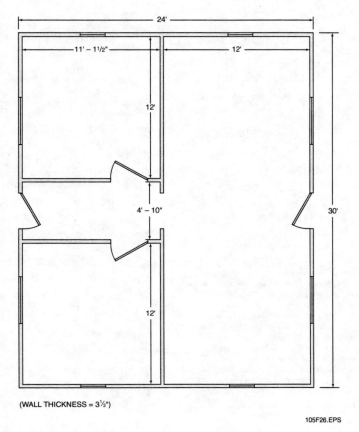

(WALL THICKNESS = 3½")

105F26.EPS

Figure 26 ◆ Sample floor plan.

Assign reading of Sections 8.0.0–10.1.0 for the next class session.

Classroom

Explain the method for estimating framing materials using the example in the Trainee Module.

Audiovisual

Show Transparency 30 (Figure 26).

Soleplates and Top Plates

Assume that the plate material is ordered in 12' lengths.

Step 1 Determine the length of the walls in feet. For loadbearing walls, multiply this number by 3 to account for the soleplates and double top plates (remember that a double top plate is made of two 2 × 4s stacked together).

Step 2 Divide the result by 12 to get the number of 12' pieces. Round up to the next full number, and allow for waste.

Example:

For the exterior walls, add the lengths of the two long walls, plus the two narrow walls:

$$30' + 30' + 24' + 24' = 108'$$

Also add the length of the interior partitions:

$$30' + 12' + 12' = 54'$$

Add those numbers together:

$$108' + 54' = 162'$$

Then multiply by 3 to account for the soleplate and double top plate:

$$162' \times 3 = 486'$$

Divide by 12 to get the number of 12' pieces needed:

$$486' \div 12 = 40\tfrac{1}{2} \text{ pieces}$$

Since you can buy only full pieces, you will need to round up to the next whole number, which is 41. So, for all of the soleplates, top plates, and double top plates in this building, you need 41 pieces of 12' 2 × 4s.

> **Note**
>
> The example estimate here for soleplate and top plate material is a very simple one and works for this small, simple structure. Actually, only the center loadbearing partition (and the exterior walls, of course) will require double top plates. But the extra amount of lumber from this simpler estimate is small, and you need to figure more lumber for waste anyway. In a larger, more complex structure, you need to calculate the different quantities of lumber for loadbearing and non-bearing partitions to get a more precise estimate.

Studs

Step 1 Determine the length in feet of all the walls.

Step 2 The general industry standard is to allow one stud for each foot of wall length, even when you are framing 16" on center. This should cover any additional studs that are needed for openings, corners, partition Ts, and blocking.

When you were estimating for soleplate and top plate material, you found that there were 162' of exterior walls and interior partitions. You will need one stud for every foot of wall, or 162 studs.

Headers

Let's assume that you'll use 2 × 12 headers with plywood spacers. This might cost a little more, but you won't need to lay out, cut, and assemble cripple studs above the headers.

Step 1 Use the architect's drawings (to be safe, check both the floor plans and the door and window schedules to make sure they agree) to find the number and size of each window and door.

Step 2 Add 5¼" to the finish width of each door and window.

Step 3 Double the length of each header.

Step 4 Combine the lengths obtained in Step 3 into convenient lengths for ordering.

Step 5 Order enough ½" plywood for spacers.

Example:

In *Figure 26*, there are four 4830 windows (each 48" wide):

$$48" + 5\tfrac{1}{4}" = 53\tfrac{1}{4}"; \quad 53\tfrac{1}{4}" \times 4 = 213"$$

Two 2430 windows (each 24" wide):

$$24" + 5\tfrac{1}{4}" = 29\tfrac{1}{4}"; \quad 29\tfrac{1}{4}" \times 2 = 58\tfrac{1}{2}"$$

Four 3680 doors (each 36" wide):

$$36" + 5\tfrac{1}{4}" = 41\tfrac{1}{4}"; \quad 41\tfrac{1}{4}" \times 4 = 165"$$

One 3'–0" opening with no door:

$$36" + 5\tfrac{1}{4}" = 41\tfrac{1}{4}"$$

Add those numbers together:

$$213" + 58\tfrac{1}{2}" + 165" + 41\tfrac{1}{4}" = 477\tfrac{3}{4}"$$

Point out that only exterior walls and interior loadbearing walls require a double top plate.

Instructor's Notes:

_____

_____

_____

_____

_____

_____

_____

_____

Then double that total:

$$477\tfrac{3}{4}" \times 2 = 955\tfrac{1}{2}"$$

Divide by 12 to get feet:

$$955\tfrac{1}{2}" \div 12 = 79'\text{–}7\tfrac{1}{2}"$$

Round that up to 80'.

Diagonal Bracing
You need to install diagonal bracing at each end of all exterior walls. You can use metal strips or 1 × 4 lumber. Braces run from the top plate to the sole-plate at a 45-degree angle. The walls for this building are 8' high, so they would need a 12' brace. (Review the Core module, *Introduction to Con-*struction Math, to determine how to find the length of the brace using the formula for the hypotenuse of a right triangle.)

Step 1 Determine the number of outside corners.

Step 2 Figure the length of each brace based on the height of the walls, then multiply by the number of corners.

Example:
This part is pretty easy. *Figure 26* shows that this building is a rectangle with four corners. Each corner has two legs, so there are eight places where you would install diagonal braces. Since you need

Explain why smaller buildings require an additional 15% to be added to the estimate as a safety factor.

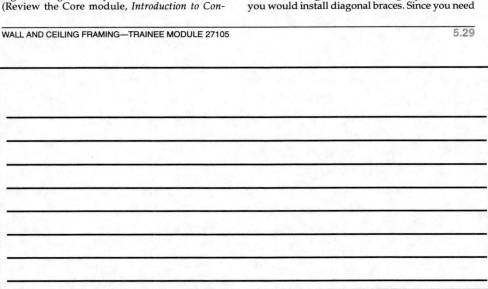

Classroom

Provide example estimating problems and have the trainees solve them.

Describe the methods of furring and framing for masonry walls with door and window openings. Emphasize that pressure-treated lumber must be used.

Audiovisual

Show Transparencies 31 through 33 (Figures 27 through 29).

Teaching Tip

Take the trainees to a construction site to observe masonry wall framing in progress.

12' of bracing at each location, 8 × 12' = 96' of bracing altogether.

Remember that let-in or diagonal bracing is typically applied to the interior side of the exterior walls.

Ceiling Joists

As mentioned previously, it is more likely that carpenters will install roof trusses instead of joists in a one-story building. You will learn more about roof trusses and modern roof framing methods in later modules in this book. For now, let's again focus on the older, simpler way of framing with ceiling joists.

Step 1 Determine the span of the building.

Step 2 Figure the number of joists based on the spacing (remember that you'll place joists 16" on center), then add one for the end joist.

Step 3 Multiply the span by the number of joists to get total length. Add 6" per joist per splice where joists will be spliced at bearing walls, partitions, and girders.

Example:

The building in *Figure 26* is 30' by 24'. Let's take the short dimension (24'). Fortunately, there is a partition down the long dimension of the building. That breaks the 24' span into two 12' spans.

Divide the long dimension of the building (30') by 16" to determine the number of joist locations. First, convert 30' to inches:

$$30' \times 12" = 360"$$

Divide that by 16":

$$360" \div 16" = 22\frac{1}{2} \text{ (round up to 23)}$$

Then add one for the end joist:

$$23 + 1 = 24$$

The 24' width is too big to span with one long joist, so you'll span it with two 12' joists. Multiply the last number by two:

$$24 \times 2 = 48$$

Now calculate the total length, adding 6" for each joist at each splice.

$$48 \times 12'\text{-}6" = 600'$$

So how much lumber do you order? So far, we've worked mostly with 16' lengths of lumber. If you divide 600' by 16', you get 37½ pieces, which you'd round up to 38.

However, it's not that simple. If you cut a 12'-6" joist out of a 16' piece of lumber, you have only 3'-6" left over. That's not long enough to

span 12' (half of the total 24' span). Instead, you would most likely order 48 pieces of 14' 2 × 6s. You'd end up with 1'-6" of waste from each piece of lumber, but you can use the waste for bridging and blocking.

9.0.0 ◆ WALL FRAMING IN MASONRY

You must be aware of the methods used in furring masonry walls in order to install the interior finish. As a general rule, furring of masonry walls should be done on 16" centers. Some contractors will apply 1 × 2 furring strips 24" OC. This may save material, but it does not provide the same quality as a wall constructed on 16" centers. Carpenters have little control over the quality of the masonry structure. It is important that the walls are put up square and plumb. If a structure starts off level and plumb, little difficulty will be encountered with measuring, cutting, and fitting. If the walls or floor are not plumb and level, problems can be expected throughout the structure.

Furring is applied to a masonry wall with masonry nails measuring 1¼" to 1¾" long. In addition to the furring strips, 1 × 4 and 1 × 6 stock is used. Remember, all material that comes in contact with concrete or masonry must be pressure-treated.

Backing for partitions against a masonry block wall is done using one of the following methods. In the first method, locate where the partition is going to be, then nail a 1 × 6 board, centered on the center of the partition (i.e., locate the center of the partition at the bottom corner of the wall, move back 2¾", and mark the wall by plumbing with a straightedge and level). Attach the 1 × 6 to the wall with one edge on the mark. This will allow an even space on either side of the partition to receive the drywall, as shown in *Figure 27*.

In the second method, secure the partition to the block wall and install a furring strip on each side of the partition. This will cause a problem when the drywall is installed; i.e., the partition stud is 1½", and by adding a ¾" furring strip to the wall we have only ¾" of stud left to work with. Nailing the ½" drywall to the furring strip on the wall will leave only ¼" of nailing on the stud. This does not allow enough room to nail the ½" drywall, so it may become necessary to install a 16" 2 × 4 block to the bottom, center, and top of the partition, as shown in *Figure 28*.

The proper nailing surface for drywall is ¾" to 1". In preparing the corners of the block wall to receive the furring strips, enough space should be allowed for the drywall to slip by the furring strips. Come away from the corner ⅝" in either direction with the strips (*Figure 29*).

Instructor's Notes:

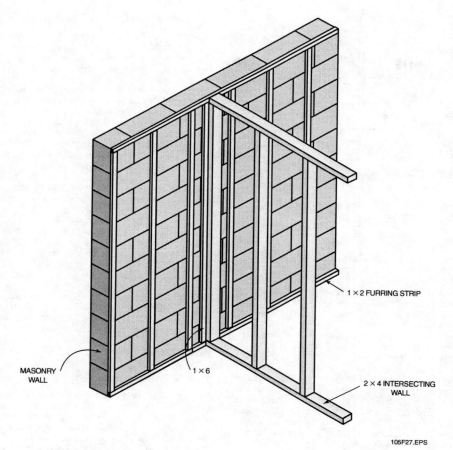

Figure 27 ◆ Partition backing using a 1 × 6.

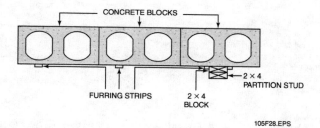

Figure 28 ◆ Partition backing using 2 × 4 blocks.

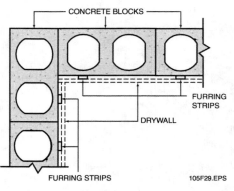

Figure 29 ◆ Placement of furring strips at corners of masonry walls.

A 1 × 4 is used at floor level to receive the baseboard. Either a narrow or wide baseboard can be used. Some carpenters will install a simple furring strip at floor level and depend on the vertical strips for baseboard nailing. Once the drywall has been installed, it is difficult to find the strips when nailing the baseboard.

A sequence of installation should be established for nailing the furring strips to the wall. Either start from the right and work to the left or work from left to right. This sequence will allow the person doing the trim work to locate the furring strips. When applying the furring strips to the wall, remember that the strips must be placed 16" OC for the drywall to be nailed properly. Start the first strip at 15¼" and then lay out the second strip 16" from the first mark. Laying out furring strips is done in the same way as laying out wall studs.

9.1.0 Framing Door and Window Openings in Masonry

In modern construction, metal door frames and metal window frames or channels are used almost exclusively with masonry walls. However, in rare instances, wood framing is used. Each installation is unique and a complete examination of this subject is beyond the scope of this module. There are two general rules, however:

- Use treated wood in contact with masonry.
- Use cut nails, expansion bolts, anchor bolts, or other fasteners to attach the wood to the masonry securely.

10.0.0 ◆ METAL STUDS IN FRAMING

Depending on the gauge, steel studs are typically stronger, lighter, and easier to handle than wood. Unlike wood studs, steel studs will not split, warp, swell, or twist. Furthermore, steel studs will not burn as wood studs would. Steel studs are currently more expensive than wood studs, but it is expected that as lumber prices continue to rise, the costs will begin to equalize.

Steel studs are prepunched to permit quick installation of electrical distribution systems.

Metal studs have become popular in residential, commercial, and industrial construction. Metal studs may be spaced 16" or 24" OC. On a non-bearing wall, spacing is determined by drywall type and thickness. Unlike wood (which has defects), metal studs are consistent in material composition.

There are three types of metal studs. The first is used for non-bearing walls that have facings to accept drywall. The second will accept lath and plaster on both interior and exterior walls. The third type is a wide-flange metal stud, which is used for both loadbearing and non-bearing walls.

A wide variety of accessories are available for metal studs. There are tracks for floors and ceilings to which the studs are fastened. Tracks are also available for sills, fascia, and joint-end enclosures. Other accessories include channels, angles, and clips. For residential construction, steel trusses are also available.

10.1.0 Fabrication

For layout, metal studs are marked to the centerline rather than the edge. The open side of the stud should always face the beginning point of the layout. The bottom channel is fastened to the concrete floor with small powder-actuated fasteners. A metal self-tapping screw is used to fasten the studs to the track (*Figure 30*). The electric screwgun takes the place of the hammer. The studs may also be welded instead of screwed.

When constructing an opening, two studs are put back to back and screwed or welded together. The stud that will act as the trimmer stud will be cut to the height specified to receive the header. A section of floor track can be used for the bottom part of the header, with short pieces of studs put in place over the header and secured in place. Blocking may be required to fasten millwork.

Summary

This module covered how to identify and use all the wall and ceiling components; how to lay out

Instructor's Notes:

_____

_____

_____

_____

_____

_____

_____

_____

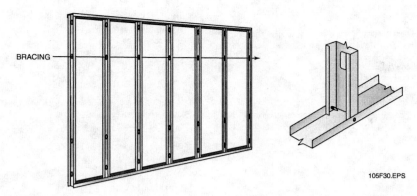

BRACING

105F30.EPS

Figure 30 ◆ Metal framing.

walls and ceilings; how to measure and cut lumber; how to assemble, erect, plumb, brace, and sheath walls; how to lay out and build doors, windows, and other wall openings; how to install ceilings; how to estimate materials; how to work with masonry walls; how to construct walls with metal studs; and much more.

The most important skill in wall and ceiling framing is the need to be accurate. For example, in framing walls you must make sure the walls are straight, plumb, and square. Precise layout and measuring of studs and headers is also critical.

Even a very small error can cause big problems. For example, over a span of 20', an error of only $\frac{1}{16}$" at one end of the span could become $1\frac{1}{4}$" at the other end. The closer you get to the end of the wall, the more patching and fitting you will have to do to get the sheathing, windows, and exterior doors to fit.

In other words, saving a little time at the beginning can cost you a lot of time at the end. Remember this as you move forward in your training.

Classroom

Have the trainees complete the Review Questions and go over the answers prior to administering the Module Examination.

Examination

Administer the Module Examination and Performance Test. Be sure to record the results on Craft Training Report Form 200 and submit the form to the Training Program Sponsor.

Performance Profile Test

Ensure that all Performance Profile Tests have been completed and Performance Profile Sheets for each trainee are filled out. Be sure to record the results of the Testing on Craft Training Report Form 200 and submit the results to the Training Program Sponsor. Answers can be found on the key in the Test Booklet.

1. A short framing member that fills the space between the rough sill and the soleplate is a _____.
   a. spacer
   b. cripple stud
   c. trimmer stud
   d. top plate

2. The framing member that forms the side of a rough opening for a window or door is the _____.
   a. trimmer stud
   b. header
   c. soleplate
   d. cripple

3. The straightest, least-defective studs are normally used for _____.
   a. window and door trimmers
   b. regular studs
   c. corners
   d. partition Ts

4. The framing member that distributes the weight of the structure around a door or window opening is the _____.
   a. trimmer stud
   b. cripple stud
   c. top plate
   d. header

5. The width of a header is equal to the _____.
   a. width of the rough opening
   b. width of the opening plus the width of two trimmer studs plus the width of two common studs
   c. width of the opening plus the width of two trimmer studs
   d. width of the trimmer plus the width of two cripple studs

6. The length of a cripple stud above a window opening equals the length of a common stud less the height of the trimmer, combined with the thickness of the _____.
   a. soleplate
   b. rough sill
   c. top plate
   d. header

7. When calculating the length of a regular stud for a wood frame floor, the thickness of the _____ is not considered.
   a. soleplate
   b. subfloor
   c. underlayment
   d. double top plate

8. Pieces of 2 × 4 placed horizontally between each pair of studs are used to _____.
   a. brace the frame
   b. retard the spread of fire
   c. provide nailers for siding
   d. provide a place for carpenters to stand when installing ceiling joists

9. A long 2 × 4 with a ¾" standoff block nailed to each end is used as a _____.
   a. wall brace
   b. straightedge
   c. gauge block
   d. trimmer stud

10. When the sheathing acts as the finish surface, you should use _____ sheets.
    a. ⅜"
    b. ½"
    c. ¾"
    d. 1"

11. When you are installing APA-rated sheathing material, you should space the nails _____ apart at the panel edges.
    a. 6"
    b. 12"
    c. 16"
    d. 24"

12. A ribband is a(n) _____.
    a. brace placed between studs
    b. L-shaped joist brace
    c. type of joist hanger
    d. 1 × 4 strip used to prevent joists from twisting or bowing

13. Fabricated headers normally use _____ as spacers.
    a. furring strips
    b. ½" plywood
    c. ¾" plywood
    d. 2 × 4 blocks

14. The purpose of prepunched holes in metal studs is to _____.
    a. make them lighter
    b. save material
    c. provide runs for piping and wiring
    d. make it easier to attach drywall

15. Metal studs are attached to the bottom channel using _____.
    a. wood screws
    b. powder-actuated fasteners
    c. construction adhesive
    d. self-tapping metal screws

Instructor's Notes:

_____

_____

_____

_____

_____

_____

_____

_____

David Scalf
Estimator and Instructor
Viox Services, Cincinnati, Ohio

David Scalf began his career in 1991 as a carpenter with Viox Services, the largest Ohio-based facility maintenance and construction company. Today he is an estimator and primary CAD (computer-aided design) operator.

How did you first get interested in carpentry?
My father was a carpenter, so I guess it was in the genes. My first paying job was building houseboats in Somerset, Kentucky. I worked on interior paneling and trim.

Did you begin formal training after starting at Viox, and how has it prepared you for your job?
Yes, my training through our local Associated Builders & Contractors (ABC) chapter began immediately. Even though you learn best by doing, I found that you need someone to guide you in the right direction. Classroom hours offer you that direction and help you to understand that you're doing it right. I learned how to do what I already knew faster, better, or in a different way.

What's been your career path over the past seven years?
Throughout the four-year apprenticeship program, I worked in the field. We were required to have 8,000 hours on the job to graduate. I went from crew leader to lead carpenter and then moved into the estimating department as an assistant estimator. Today I'm the primary CAD operator, which sometimes surprises me because five years ago I didn't even know how to turn a computer on. Now I'm using different types of software programs each day and spend about 80% of my time working on the computer. In addition to providing CAD training for all Viox employees, I'm also responsible for new associate safety training and will be teaching fourth-year apprentices for ABC.

What was your typical day like in the field, and what's it like now that you're in the office?
When I began, I mostly worked in remodeling of existing structures. I especially liked working on small projects where there was close contact with clients and you got to see the project from start to finish.

Now I spend the majority of my day in the office. I'm currently involved in a lot of design-build work.

We get involved even before the architects are hired and work directly with the clients. Working with us instead of an architect in the initial phases of layout can save the client a significant amount of money. Usually we carry the client through the entire process and often select and work with the architects once they are brought in.

As an apprentice, you came in second place in ABC's National Craft Olympics. What was that like?
That was probably one of the highlights of my apprenticeship years. I competed locally in late 1995, won, and then went on to the nationals in Orlando. We had to take a written test and also demonstrate hands-on ability. We were judged both on performance and time. I still remember what I had to do: build a plastic laminate countertop, tie rebar, shoot elevations with a transit, and build a metal stud wall.

What did you win?
New power tools, $500, plus the all-expense-paid trip to Orlando.

Why have you been so successful?
I think it's mostly tied to my original commitment to endure the four-year apprenticeship program. It takes a real effort to stay with it. Now, it's because I'm constantly pushing myself and my employer to do more. I'm never really content; I'm always looking to the next level.

As an instructor now, what advice do you give your students?
The most important thing is to find the right employer—one that provides the opportunity to be trained. You're not looking for a summer job, but a career. There needs to be long-term commitment both ways.

Trade Terms Introduced in This Module

Blocking: A wood block used as a filler piece and a support between framing members.

Cripple stud: In wall framing, a short framing stud that fills the space between a header and a top plate or between the sill and the soleplate.

Double top plate: A plate made of two members to provide better stiffening of a wall. It is also used for connecting splices, corners, and partitions that are at right angles (perpendicular) to the wall.

Drying-in: Applying sheathing, windows, and exterior doors to a framed building.

Furring strip: Narrow wood strips nailed to a wall or ceiling as a nailing base for finish material.

Gable roof: A roof with two slopes that meet at a center ridge.

Header: A horizontal structural member that supports the load over a window or door.

Hip roof: A roof with four sides or slopes running toward the center.

Jamb: The top (head jamb) and side members of a door or window frame that come into contact with the door or window.

Let-in: A notch cut into a joist, stud, block, etc., to "let in" another piece such as a brace.

Ribband: A 1 × 4 nailed to the ceiling joists at the center of the span to prevent twisting and bowing of the joists.

Sill: The lower framing member attached to the top of the lower cripple studs to form the base of a rough opening for a window.

Soleplate: The lowest horizontal member of a wall or partition to which the studs are nailed. It rests on the rough floor.

Strongback: An L-shaped arrangement of lumber used to support ceiling joists and keep them in alignment.

Top plate: The upper horizontal framing member of a wall used to carry the roof trusses or rafters.

Trimmer stud: A vertical framing member that forms the sides of rough openings for doors and windows. It provides stiffening for the frame and supports the weight of the header.

Instructor's Notes:

_____

_____

_____

_____

_____

_____

_____

_____

Answers to Review Questions

Answer	Section
1. b	2.0.0
2. a	2.0.0
3. c	2.1.0
4. d	2.3.0
5. c	2.3.0
6. d	4.0.0
7. b	4.0.0
8. b	5.1.0
9. b	6.1.0
10. c	6.1.2
11. a	6.1.2
12. d	7.1.0
13. b	2.3.1
14. c	10.0.0
15. d	10.1.0

Additional Resources

This module is intended to present thorough resources for task training. The following reference works are suggested for further study. These are optional materials for continuing education rather than for task training.

Carpentry. Homewood, IL: American Technical Publishers.

Carpentry. Albany, NY: Delmar Publishers.

Modern Carpentry. Tinley Park, IL: The Goodheart-Willcox Co. Inc.

Instructor's Notes:

_____

_____

_____

_____

_____

_____

_____

_____

Section 4.0.0   *Think About It – Determining Stud Length*

The answers to these questions depend on the ceiling thickness and, if used, the floor underlayment thickness. Assuming the same ceiling and underlayment thicknesses shown in *Figure 16*, the answers are:

1. Stud length = 9'-6" + (underlayment and ceiling thickness) − (sole and double top plate thickness).
    Substituting a ceiling thickness of ½", an underlayment thickness of ⅜", and 1½" for the sole and each plate thickness yields:

    Stud length = 9'-6" + (½" + ⅜") − (1½" + 3") = 9'-2⅝".

2. Cripple length = 32" + (underlayment thickness) − (sole and sill plate thickness).
    Substituting an underlayment thickness of ⅜" and 1½" for the sole and sill thicknesses yields:

    Cripple length = 32" + ⅜" − (1½" + 1½") = 29⅜".

If desired, have the trainees determine appropriate stud and cripple lengths by substituting the values used locally for underlayment and finish ceiling thicknesses.

The NCCER makes every effort to keep these textbooks up-to-date and free of technical errors. We appreciate your help in this process. If you have an idea for improving this textbook, or if you find an error, a typographical mistake, or an inaccuracy in the NCCER's Craft Training textbooks, please write us, using this form or a photocopy. Be sure to include the exact module number, page number, a detailed description, and the correction, if applicable. Your input will be brought to the attention of the Technical Review Committee. Thank you for your assistance.

Instructors – If you found that additional materials were necessary in order to teach this module effectively, please let us know so that we may include them in the Equipment/Materials list in the Instructor's Guide.

Write: Curriculum Revision and Development Department
National Center for Construction Education and Research
P.O. Box 141104, Gainesville, FL 32614-1104

Fax: 352-334-0932

E-mail: curriculum@nccer.org

Craft

Module Name

Copyright Date

Module Number

Page Number(s)

Description

(Optional) Correction

(Optional) Your Name and Address

Roof Framing

27106-01

MODULE OVERVIEW

This module introduces the carpentry trainee to the methods and procedures used in roof framing.

PREREQUISITES

Please refer to the Course Map in the Trainee Module. Prior to training with this module, it is suggested that the trainee shall have successfully completed the following modules:

Core Curriculum; Carpentry Level One, Modules 27101 through 27105

LEARNING OBJECTIVES

Upon completion of this module, the trainee will be able to:

1. Understand the terms associated with roof framing.
2. Identify the roof framing members used in gable and hip roofs.
3. Identify the methods used to calculate the length of a rafter.
4. Identify the various types of trusses used in roof framing.
5. Use a rafter framing square, speed square, and calculator in laying out a roof.
6. Identify various types of sheathing used in roof construction.
7. Frame a gable roof with vent openings.
8. Frame a roof opening.
9. Construct a frame roof, including hips, valleys, commons, jack rafters, and sheathing.
10. Erect a gable roof using trusses.
11. Estimate the materials used in framing and sheathing a roof.

PERFORMANCE OBJECTIVES

Under the supervision of the instructor, the trainee should be able to:

1. Lay out rafter locations on a top plate.
2. Lay out, cut, and erect rafters for a gable roof.
3. Frame a gable end with vent openings.
4. Frame an opening in a roof.
5. Apply roof sheathing.
6. Lay out, cut, and erect rafters for an intersecting hip roof with valley.
7. Erect trusses for a gable roof.

NCCER STANDARDIZED CRAFT TRAINING PROGRAM

The National Center for Construction Education and Research (NCCER) provides a standardized national program of accredited craft training. Key features of the program include instructor certification, competency-based training, and performance testing. The program provides trainees, instructors, and companies with a standard form of recognition through a National Craft Training Registry. The program is described in full in the *Guidelines for Accreditation*, published by the NCCER. For more information on standardized craft training, contact the NCCER at P.O. Box 141104, Gainesville, FL 32614-1104, 352-334-0911, visit our Web site at www.NCCER.org, or e-mail info@NCCER.org.

HOW TO USE THIS ANNOTATED INSTRUCTOR'S GUIDE

Each page presents two sections of information. The larger section displays each page exactly as it appears in the Trainee Module. The narrow column ties suggested trainee and instructor actions to each page and provides icons to call your attention to material, safety, audiovisual, or testing requirements. The bottom of each page includes space for your notes.

 **Teaching Tip**   If you see the Teaching Tip icon, that means there is a teaching tip associated with this section. Also refer to any suggested teaching tips at the end of the module.

SAFETY CONSIDERATIONS

Ensure that the trainees are equipped with appropriate personal protective equipment.

PREPARATION

Before teaching this module, you should review the Module Outline, the Learning and Performance Objectives, and the Materials and Equipment List. Be sure to allow ample time to prepare your own training or lesson plan and gather all required equipment and materials.

MATERIALS AND EQUIPMENT LIST

Materials:

Transparencies
Markers/chalk
8d common nails
8d box nails
16d box nails
16d common nails
Roof framing plan
2 × 4 or 2 × 6 framing lumber for rafters and ridgeboards
Joist and header material for roof opening
½" CD plywood or other sheathing material
Nails for sheathing
H-clips
Roof trusses
1 × 6 lumber or plywood for catwalk
2 × 4 lumber for braces and stakes
Sample blueprints
Copies of Job Sheets 1 through 7*
Module Examinations*
Performance Profile Sheets*
Videotapes (optional) *Roof Framing, Parts One and Two*

Equipment:

Overhead projector and screen
Whiteboard/chalkboard
Appropriate personal protective equipment
Chalkline
String line
Steel tape with markings at 16" OC
Framing hammer
Claw hammer
Spreader for lifting trusses (if applicable)
Crane for lifting trusses (if applicable)
Rafter framing square
Sawhorses
Speed square and booklet
Circular saw
Extension cord
Handsaw
4' level
6' stepladders
Plumb bob and line
Scientific calculator
Videocassette recorder (VCR)/TV set (optional)

*Packaged with this Annotated Instructor's Guide.

ADDITIONAL RESOURCES

This module is intended to present thorough resources for task training. The following reference works are suggested for both instructors and motivated trainees interested in further study. These are optional materials for continued education rather than for task training.

Carpentry, Leonard Koel. Homewood, IL: American Technical Publishers, 1997.

Carpentry, Gasper J. Lewis. Albany, NY: Delmar Publishers, 2000.

Modern Carpentry, Willis H. Wagner and Howard Bud Smith. Tinley Park, IL: The Goodheart-Willcox Company, Inc., 2000.

Roof Framing, Parts One and Two, videotape. Gainesville, FL: The National Center for Construction Education and Research.

TEACHING TIME FOR THIS MODULE

An outline for use in developing your lesson plan is presented below. Note that each Roman numeral in the outline equates to one session of instruction. Each session has a suggested time period of 2½ hours. This includes 10 minutes at the beginning of each session for administrative tasks and one 10-minute break during the session. Approximately 37½ hours are suggested to cover *Roof Framing*. You will need to adjust the time required for hands-on activity and testing based on your class size and resources.

Topic   **Planned Time**

Session I. Introduction; Types of Roofs; Basic Roof Layout

  A. Introduction   _____

  B. Types of Roofs   _____

  C. Basic Roof Layout   _____

    1. Rafter Framing Square   _____

    2. Basic Rafter Layout   _____

      a. Laying Out Rafter Locations   _____

      b. Determining the Length of a Common Rafter   _____

      c. Laying Out a Common Rafter   _____

Session II. Laboratory

  A. Laboratory   _____

  Hand out Job Sheet 27106-1. Under your supervision, have the trainees perform the tasks on the Job Sheet. Note the proficiency of each trainee on his or her Job Sheet and Skill Test Record.

Session III. Erecting a Gable Roof; Laboratory

  A. Erecting a Gable Roof

    1. Installing Rafters   _____

  B. Laboratory   _____

  Hand out Job Sheet 27106-2. Under your supervision, have the trainees perform the tasks on the Job Sheet. Note the proficiency of each trainee on his or her Job Sheet and Skill Test Record.

Session IV. Framing the Gable Ends; Framing a Gable Overhang; Laboratory

  A. Framing the Gable Ends   _____

  B. Framing a Gable Overhang   _____

  C. Laboratory   _____

  Hand out Job Sheet 27106-3. Under your supervision, have the trainees perform the tasks on the Job Sheet. Note the proficiency of each trainee on his or her Job Sheet and Skill Test Record.

Session V. Framing an Opening in the Roof; Laboratory

  A. Framing an Opening in the Roof   _____

  B. Laboratory   _____

  Hand out Job Sheet 27106-4. Under your supervision, have the trainees perform the tasks on the Job Sheet. Note the proficiency of each trainee on his or her Job Sheet and Skill Test Record.

Session VI. Laying Out and Erecting Hips and Valleys

  A. Laying Out and Erecting Hips and Valleys

    1. Hip Rafters   _____

    2. Valley Layout   _____

    3. Jack Rafter Layout   _____

Session VII. Laboratory

   A. Laboratory   _____

Hand out Job Sheet 27106-5. Under your supervision, have the trainees perform the tasks on the Job Sheet. Note the proficiency of each trainee on his or her Job Sheet and Skill Test Record.

Session VIII. Installing Sheathing; Laboratory

   A. Installing Sheathing   _____

   B. Laboratory   _____

Hand out Job Sheet 27106-6. Under your supervision, have the trainees perform the tasks on the Job Sheet. Note the proficiency of each trainee on his or her Job Sheet and Skill Test Record.

Session IX. Rafter Layout Using a Speed Square

   A. Rafter Layout Using a Speed Square   _____

     1. Procedure for Laying Out Common Rafters   _____

     2. Laying Out a Hip Rafter with a Speed Square   _____

Session X. Rafter Layout Using a Calculator

   A. Rafter Layout Using a Calculator   _____

     1. Using Trig Functions on a Calculator   _____

Session XI. Truss Construction

   A. Truss Construction   _____

     1. Truss Installation   _____

     2. Bracing of Roof Trusses   _____

Session XII. Truss Construction; Laboratory

   A. Truss Construction   _____

   B. Laboratory   _____

Hand out Job Sheet 27106-7. Under your supervision, have the trainees perform the tasks on the Job Sheet. Note the proficiency of each trainee on his or her Job Sheet and Skill Test Record.

Session XIII. Determining Quantities of Materials

   A. Determining Quantities of Materials   _____

     1. Determine Materials Needed for a Gable Roof   _____

Session XIV. Dormers; Plank-and-Beam Roof Construction

   A. Dormers   _____

   B. Plank-and-Beam Roof Construction   _____

Session XV. Metal Roof Framing; Module Examination and Performance Testing

   A. Metal Roof Framing   _____

   B. Summary   _____

     1. Summarize module.

     2. Answer questions.

   C. Module Examination   _____

     1. Trainees must score 70% or higher to receive recognition from the NCCER.

     2. Record the testing results on Craft Training Report Form 200 and submit the results to the Training Program Sponsor.

   D. Performance Testing   _____

     1. Trainees must perform each task to the satisfaction of the instructor to receive recognition from the NCCER.

     2. Record the testing results on Craft Training Report Form 200 and submit the results to the Training Program Sponsor.

Module 27106-01

Roof Framing

Instructor's Notes:

_____

_____

_____

_____

_____

_____

_____

_____

Course Map

This course map shows all of the modules in the first level of the Carpentry curriculum. The suggested training order begins at the bottom and proceeds up. Skill levels increase as a trainee advances on the course map. The training order may be adjusted by the local Training Program Sponsor.

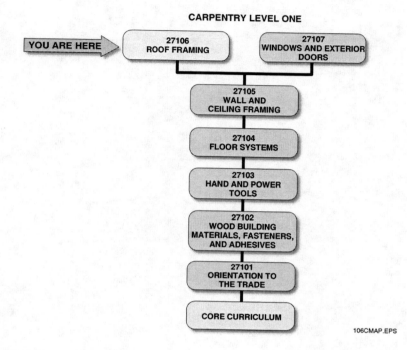

CARPENTRY LEVEL ONE

YOU ARE HERE →

27106
ROOF FRAMING

27107
WINDOWS AND EXTERIOR DOORS

27105
WALL AND CEILING FRAMING

27104
FLOOR SYSTEMS

27103
HAND AND POWER TOOLS

27102
WOOD BUILDING MATERIALS, FASTENERS, AND ADHESIVES

27101
ORIENTATION TO THE TRADE

CORE CURRICULUM

106CMAP.EPS

Assign reading of Module 27106.

1.0.0   INTRODUCTION 6.1

2.0.0   TYPES OF ROOFS 6.1

3.0.0   BASIC ROOF LAYOUT 6.3

    3.1.0   Rafter Framing Square 6.4

    3.2.0   Basic Rafter Layout 6.4

    3.2.1   Laying Out Rafter Locations 6.5

    3.2.2   Determining the Length of a Common Rafter 6.5

    3.2.3   Laying Out a Common Rafter 6.9

    3.3.0   Erecting a Gable Roof 6.11

    3.3.1   Installing Rafters 6.11

    3.3.2   Framing the Gable Ends 6.11

    3.3.3   Framing a Gable Overhang 6.13

    3.3.4   Framing an Opening in the Roof .6.14

4.0.0   LAYING OUT AND ERECTING HIPS AND VALLEYS 6.15

    4.1.0   Hip Rafters 6.15

    4.2.0   Valley Layout 6.18

    4.3.0   Jack Rafter Layout 6.18

5.0.0   INSTALLING SHEATHING 6.19

6.0.0   RAFTER LAYOUT USING A SPEED SQUARE 6.23

    6.1.0   Procedure for Laying Out Common Rafters 6.24

    6.2.0   Laying Out a Hip Rafter with a Speed Square 6.24

7.0.0   RAFTER LAYOUT USING A CALCULATOR 6.27

    7.1.0   Using Trig Functions on a Calculator 6.28

8.0.0   TRUSS CONSTRUCTION 6.31

    8.1.0   Truss Installation 6.34

    8.2.0   Bracing of Roof Trusses 6.34

9.0.0   DETERMINING QUANTITIES OF MATERIALS 6.36

    9.1.0   Determine Materials Needed for a Gable Roof 6.37

10.0.0 DORMERS 6.37

11.0.0 PLANK-AND-BEAM ROOF CONSTRUCTION 6.38

12.0.0 METAL ROOF FRAMING 6.38

SUMMARY 6.39

REVIEW QUESTIONS 6.40

GLOSSARY 6.43

ANSWERS TO REVIEW QUESTIONS 6.44

ADDITIONAL RESOURCES 6.45

Figures

Figure 1   Types of roofs 6.2
Figure 2   Roof framing members 6.3
Figure 3   Roof layout factors 6.4
Figure 4   Rafter tables on a framing square 6.5
Figure 5   Application of the rafter square 6.6
Figure 6   Parts of a rafter 6.7
Figure 7   Marking rafter locations 6.8
Figure 8   Determining rafter length with the framing square 6.9
Figure 9   Framing square step-off method 6.9
Figure 10   Marking the rafter cuts 6.10
Figure 11   Rafter installation 6.11
Figure 12   Bracing a long roof span 6.12
Figure 13   Gable end vents and frame 6.12
Figure 14   Gable stud 6.13
Figure 15   Framing a gable overhang 6.13
Figure 16   Framing an opening in a roof 6.14
Figure 17   Example of intersecting roof sections 6.15
Figure 18   Overhead view of a roof layout 6.16
Figure 19   Layout of rafter locations on the top plate 6.16
Figure 20   Using the framing square to determine the length
            of a hip rafter 6.17
Figure 21   Hip rafter position 6.17
Figure 22   Dropped bird's mouth cut 6.18
Figure 23   Valley layout 6.18
Figure 24   Valley rafter layout 6.19
Figure 25   Jack rafter locations 6.19
Figure 26   Hip and jack rafter layout 6.20
Figure 27   Use of H-clips 6.21
Figure 28   Underlayment installation 6.23
Figure 29   Speed square 6.23
Figure 30   Laying out a common rafter 6.25
Figure 31   Hip rafter layout 6.26
Figure 32   Roof framing angles 6.27
Figure 33   Sides of a triangle 6.28
Figure 34   Calculating hip and valley rafters 6.29
Figure 35   First practical use of trigonometry 6.29
Figure 36   Angle-side relationships 6.30
Figure 37   Roof framing example 6.30
Figure 38   Components of a truss 6.32
Figure 39   Types of trusses 6.32

Instructor's Notes:

_____

_____

_____

_____

_____

_____

_____

_____

Figure 40   Erecting trusses 6.33
Figure 41   Use of a tiedown to secure a truss 6.34
Figure 42   Example of permanent bracing specification 6.35
Figure 43   Example of temporary bracing of trusses 6.35
Figure 44   Shed dormer 6.37
Figure 45   Gable dormer framing 6.38
Figure 46   Example of post-and-beam construction 6.39

Roof Framing

Materials

Ensure you have everything required to teach the course. Check the Materials and Equipment List at the front of this Instructor's Guide.

Audiovisual

Show Transparency 1, Course Objectives.

Show Transparency 2, Performance Profile Tasks.

Objectives

Upon completion of this module, the trainee will be able to:

1. Understand the terms associated with roof framing.
2. Identify the roof framing members used in gable and hip roofs.
3. Identify the methods used to calculate the length of a rafter.
4. Identify the various types of trusses used in roof framing.
5. Use a rafter framing square, speed square, and calculator in laying out a roof.
6. Identify various types of sheathing used in roof construction.
7. Frame a gable roof with vent openings.
8. Frame a roof opening.
9. Construct a frame roof, including hips, valleys, commons, jack rafters, and sheathing.
10. Erect a gable roof using trusses.
11. Estimate the materials used in framing and sheathing a roof.

Prerequisites

Successful completion of the following Task Modules is recommended before beginning study of this Task Module: Core Curriculum; Carpentry Level One, Modules 27101 through 27105.

Required Trainee Materials

1. Trainee Task Module
2. Appropriate personal protective equipment

1.0.0 ◆ INTRODUCTION

Roof framing is the most demanding of the framing tasks. Floor and wall framing generally involves working with straight lines. Residential roofs are usually sloped in order to shed water from rain or melting snow. In areas where there is heavy snowfall, the roof must be constructed to bear extra weight. Because a roof is sloped, laying out a roof involves working with precise angles in addition to straight lines.

In this module, you will learn about the different types of roofs used in residential construction. You will also learn how to lay out and frame a roof.

2.0.0 ◆ TYPES OF ROOFS

The most common types of roofs used in residential construction are shown in *Figure 1* and described below.

- *Gable roof*—A gable roof has two slopes that meet at the center (ridge) to form a gable at each end of the building. It is the most common type of roof because it is simple, economical, and can be used on any type of structure.
- *Hip roof*—A hip roof has four sides or slopes running toward the center of the building. Rafters at the corners extend diagonally to meet at the ridge. Additional rafters are framed into these rafters.
- *Gable and valley roof*—This roof consists of two intersecting gable roofs. The part where the two roofs meet is called a *valley*.
- *Mansard roof*—The mansard roof has four sloping sides, each of which has a double slope. As compared with a gable roof, this design provides more available space in the upper level of the building.
- *Gambrel roof*—The gambrel roof is a variation on the gable roof in which each side has a break, usually near the ridge. The gambrel roof provides more available space in the upper level.
- *Shed roof*—Also known as a *lean-to roof*, the shed roof is a flat, sloped construction. It is common on high-ceiling contemporary construction, and is often used on additions.

Classroom

Discuss the importance of proper roof framing.

Describe the six types of roofs. Point out that roofs can be stick-built or truss-built. Stick-built roofs usually require interior loadbearing walls for the ceiling joists.

Explain that terms shown in bold (blue) are defined in the Glossary at the back of this module.

Audiovisual

Show Transparency 3 (Figure 1).

Teaching Tip

Acquire photographs or drawings from architectural firms or building material manufacturers that show specific types of building construction in various stages of completion. Pass them around for the trainees to examine.

ROOF FRAMING—TRAINEE MODULE 27106   6.1

Mansard Roof

Mansard roofs are popular in the design of many fast-food restaurants and small office buildings.

106P0601.EPS

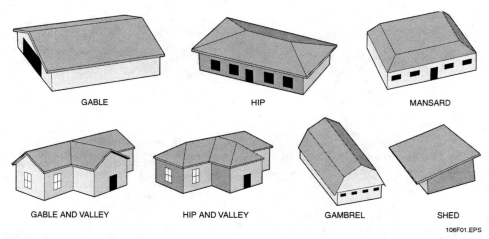

GABLE   HIP   MANSARD

GABLE AND VALLEY   HIP AND VALLEY   GAMBREL   SHED

106F01.EPS

Figure 1 ◆ Types of roofs.

There are two basic roof framing systems. In stick-built framing, ceiling joists and rafters are laid out and cut by carpenters on the site and the frame is constructed one stick at a time.

In truss-built construction, the roof framework is prefabricated off site. The truss contains both the rafters and the ceiling joist. Trusses and truss construction will be discussed later in this module.

Take the trainees to local sites to observe various types of completed roofs as well as stick-built and truss-built roofs under construction.

6.2   CARPENTRY LEVEL ONE—TRAINEE MODULE 27106

Instructor's Notes:

3.0.0 ◆ BASIC ROOF LAYOUT

Rafters and ceiling joists provide the framework for all roofs. The main components of a roof are shown in *Figure 2* and described below.

- *Ridge (ridgeboard)*—The highest horizontal roof member. It helps to align the rafters and tie them together at the upper end. The ridgeboard is one size larger than the rafters.
- *Common rafter*—A structural member that extends from the top plate to the ridge in a direction perpendicular to the wall plate and ridge. Rafters often extend beyond the roof plate to form the overhang (eaves) that protect the side of the building.
- *Hip rafter*—A roof member that extends diagonally from the corner of the plate to the ridge.

- *Valley rafter*—A roof member that extends from the plate to the ridge along the lines where two roofs intersect.
- *Jack rafter*—A roof member that does not extend the entire distance from the ridge to the top plate of a wall. The hip jack and valley jack are shown in *Figure 2*. A rafter fitted between a hip rafter and a valley rafter is called a *cripple jack*. It touches neither the ridge nor the plate.
- *Plate*—The wall framing member that rests on top of the wall studs. It is sometimes called the *rafter plate* because the rafters rest on it. It is also referred to as the *top plate*.

As you can see in *Figure 3*, on any pitched roof, rafters rise at an angle to the ridgeboard. Therefore, the length of the rafter is greater than the horizontal distance from the plate to the ridge. In

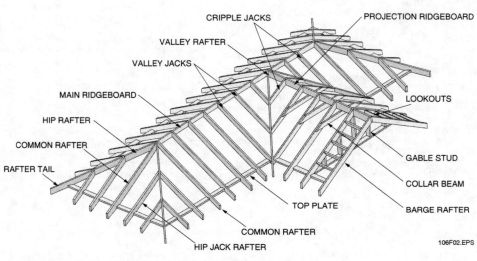

CRIPPLE JACKS — PROJECTION RIDGEBOARD
VALLEY RAFTER
VALLEY JACKS
MAIN RIDGEBOARD
HIP RAFTER
COMMON RAFTER
RAFTER TAIL
LOOKOUTS
GABLE STUD
COLLAR BEAM
BARGE RAFTER
TOP PLATE
COMMON RAFTER
HIP JACK RAFTER
COMMON RAFTER

106F02.EPS

Figure 2 ◆ Roof framing members.

Pitch and Slope

Carpenters may use the terms *pitch* and *slope* interchangeably on the job site, but the two terms actually refer to two different concepts. Slope is the amount of rise per foot of run and is always referred to as a *number in 12*. For example, a roof that rises 6" for every foot of run has a 6 in 12 slope (the 12 simply refers to the number of inches in a foot). Pitch, on the other hand, is the ratio of rise to the span of the roof and is expressed as a fraction. For example, a roof that rises 8' over a 32' span is said to have a pitch of ¼ (⁸⁄₃₂ = ¼).

order to calculate the correct rafter length, the carpenter must factor in the slope of the roof. Here are some additional terms you will need to know in order to lay out rafters:

- *Span*—The horizontal distance from the outside of one exterior wall to the outside of the other exterior wall.
- *Run*—The horizontal distance from the outside of the top plate to the center line of the ridgeboard (usually equal to half of the span).
- *Rise*—The total height of the rafter from the top plate to the ridge. This is stated in inches per foot of run.
- *Pitch*—The angle or degree of slope of the roof in relation to the span. Pitch is expressed as a fraction; e.g., if the total rise is 6' and the span is 24', the pitch would be ¼ (6 over 24).
- *Slope*—The inclination of the roof surface expressed as the relationship of rise to run. It is stated as a unit of rise to so many horizontal units; e.g., a roof that has a rise of 5" for each foot of run is said to have a 5 in 12 slope (*Figure 3*). The roof slope is sometimes referred to as the *roof cut*.

The first step in determining the correct length of a rafter is to find the unit rise, which is usually shown on the building's elevation drawing. The unit rise is the number of inches the rafter rises vertically for each foot of run. The greater the rise per foot of run, the greater the slope of the roof.

There are several ways to calculate the length of a rafter. It can be done with a framing (rafter) square or speed square, or it can be done using a calculator. We will show the framing square

method now, then discuss the speed square and calculator approaches later.

3.1.0 Rafter Framing Square

The rafter framing square is a special carpenter's square that is calibrated to show the length per foot of run for each type of rafter (*Figure 4*). Note that the *tongue* is the short (16") section of the square and the *blade* (or body) is the long (24") section. The corner is known as the *heel*. Rafter tables are normally provided on the back of the square. The rafter tables usually give the rafter dimensions in length per foot of run, but some give length per given run.

The framing square is used to determine the rafter length and to measure and mark the cuts that must be made in the rafter (*Figure 5*). As you can see, you can relate the pitch and slope to the rise per foot of run. The rise per foot of run is always the same for a given pitch or slope. For example, a pitch of ½, which is the same as a 12 in 12 slope, equals 12" of rise per foot of run. Framing squares are discussed in more detail later in the module.

3.2.0 Basic Rafter Layout

Laying out the framing for a roof involves four tasks:

- Marking off the rafter locations on the top plate
- Determining the length of each rafter
- Making the plumb cuts at the ridge end and tail end of each rafter
- Making the bird's mouth cut in each rafter

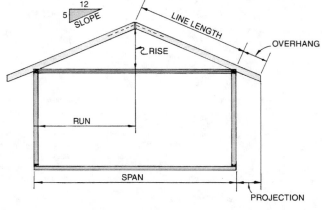

Figure 3 ◆ Roof layout factors.

106F03.TIF

CARPENTRY LEVEL ONE—TRAINEE MODULE 27106

Instructor's Notes:

_____

_____

_____

_____

_____

_____

_____

_____

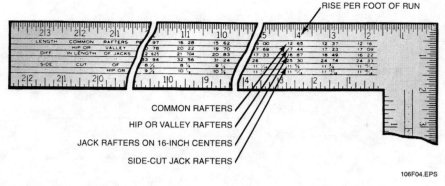

RISE PER FOOT OF RUN

COMMON RAFTERS
HIP OR VALLEY RAFTERS
JACK RAFTERS ON 16-INCH CENTERS
SIDE-CUT JACK RAFTERS

106F04.EPS

Figure 4 ◆ Rafter tables on a framing square.

Rafters must be laid out and cut so that the ridge end will fit squarely on the ridge and the tail end will present a square surface for the false fascia (sub fascia) board. In addition, a bird's mouth cut (*Figure 6*) must be made at the correct location and angle for the rafter to rest squarely on the plate.

3.2.1 Laying Out Rafter Locations

The following is a basic procedure for marking the locations of the rafters on the top plate (*Figure 7*) for 24" on center (OC) construction. Keep in mind that in most cases, the ceiling joists would be in place and the rafter locations would have already been marked as described in a previous module.

Step 1 Locate the first rafter flush with one end of the top plate.

Step 2 Measure the width of a rafter and mark and square a line the width of a rafter in from the end.

Step 3 Use a measuring tape to space off and mark the rafter locations every 24" all the way to the end. Square the lines.

Note

The distance between the last two rafters may be less than 24", but not more than 24".

Step 4 Repeat the process on the opposite top plate, starting from the same end as before.

Step 5 Cut the ridgeboard to length, allowing for a barge rafter at each end, if required.

Step 6 Place the ridgeboard on the top plate and mark it for correct position. Then measure and mark the rafter locations with an R on the ridgeboard.

To determine the number of rafters you need, count the marks on both sides of the ridge or top plate.

3.2.2 Determining the Length of a Common Rafter

The following is an overview of the procedure for laying out the rafters and joists of a gable roof.

Cutting Sequence

Some carpenters prefer to make the tail end plumb cut while laying out the rafter. Others prefer to tail cut all the rafters at once after the rafters are installed. One method isn't necessarily better than the other and is simply a matter of preference.

Show Transparencies 8 and 9 (Figures 6 and 7).

Demonstration

Demonstrate the layout of rafter locations. Point out that rafters can also be positioned on 16" centers to support high roof loads caused by ice or snow accumulation.

Teaching Tip

For roof framing demonstrations or laboratory exercises, you may wish to use half-size (half-scale) layout dimensions and lumber (for example, 1 × 2s = 2 × 4s, 1 × 4s = 2 × 8s, etc.).

If desired, the video *Roof Framing, Part One* may be used to demonstrate rafter erection for a gable roof.

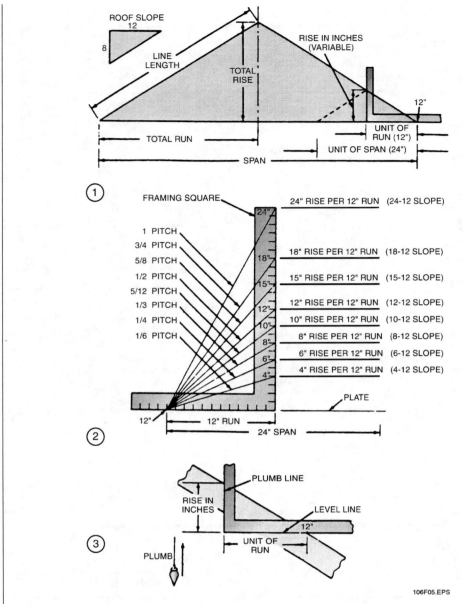

Figure 5 ◆ Application of the rafter square.

CARPENTRY LEVEL ONE—TRAINEE MODULE 27106

106F05.EPS

Instructor's Notes:

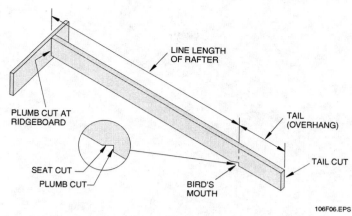

Figure 6 ◆ Parts of a rafter.

Discuss the "Think About It." (See the answers at the end of this module.)

Hand out Job Sheet 27106-1. Under your supervision, have the trainees perform the tasks on the Job Sheet. Note the proficiency of each trainee.

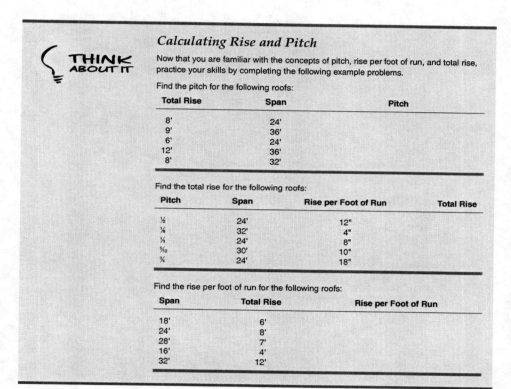

THINK ABOUT IT

Calculating Rise and Pitch

Now that you are familiar with the concepts of pitch, rise per foot of run, and total rise, practice your skills by completing the following example problems.

Find the pitch for the following roofs:

Total Rise	Span	Pitch
8'	24'	
9'	36'	
6'	24'	
12'	36'	
8'	32'	

Find the total rise for the following roofs:

Pitch	Span	Rise per Foot of Run	Total Rise
½	24'	12"	
⅙	32'	4"	
⅓	24'	8"	
⁵⁄₁₂	30'	10"	
¾	24'	18"	

Find the rise per foot of run for the following roofs:

Span	Total Rise	Rise per Foot of Run
18'	6'	
24'	8'	
28'	7'	
16'	4'	
32'	12'	

ROOF FRAMING—TRAINEE MODULE 27106

6.7

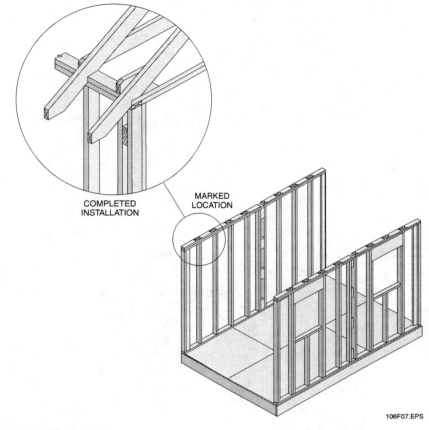

Figure 7 ◆ Marking rafter locations.

Explain the rafter table method of determining rafter length.

Show Transparencies 10 and 11 (Figures 8 and 9).

Here is an easy way of determining the required length of a rafter:

Step 1 Start by measuring the building span, then divide that in half to determine the run.

Step 2 Determine the rise. This can be done in either of the following ways:
• Calculate the total rise by multiplying the span by the pitch (e.g., 40' span × ¼ pitch = 10' rise).
• Look for it on the slope diagram on the roof plan as discussed previously.

Step 3 Divide the total rise (in inches) by the run (in feet) to obtain the rise per foot of run.

Step 4 Look up the required length on the rafter tables on the framing square.

For example, if the roof has a span of 20', the run would be 10'. Then, assuming that the blueprint shows the rise per foot of run to be 8", the correct rafter length would be 14.42" (14$\frac{5}{12}$") per foot of run. Since you have 10' of run, the rafter length would be 144⅛" (*Figure 8*). If an overhang is used, the overhang length must be added to the rafter length.

6.8

Instructor's Notes:

_____

_____

_____

_____

_____

_____

_____

_____

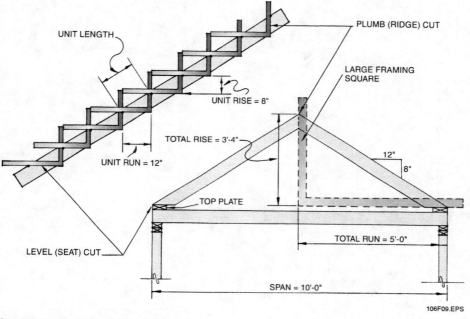

Figure 8 ◆ Determining rafter length with the framing square.

ridge plumb cut is determined, then the framing square is stepped once for every foot of run. The final step marks the plumb cut for the bird's mouth. This is not the preferred method because it is not as precise as other methods.

In a later section of this module, we will cover a method of determining the length of a rafter using a calculator.

3.2.3 Laying Out a Common Rafter

The following is an overview of the procedure for laying out and cutting a common rafter:

Step 1 Start with a piece of lumber a little longer than the required length of the rafter, including the tail. If the lumber has a crown or bow, it should be at the top of the rafter. Lay the rafter on sawhorses with the crown (if any) at the top.

Step 2 Start by marking the ridge plumb cut using the framing square (*Figure 10*). Be sure to subtract half the thickness of the ridgeboard. Make the cut.

Another way to calculate rafter length is to measure the distance between 8 on the framing square blade and 12 on the tongue, as shown in *Figure 5*.

Yet another method of determining the approximate length of a rafter is the *framing square step-off method* shown in *Figure 9*. In this procedure, the

Figure 9 ◆ Framing square step-off method.

Discuss the "Think About It." (See the answers at the end of this module.)

Performance Profile Test

Have each trainee complete to your satisfaction **Performance Profile Test 1.** Fill out Performance Profile Sheets for each trainee.

Homework

Assign reading of Sections 3.3.0–3.3.1 for the next class session.

Calculating Common Rafter Lengths

Now that you are familiar with the method of arriving at the length of a rafter using the square, find the lengths of common rafters for the following spans:

Span	Run	Rise per Foot of Run	Rafter Length
26'	13'	6"	
24'	12'	4"	
28'	14'	8"	
32'	16'	12"	
30'	15'	7"	

Note

12 is a factor used to obtain a value in feet. Be sure to reduce or convert to the lowest terms.

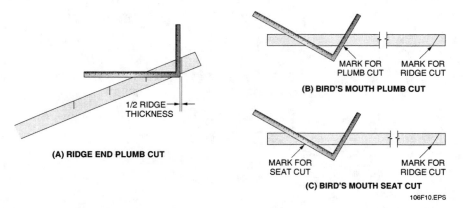

(A) RIDGE END PLUMB CUT

(B) BIRD'S MOUTH PLUMB CUT

(C) BIRD'S MOUTH SEAT CUT

106F10.EPS

Figure 10 ◆ Marking the rafter cuts.

Step 3 Measure the length of the rafter from the plumb cut mark to the end (excluding the tail) and mark another plumb cut for the bird's mouth. Reposition the framing square and mark the bird's mouth seat.

Step 4 Make the end plumb cut, then cut out the bird's mouth. Cut the bird's mouth partway with a circular saw, then use a hand saw to finish the cuts.

Step 5 Use the first rafter as a template for marking the remaining rafters. As the rafters are cut, stand them against the building at the joist locations.

Instructor's Notes:

_____

_____

_____

_____

_____

_____

_____

_____

3.3.0 Erecting a Gable Roof

This section contains an overview of the procedure for erecting a gable roof.

 WARNING!
Be sure to follow applicable fall protection procedures.

3.3.1 Installing Rafters

Rafters are installed using the following procedure.

 Note
It is a good idea to mark a 2 × 4 in advance with the total rise and use it as a guide for the height of the ridgeboard.

Step 1 Start by placing boards over the joists to walk on. Nail a rafter at each end of the ridgeboard, then lift the ridgeboard to a temporary position, secure it, and nail the bird's mouth of each rafter to the joists (*Figure 11*). Nail the rafters in pairs.

Step 2 On the opposite side, start by nailing the bird's mouth to the joist, then toenail the plumb cut into the ridgeboard. Once this is done, use a temporary brace to hold the ridgeboard in place while installing the remaining rafters. Remember to keep the ridgeboard straight and the rafters plumb.

Step 3 Run a line and trim the rafter tails.

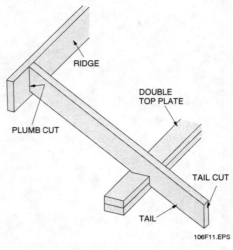

Figure 11 ◆ Rafter installation.

If the rafter span is long, additional support will be required. *Figure 12* shows the use of strongbacks, purlins, braces, and collar beams (collar ties) for this purpose. Two by six collar ties are installed at every second rafter. Two by four diagonal braces are notched into the purlins. Strongbacks are L-shaped members that run the length of the roof. They are used to straighten and strengthen the ceiling joists.

3.3.2 Framing the Gable Ends

Attics must be vented to allow heat that rises from the lower floors of the building to escape to the

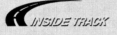 **Rafter Marks**

When it is time to place the rafters into position on the top plate, make sure they are placed on the correct side of your marks. Start installing the rafters from one end of the roof and work towards the other end. Double-check each rafter position for consistency before nailing it into place. Roof framing mistakes are time consuming to fix and must be avoided.

 **Classroom**

Provide an overview of the procedure for erecting a gable roof.

 **Audiovisual**

Show Transparencies 13 and 14 (Figures 11 and 12).

 **Demonstration**

With the help of several trainees, demonstrate the erection of a portion of a gable roof. Emphasize the use of fall protection when erecting roofs. Point out that long rafter spans will require extra support.

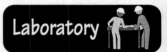 **Laboratory**

Hand out Job Sheet 27106-2. Under your supervision, have the trainees perform the tasks on the Job Sheet. Note the proficiency of each trainee.

 **Performance Profile Test**

Have each trainee complete to your satisfaction Performance Profile Test 2 on rafters for a gable roof. Fill out Performance Profile Sheets for each trainee.

 **Homework**

Assign reading of Sections 3.3.2–3.3.3 for the next class session.

outdoors. There is also a need to vent moisture that accumulates in an attic due to condensation that occurs when rising heat from below meets the cooler air in the unheated attic space. Several methods are used to vent roofs. *Figure 13* shows two types of gable end vents.

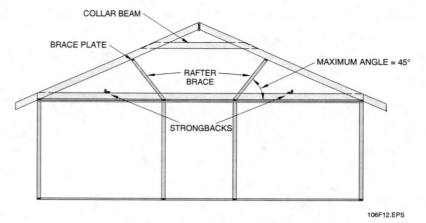

106F12.EPS

Figure 12 ◆ Bracing a long roof span.

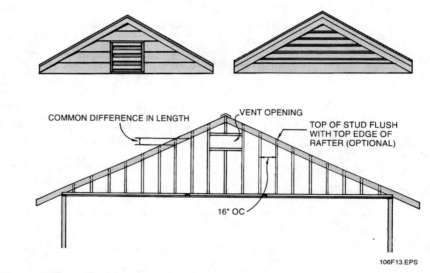

106F13.EPS

Figure 13 ◆ Gable end vents and frame.

Instructor's Notes:

_____

_____

_____

_____

_____

_____

_____

_____

Notice that the lengths of the studs decrease as they approach the sides. Each pair of studs must be measured and cut to fit.

Step 1 Start by plumbing down from the center of the rafter to the top plate, then mark and square a line on the top plate at this point.

Step 2 Lay out the header and sill for the vent opening by measuring 8" (in this case) on either side of the plumb line.

Step 3 Mark the stud locations above the wall studs, then stand a stud upright at the first position, plumb it, and mark the diagonal cut for the top of the stud.

Step 4 Measure and mark the next stud in the sequence. The difference in length between the first and second studs is the common difference, which can be applied to all remaining studs.

Step 5 Cut and install the studs as shown in *Figure 14*. Toenail or straight nail the gable stud to the double top plate. The notch should be as deep as the rafter.

Step 6 When studs are in place, cut and install the header and sill for the vent opening. Then lay out, cut, and install cripple studs above and below the vent using the same method as before.

3.3.3 Framing a Gable Overhang

If an extended gable overhang (*rake*) is required, the framing must be done before the roof framing is complete. *Figure 15* shows two methods of framing overhangs. In the view on the right, 2 × 4 lookouts are laid into notched rafters. In the view on the left, which can be used for a small overhang, a 2 × 4 barge rafter and short lookouts are used.

Note
Install a brace across the rafters to be cut. It will temporarily hold the framing in place until the headers are installed.

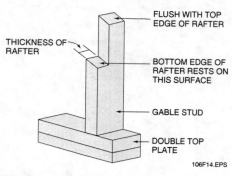

Figure 14 ◆ Gable stud.

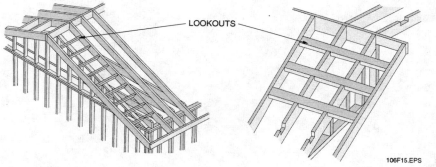

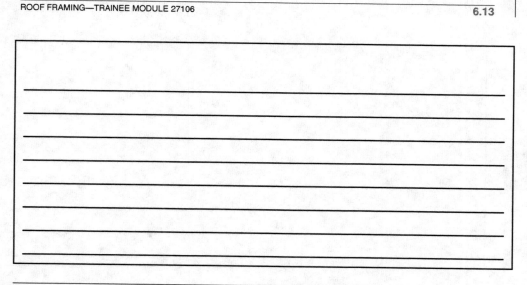

Figure 15 ◆ Framing a gable overhang.

Demonstrate how to frame a portion of a gable end and gable overhang.

Hand out Job Sheet 27106-3. Under your supervision, have the trainees perform the tasks on the Job Sheet. Note the proficiency of each trainee.

Have each trainee complete to your satisfaction Performance Profile Test 3 on framing a gable end. Fill out Performance Profile Sheets for each trainee.

Assign reading of Section 3.3.4 for the next class session.

Classroom

Provide an overview of the procedure for framing a roof opening.

Audiovisual

Show Transparency 18 (Figure 16).

Demonstration

Demonstrate how to frame a roof opening. Point out that a brace should be nailed across all rafters to be cut, and to adjacent uncut rafters to hold the cut rafters in place until the headers are installed.

Laboratory

Hand out Job Sheet 27106-4. Under your supervision, have the trainees perform the tasks on the Job Sheet. Note the proficiency of each trainee.

Performance Profile Test

Have each trainee complete to your satisfaction Performance Profile Test 4 on framing an opening in a roof. Fill out Performance Profile Sheets for each trainee.

Homework

Assign reading of Sections 4.0.0–4.3.0 for the next class session.

Barge Rafter and Lookouts

A gable overhang consists of a barge (fly) rafter on the end and lookouts that connect the barge to a common rafter. The lookouts provide structural support for the overhang and a solid foundation for a soffit or other decorative finish.

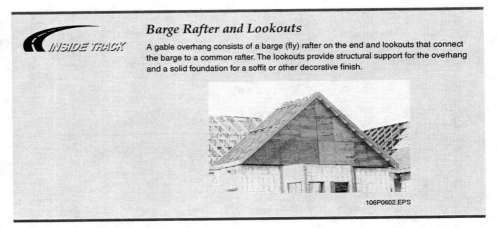

106P0602.EPS

3.3.4 Framing an Opening in the Roof

It is sometimes necessary to make an opening in a roof for a chimney, skylight, or roof window (*Figure 16*). The following is a general procedure for framing such an opening.

Step 1 Lay out the opening on the floor beneath the opening, then use a plumb bob to transfer the layout to the roof. If you are framing a chimney, be sure to leave adequate clearance. If the opening is large, allow for double headers.

Step 2 Cut the rafters per the layout. Install the headers, then install a double trimmer rafter on either side of the opening, as shown in *Figure 16*.

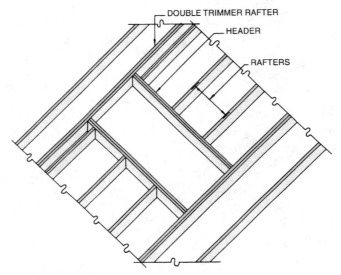

DOUBLE TRIMMER RAFTER
HEADER
RAFTERS

Figure 16 ◆ Framing an opening in a roof.

CARPENTRY LEVEL ONE—TRAINEE MODULE 27106

Instructor's Notes:

CARPENTRY LEVEL ONE—INSTRUCTOR'S GUIDE MODULE 27106

Framing Roof Openings

Before you cut an opening in a roof, check with the site engineer to make sure that you are proceeding according to specifications. If you don't double-check with the engineer, you may cut an incorrect opening and new rafters or trusses may be required. This type of error is very time consuming and expensive to fix. Also, you should always be sure to check local fire codes for the proper clearance around a chimney and other roof openings. If the framing for the opening is too close to a chimney, it could cause a fire.

Discuss roof projections and hip roofs. Identify the various framing members used.

Show Transparencies 19 through 21 (Figures 17 through 19).

Discuss hip rafter layout. Explain the 17" rule.

4.0.0 ◆ LAYING OUT AND ERECTING HIPS AND VALLEYS

An intersecting roof contains two or more sections sloping in different directions. Examples are the connection of two gable sections or a gable and hip combination such as that shown in *Figure 17*. *Figure 18* shows an overhead view of the same layout.

A valley occurs wherever two gable or hip roof sections intersect. Valley rafters run at a 45° angle to the outside walls of the building.

The material that follows provides an overview of the procedures for laying out hip and valley sections and the various types of rafters used in framing these sections. The first step, as always, is to lay out the rafter locations on the top plate (*Figure 19*). The layout of the common rafters for a hip roof is the same as that for a gable roof. The next step would be to lay out, cut, and install the common rafters and the main and projection ridgeboards, as described earlier in this module. At that point, you are ready to lay out the hip or valley section.

4.1.0 Hip Rafters

A hip rafter is the diagonal of a square formed by the walls and two common rafters (*Figure 18*). Because they travel on a diagonal to reach the ridge, the hip rafters are longer than the common rafters. The unit run is 17" (16.97" rounded up), which is the length of the diagonal of a 12" square. You can see this on the top line of the rafter tables on the framing square. There are two hip rafters in every hip section.

For every hip roof of equal pitch, for every foot of run of common rafter, the hip rafter has a run of 17". This is a very important fact to remember. It can be said that the run of a hip rafter is 17 divided by 12 times the run of a common rafter. The total rise of the hip rafter is the same as that of a common rafter. For example, if a common rafter has an 8" rise per foot of run, the hip rafter would also have a rise of 8" per 17" run. Therefore, the rise of a hip rafter would be the same as a common rafter at any given corresponding point.

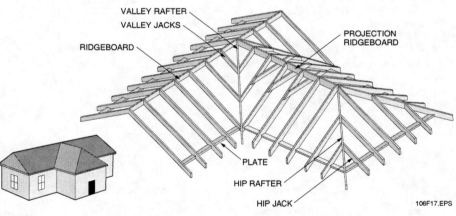

Figure 17 ◆ Example of intersecting roof sections.

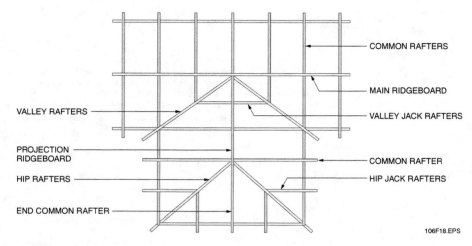

Figure 18 ◆ Overhead view of a roof layout.

COMMON RAFTERS

MAIN RIDGEBOARD

VALLEY RAFTERS

VALLEY JACK RAFTERS

PROJECTION RIDGEBOARD

COMMON RAFTER

HIP RAFTERS

HIP JACK RAFTERS

END COMMON RAFTER

106F18.EPS

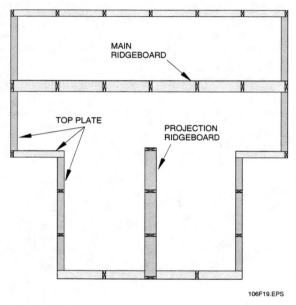

MAIN RIDGEBOARD

TOP PLATE

PROJECTION RIDGEBOARD

106F19.EPS

Figure 19 ◆ Layout of rafter locations on the top plate.

To find the length of a hip rafter, first find the rise per foot of run of the roof, then locate that specific number on the inch scale line of the framing square. Find the corresponding numbers on the second line of the framing square. For example, assume a roof has a rise per foot of run of 8" (8 in 12). The span of the building is 10'. The second line under the 8 on the blade of the rafter (framing)

Instructor's Notes:

_____

_____

_____

_____

_____

_____

_____

_____

square is 18.76. You have now established that the unit length of the hip rafter is 18.76" or 18¾" for every foot of the common rafter. This unit length must be multiplied by the run of the structure, which is 5'. The sum of the two factors multiplied together must then be divided to find the length of the hip rafter. For example, 18.76" × 5 = 93.8". 93.8" ÷ 12 = 7.81" or 7'-9¾". Therefore, the total length of the hip rafter would be 7'-9¾".

No matter what the rise per foot of run is, the length of the common and hip rafters is determined in the same fashion. The method of laying out and marking a hip rafter is very similar to the layout of a common rafter, with one exception. A common rafter is laid out using 12 on the tongue and the rise per foot of run on the blade. A hip rafter is laid out by using 17 on the blade and the rise per foot of run on the tongue (*Figure 20*).

The basic procedure for laying out a hip rafter is as follows:

Step 1 Determine the length of the hip rafters using the framing square (*Figure 20*) or rafter tables and add the overhang.

Step 2 Mark the plumb cuts on the hip rafters. The hip rafters must be shortened by half the diagonal thickness of the ridgeboard. Also, they must be cut on two sides to fit snugly between the common rafter and the ridgeboard (*Figure 21*). These cuts are known as *side cuts* or *cheek cuts*.

Show Transparencies 22 through 24 (Figures 20 through 22).

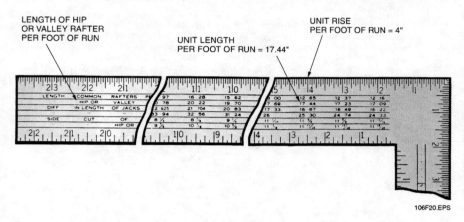

Figure 20 ◆ Using the framing square to determine the length of a hip rafter.

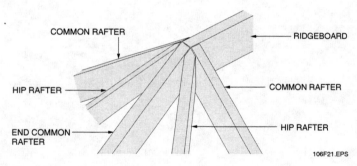

Figure 21 ◆ Hip rafter position.

Demonstrate how to lay out a hip rafter.

Discuss valley rafter layout. Point out the difference between hip and valley rafter seat and tail cuts.

Show Transparencies 25 and 26 (Figures 23 and 24).

Discuss jack rafter layout. Point out the difference between hip and valley jack rafters.

Show Transparencies 27 and 28 (Figures 25 and 26).

Step 3 The bird's mouth plumb and seat cuts are determined in the same way as for a common rafter. However, the seat cut is dropped half the thickness of the rafter to align it with the top plane of the common rafter (distance A, *Figure 22*). This drop is necessary because the corners of the rafters would otherwise be higher than the plane on which the roof surface is laid. Another way to accomplish this is to chamfer the top edges of the rafter. This procedure is known as *backing*. Most carpenters use the dropping method because it is faster.

4.2.0 Valley Layout

Each valley requires a valley rafter and some number of valley jack rafters (*Figure 23*). The layout of a valley rafter is basically the same as that of the hip rafter, with 17" used for the unit run. The only difference in layout between the hip and valley rafters is in the seat and tail cuts.

For a valley rafter, the bird's mouth plumb cut must be angled to allow the rafter to drop down into the inside corner of the building (*Figure 24*). In addition, the tail end cuts must be made so that corner made by the valley will line up with the rest of the roof overhang. Like the hip rafter, the valley rafter must be aligned with the plane of the common rafter. Cheek cuts are also required on valley rafters to allow them to fit between the two ridgeboards.

The layout of valley jacks is the same as that for hip jacks, with the exception that the valley jacks usually run in the opposite direction; i.e., toward the ridge (*Figure 23*).

4.3.0 Jack Rafter Layout

Hip jack rafters run from the top plate to the hip rafters. Valley jacks run from the valley rafter to the ridge (*Figure 17*). Notice that layout of valley jack rafters usually starts at the building line and moves toward the ridge.

As with gable end studs, there is a common difference from one jack rafter to the next (*Figure 25*).

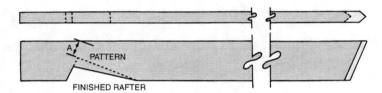

Figure 22 ◆ Dropped bird's mouth cut.

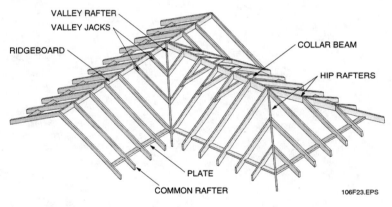

Figure 23 ◆ Valley layout.

CARPENTRY LEVEL ONE—TRAINEE MODULE 27106

Instructor's Notes:

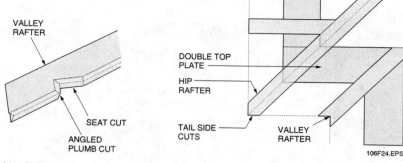

Figure 24 ◆ Valley rafter layout.

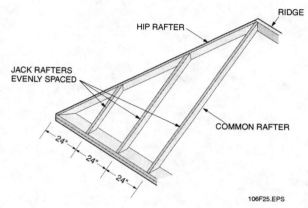

Figure 25 ◆ Jack rafter locations.

Here is an overview of the jack rafter layout process.

Step 1 The third and fourth lines of the framing square have the information you need to determine the lengths of jack rafters for 16" and 24" OC construction, respectively. The number you read from the framing square is the difference in calculated length between the common rafter and the jack rafter (*Figure 26*). The longest jack rafter is referred to on the plans as the *#1 jack rafter*. Jack rafters must be cut and installed in pairs to prevent the hip or valley from bowing.

Step 2 Use the common rafter to mark the bird's mouth.

Step 3 To lay out the additional jack rafters, subtract the common difference from the last jack rafter you laid out. For each jack rafter with a cheek cut on one side, there must be one of equal length with the cheek cut on the opposite side.

5.0.0 ◆ INSTALLING SHEATHING

The sheathing should be applied as soon as the roof framing is finished. The sheathing provides additional strength to the structure, and provides a base for the roofing material.

Some of the materials commonly used for sheathing are plywood, OSB, waferboard, shiplap, and common boards. When composition shingles are used, the sheathing must be solid. If

Demonstrate how to lay out a jack rafter.

Hand out Job Sheet 27106-5. Under your supervision, have the trainees perform the tasks on the Job Sheet. Note the proficiency of each trainee.

Have each trainee complete to your satisfaction Performance Profile Test 6. Fill out Performance Profile Sheets for each trainee.

Assign reading of Section 5.0.0 for the next class session.

Discuss the importance of sheathing.

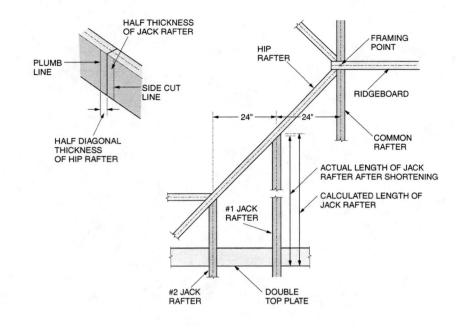

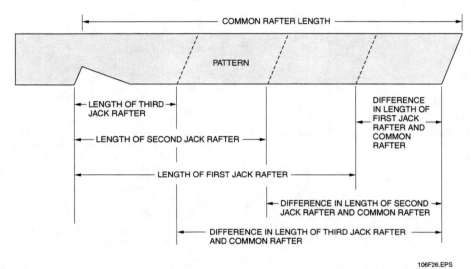

Figure 26 ◆ Hip and jack rafter layout.

106F26.EPS

Instructor's Notes:

_____

_____

_____

_____

_____

_____

_____

_____

Calculating Jack Rafter Lengths

Use the rafter tables to find the lengths of the following jack rafters at 16" OC when a common rafter is 14'-6" long.

Rise per Foot of Run	Jack Rafter	Length
6"	2nd	
4"	4th	
9"	5th	
12"	3rd	
5"	4th	

Rafter Tails

Prior to installing the sheathing, check the rafter tails to make sure they form a straight line. If you made accurate measurements and cuts, the rafters should all be the same length. If they are not, identify and measure the shortest rafter tail. Then mark the same distance on the rafter tails at each end of the roof span. Use these marks to snap a chalkline across the entire span of rafter tails, then trim each end to the same length. This will avoid an unsightly, crooked roof edge and provide a solid nailing base for the fascia.

Crane Delivery

If a crane is used to place a stack of sheathing on a roof, a special platform must be in place to provide a level surface. The platform must be placed over a loadbearing wall so that the weight of the sheathing doesn't cause structural damage to the framing.

wood shakes are used, the sheathing boards may be spaced. When solid sheathing is used, leave a ⅛" space between panels to allow for expansion.

The following is an overview of roof sheathing requirements using plywood or other 4 × 8 sheet material.

Step 1 Start by measuring up 48¼" from where the finish fascia will be installed. Chalk a line at that point, then lay the first sheet down and nail it. Install H-clips midway between the rafters or trusses before starting the next course (*Figure 27*). These clips eliminate the need for tongue-in-groove panels.

Step 2 Apply the remaining sheets. Stagger the panels by starting the next course with a half sheet. Let the edges extend over the hip, ridge, and gable end. Cut the extra sheathing off with a circular saw.

Once the sheathing has been installed, an underlayment of asphalt-saturated felt or other specified material must be installed to keep moisture out until the shingles are laid. For roofs with a slope of 4" or more, 15-pound roofer's felt is commonly used.

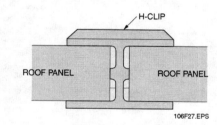

Figure 27 ◆ Use of H-clips.

Discuss the "Think About It." (See the answers at the end of this module.)

Provide an overview of the procedure for applying roof sheathing.

Show Transparency 29 (Figure 27).

Demonstrate the application of roof sheathing. Point out that sheathing must be properly stacked on a roof, sheathing thickness may be specified by local codes, and sheathing joints must be staggered and spaced for expansion.

Hand out Job Sheet 27106-6. Under your supervision, have the trainees perform the tasks on the Job Sheet. Note the proficiency of each trainee.

Emphasize the use of
proper fall protection and
footwear when applying
roof sheathing.

Sheathing Safety

When installing sheathing on an inclined roof, always be sure to use a toeboard and safety harness. Battling gravity is impossible without the proper equipment. You can gain additional footing by wearing skid-resistant shoes. Falling from a roof can obviously result in serious injury or death. Taking the time to follow the proper precautions will prevent you from slipping or falling. The picture shows a safety harness anchor and toeboard.

106P0603.EPS

Sheathing

If the trusses or rafters are on 24" centers, use ⅝" sheathing. With 16" centers, ½" sheathing may be used. This may vary depending on the local codes. Before starting construction, be sure to check the sheathing and fastener code requirements.

Felt Installation

The felt underlayment should be applied to the installed sheathing as soon as possible. It is important that both the sheathing and felt be dry and smooth at the time of installation. If the roof is damp or wet, wait a couple of days for it to dry completely before installing the felt. Moisture will cause long-term damage to both the sheathing and underlayment.

Show Transparency 30
(Figure 28).

Material such as coated sheets or heavy felt that could act as a vapor barrier should not be used. They can allow moisture to accumulate between the sheathing and the underlayment. The underlayment is applied horizontally with a 2" top lap and a 4" side lap, as shown in *Figure 28(A)*. A 6" lap should be used on each side of the centerline of hips and valleys. A metal drip edge is installed along the rakes and eaves to keep out wind-driven moisture.

In climates where snow accumulates, a waterproof underlayment, as shown in *Figure 28(B)*, should be used at roof edges and around chimneys, skylights, and vents. This underlayment has

Instructor's Notes:

_____

_____

_____

_____

_____

_____

_____

_____

Roof Openings

A 21-year-old apprentice was installing roofing on a building with six unguarded skylights. During a break, he sat down on one of the skylights. The plastic dome shattered under his weight and he fell to a concrete floor 16 feet below, suffering fatal head injuries.

The Bottom Line: Never sit or lean on a skylight. Always provide appropriate guarding and fall protection for work around skylights and other roof openings.

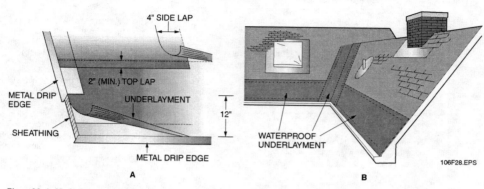

Figure 28 ◆ Underlayment installation.

an adhesive backing that adheres to the sheathing. It protects against water damage that can result from melting ice and snow that backs up under the shingles.

6.0.0 ◆ RAFTER LAYOUT USING A SPEED SQUARE

The speed square, also known as a *super square* or *quick square*, is a combination tool consisting of a protractor, try miter, and framing square. A standard speed square is a 6" triangular tool with a large outer triangle and a smaller inner triangle (*Figure 29*). The large triangle has a 6" scale on one edge, a full 90° scale on another edge, and a T-bar on the third edge. The inner triangle has a 2" square on one side. The speed square is the same on both sides. A 12" speed square is used for stair layout.

To use a square, you need to know the pitch of the roof. When you buy a speed square, it usually

comes with an instruction booklet. This booklet normally contains (among other information) tables for every pitch that show the required rafter length.

Figure 29 ◆ Speed square.

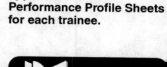
Performance Profile Test

Have each trainee complete to your satisfaction Performance Profile Test 5 on roof sheathing application. Fill out Performance Profile Sheets for each trainee.

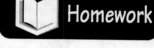
Homework

Assign reading of Sections 6.0.0–6.2.0 for the next class session.

Classroom

Discuss the scales shown on a speed square.

Audiovisual

Show Transparency 31 (Figure 30).

Demonstrate common rafter layout using a speed square.

Have the trainees perform common rafter layout using a speed square. Note the proficiency of each trainee.

Explain that hip and valley rafter layout can be simplified by using a speed square.

6.1.0 Procedure for Laying Out Common Rafters

Step 1 Choose a piece of lumber that is slightly longer than that needed for the rafter. Remember, the eaves or overhang are not included in the measurements found in the information booklet. The length of the overhang is added after you have determined the length of the rafter. Place the lumber on a pair of sawhorses with the top edge of the rafter stock facing away from you. Be sure the crown is facing away from you (the crown, if any, will be the top edge of the rafter). The rafter being cut will be for a building that is 24' wide with a 4" rise per foot of run. Therefore, the length of the rafter from the center of the ridge to the seat cut (bird's mouth) will be 12'-7¾". Any overhang must be added to this.

> **Note**
> When working with the speed square, there is no measuring line. The length is measured on the top edge of the rafter (this will be shown later in this module). Assume that the right side of the lumber will be the top of the rafter. Place the speed square against the top edge of the lumber and set the square with the 4 on the common scale. Draw a line along the edge of the speed square, as shown in *Figure 30(A)*.

Step 2 From the mark just made, measure with a steel tape along the top edge of the rafter the length required for the total length of the rafter (12'-7¾") and make a mark, as shown in *Figure 30(B)*.

Step 3 Place the speed square with the pivot point against that mark, and move the square so the 4 on the common scale is even with the edge. Draw a line along the edge of the square, as shown in *Figure 30(C)*. This will establish the vertical seat cut (bird's mouth) or plumb cut.

Step 4 To lay out the horizontal cut, reverse the speed square so that the short line at the edge of the square is even with the line previously drawn. Place the line so that the edge of the square is even with the lower edge of the rafter, as shown in *Figure 30(D)*.

Step 5 Draw a second line at a right angle to the plumb line. This line will establish the completed seat cut (bird's mouth), as shown in *Figure 30(E)*.

Step 6 The top and bottom cuts have been established for the *total length* of the rafter. If a ridgeboard is being used, be sure to deduct half of the thickness of the ridge from the top plumb cut. If any overhang is needed, it should be measured at right angles to the plumb cut of the seat cut (bird's mouth), as shown in *Figure 30(F)*.

Step 7 Once the measurement has been established, place the square against the top edge of the rafter, lining up the 4 on the common scale and marking it, as shown in *Figure 30(G)*.

Step 8 If a bottom or vertical cut is required, follow the procedure for cutting the horizontal cut of the seat cut (bird's mouth). In following the procedure described above, a pattern for a common rafter has been established. Two pieces of scrap material (1 × 2 or 1 × 4) should be nailed to the rafter, one about 6" from the top or plumb cut and one at the seat cut (bird's mouth).

6.2.0 Laying Out a Hip Rafter with a Speed Square

Assume the hip rafter to be cut is on the same building as the common rafter (24' wide with a 4" rise per foot of run). The hip rafter must be cut with a different plumb or top cut because it is at a 45° angle to the common rafter (that is, sitting diagonally at the intersection of the two wall plates).

Because of the additional length of run in the hip rafter (17" of the run to 12" of run for a common rafter), the plumb or top cut must be figured at a different angle on the speed square. The speed square has a set pattern to establish the cuts for a hip or valley rafter, as shown in *Figure 31(A)*.

The booklet that comes with the speed square indicates that the total length of a hip or valley rafter for a building 24' wide and with a 4 in 12 rise is 17'-5¼". Remember, this is the total length with no overhang figured.

Instructor's Notes:

_____

_____

_____

_____

_____

_____

_____

_____

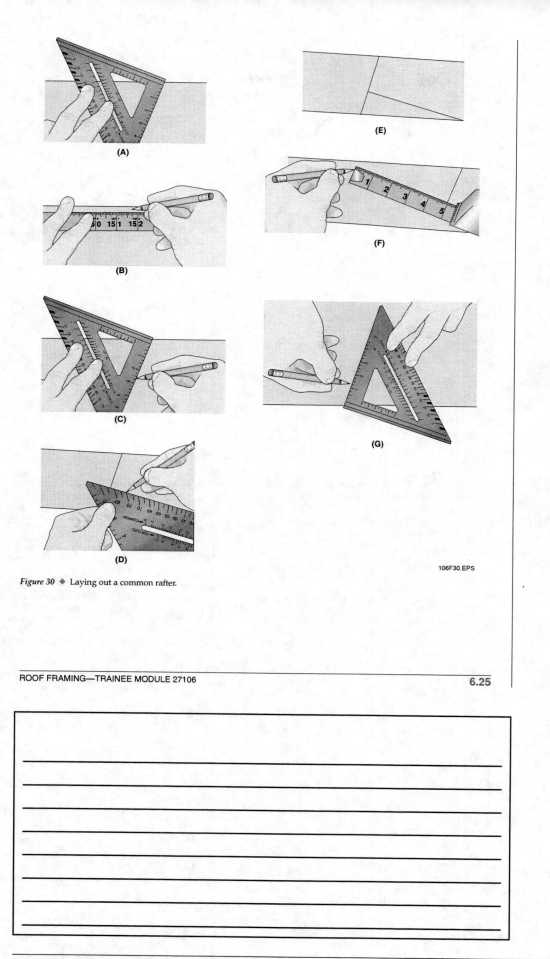

Figure 30 ◆ Laying out a common rafter.

106F30.EPS

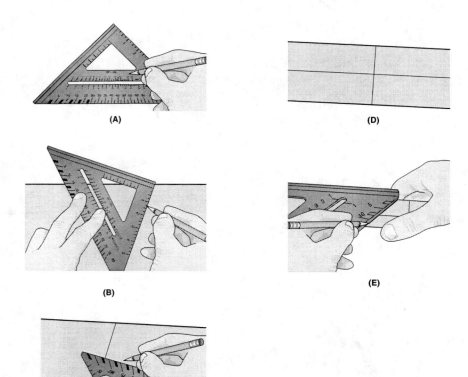

(A)

(B)

(C)

(D)

(E)

106F31.EPS

Figure 31 ◆ Hip rafter layout.

Show Transparency 32 (Figure 31).

Demonstrate hip and valley rafter layout using a speed square.

The eaves or overhang must be added to the length. Assume that the hip rafter being cut has no overhang.

Step 1 Choose a piece of lumber slightly longer than necessary. Place the lumber on a pair of sawhorses with the top edge and crown facing away from you. Assume that the right end of the lumber will be the top of the hip rafter. Place the speed square a few inches away from the top. Move the speed square so that the pivot point and the 4 on the hip valley scale are even with the outside edge, and draw a line along the edge as shown in *Figure 31(B)*. This will establish the top or plumb cut of the hip rafter.

Step 2 Measure with a tape along the top edge of the rafter the length required for the hip rafter and mark this point. Place the speed square with the pivot point on the edge of the mark and move the square until the 4 on the hip-valley scale is even with the edge of the lumber; draw a line along the edge of the square. This will establish the vertical cut of the seat cut (bird's mouth).

Step 3 To obtain the horizontal cut, reverse the square so that the small line at the pointed end of the speed square is covering the line previously drawn. Draw a line along the top edge of the speed square. This will establish the seat cut (bird's mouth), as shown in *Figure 31(C)*.

CARPENTRY LEVEL ONE—TRAINEE MODULE 27106

Instructor's Notes:

_____
_____
_____
_____
_____
_____
_____
_____

Step 4 The next step is to turn the lumber on edge and establish the center of the rafter. Draw a line on the center of the rafter so that it is over the ridge cut, as shown in *Figure 31(D)*. This is done in order to establish the double 45° angle that must be cut so that the hip rafter will sit properly at the intersection of the two common rafters.

Step 5 Deduct ¾" plus half the diagonal measure of the ridge (1¹⁄₁₆"). Add the two together and mark that distance back from the original plumb cut (ridge cut).

Note

If the ridge is 2" stock, the total measure deducted is 1¹³⁄₁₆".

Step 6 Using the speed square, draw a 45° line from the edge of the second line drawn to the center line and do the same on the opposite side of the center line, as shown in *Figure 31(E)*.

Step 7 Make the ridge cuts and the seat cut. The hip rafter is now ready to be put in place. Be sure to check that the hip rafter is not up beyond the common rafter. The common rafter may need to be dropped before nailing the hip rafter in place. Use the hip rafter as a pattern to cut the other hip rafters.

The procedure outlined for making a hip rafter is the same procedure used to cut a valley rafter. There are two alternative methods of cutting the seat cut (bird's mouth) of the hip rafter. The first is to cut the rafter plate at the intersection of the two plates. The other alternative is to cut the seat cut at a double 45° angle.

The method of cutting a valley rafter is to cut a reverse 45° cut to fit into the intersecting corners.

7.0.0 ◆ RAFTER LAYOUT USING A CALCULATOR

As you can easily see in *Figure 32*, roof framing is all about angles. Picture the rafters and ceiling joists as the sides of a triangle. When you insert a line representing the rise, you turn the large triangle into two right triangles. A right triangle is one that contains a 90° (right) angle. The height (altitude) of each triangle represents the rise, the width (base) represents the run, and the diagonal line (hypotenuse) represents a rafter.

Roof framing comes down to answering two questions:

• How long are the rafters?
• At what angle must they be cut to fit plumb to the ridgeboard and top plate?

About 2,500 years ago, a Greek mathematician named Pythagoras discovered that if you know the length of any two sides of a right triangle, you can calculate the length of the third side. For example, if you know the run (base) and rise (altitude) of a roof, you can calculate the length of a rafter (hypotenuse). The formula he developed is:

$$A^2 + B^2 = C^2$$

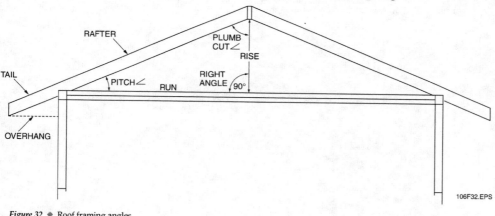

Figure 32 ◆ Roof framing angles.

Laboratory

Have the trainees perform hip and valley rafter layout using a speed square.

Homework

Assign reading of Sections 7.0.0–7.1.0 for the next class session.

Classroom

Define a right triangle. Explain the use of the Pythagorean theorem and a calculator to determine the hypotenuse of a right triangle.

Audiovisual

Show Transparencies 33 through 35 (Figures 32 through 34).

Each trainee should have a scientific calculator and know how to operate the trigonometric functions. If necessary, provide a quick demonstration on calculator use.

Give the trainees right-triangle problems representing various rises and runs. Have them solve for the hypotenuse to obtain the rafter length.

A is the rise, B is the run, and C is the rafter length (*Figure 33*). So, if you know the run and rise, you can figure the length of a rafter using the Pythagorean theorem. One minor problem you may have noticed is that the answer comes out as the square of the hypotenuse (C^2), which means you have to take its square root in order to get the actual length. Fortunately, this is easy to do using a scientific calculator. On many calculators, you simply key in the number and press the square root ($\sqrt{}$) key. (On some calculators, the square root does not have a separate key. Instead, the square root function is the inverse of the X^2 key, so you have to press INV or 2nd F, depending on your calculator, followed by X^2, to obtain the square root.)

Let us assume that you have a run of 12' and a rise of 8'.

$$A^2 + B^2 = C^2$$

Another way of representing this is:

$$C = \sqrt{A^2 + B^2}$$

or

$$
\begin{aligned}
\text{Rafter length} &= \sqrt{\text{Rise}^2 + \text{Run}^2} \\
&= \sqrt{12^2 + 8^2} \\
&= \sqrt{144 + 64} \\
&= \sqrt{208} \\
&= 14.42
\end{aligned}
$$

The square root of 208 is 14.42, or 14'-5" (.42 × 12" = 5.04"), which is the length of the rafter from the ridge to the top plate.

If the rafter has an overhang, the overhang must be calculated separately using the same formula, based on a triangle that includes the width of the overhang as the base and the distance from the soffit to the top of the rafter as the altitude (*Figure 32*). The result is then added to the common rafter length to determine the total length of the board.

Hip and valley rafters can be determined using the same formula. If you know the length of the common rafter and the span of the roof, you can calculate the length of a hip or valley rafter (*Figure 34*).

Here are some important facts you should know about triangles:

- There are two kinds of angles in a right triangle: a right angle and two acute angles. (Acute angles are less than 90°; obtuse angles are greater than 90°.)
- The sum of all three angles in a right triangle is 180°.
- Since one of the angles is (by definition) 90°, the sum of the remaining angles is 90°. Therefore, if you know one of the acute angles, you can subtract it from 90° to get the value of the remaining angle. In other words, if you know the pitch angle, you can easily figure the angle of the plumb cuts.

7.1.0 Using Trig Functions on a Calculator

Fifty years before Pythagoras, Thales (another Greek mathematician) figured out how to measure the height of the pyramids using a technique that later became known as *trigonometry* (*Figure 35*). Trigonometry recognizes that there is a relationship between the size of an angle and the lengths of the sides in a right triangle.

The sides of a triangle are referred to as *side opposite*, *side adjacent*, and *hypotenuse* with respect

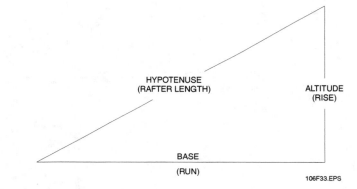

Figure 33 ◆ Sides of a triangle.

CARPENTRY LEVEL ONE—TRAINEE MODULE 27106

Instructor's Notes:

_____

_____

_____

_____

_____

_____

_____

_____

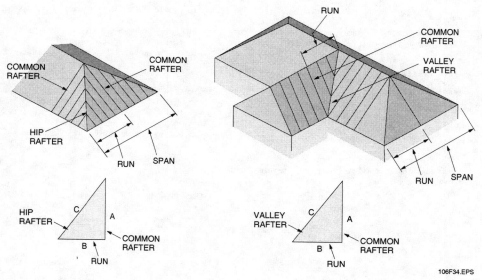

Figure 34 ◆ Calculating hip and valley rafters.

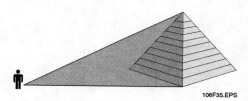

Figure 35 ◆ First practical use of trigonometry.

to either of the acute angles. For example, the side opposite angle *A* in *Figure 36* is the altitude (rise); the side adjacent to angle *A* is the base (run). A scientific calculator contains keys for sine (SIN), cosine (COS), and tangent (TAN) functions, all of which represent relationships between the sides of a right triangle. These relationships are expressed in three ways:

$$SIN\ A = \frac{\text{length of side opposite A}\quad (O)}{\text{length of hypotenuse}\qquad (H)}$$

$$COS\ A = \frac{\text{length of side adjacent to A}\ (A)}{\text{length of hypotenuse}\qquad (H)}$$

$$TAN\ A = \frac{\text{length of side opposite A}\quad (O)}{\text{length of side adjacent to A}\ (A)}$$

Here is an easy way to remember these functions:

SINE	Some Old Horse	$\frac{O}{H}$
COSINE	Caught Another Horse	$\frac{A}{H}$
TANGENT	Taking Oats Away	$\frac{O}{A}$

The formula you choose is basically a function of what you know about the triangle. If you know the lengths of any two sides, you can figure the pitch angle. If you can figure the pitch angle, you can figure the angle of the rafter plumb cuts. If you know the pitch angle and the length of one

Define the basic trigonometric functions of a triangle.

Show Transparencies 36 through 38 (Figures 35 through 37).

Discuss trigonometric solutions for determining rafter length.

side, you can figure out the lengths of the other sides.

In roof framing, you normally know the run and rise, so it is most common to use the tangent function.

Example 1:

Assume that you have a gable roof (*Figure 37*) with a run of 12' and a rise of 6', and you wish to determine the pitch angle (angle *A*):

$$TAN = \text{length of side opposite} \div \text{length of side adjacent or } TAN = \frac{O}{A}$$

In this case, the side opposite is the rise and the side adjacent is the run. Therefore, the tangent of angle *A* is:

$$TAN\ A = 6 \div 12 = .5$$

To convert the tangent to its corresponding angle, use the inverse (INV) or second function (2nd F) key, then press TAN to get the pitch angle, which is 26.56°. To determine the angle of the plumb cuts, subtract 26.56 from 90°, which yields 63.44.

Example 2:

Assume the pitch angle is 26.56°, the rafter length is 13'-5", and you want to know the rise.

In this case you would use the sine function. Since you know the angle instead of the opposite side, you need a variation of the formula:

$$SIN = \text{opposite} \div \text{hypotenuse or } SIN = \frac{O}{H}$$

becomes:

$$\text{Opposite side} = SIN \times \text{hypotenuse or } O = SIN \times H$$

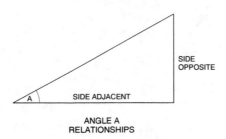
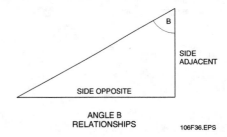

106F36.EPS

Figure 36 ◆ Angle-side relationships.

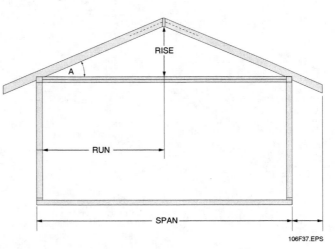

106F37.EPS

Figure 37 ◆ Roof framing example.

6.30

Instructor's Notes:

_____

_____

_____

_____

_____

_____

_____

_____

Scientific Calculators

When scientific calculators first became available in the 1960s, they cost over $400. Today, scientific calculators like the one shown here are readily available for less than $20. The first scientific calculator was known as an *electronic slide rule* because until that time, engineers carried slide rules around with them, often in holsters suspended from their belts. A slide rule is a manual device consisting of a ruler with a sliding center section, as shown below. Slide rules were used to calculate a variety of mathematical functions, including square roots and logarithms.

SCIENTIFIC CALCULATOR

106P0604.EPS

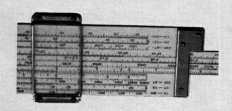

SLIDE RULE   106P0605.EPS

Key 26.56 into your calculator and press SIN; the result is 0.447 [0.447 × 13.4 = 6' (5.9898 rounded)], which is the 6' rise we started with in *Example 1*.

With practice, you should be able to do these calculations in a few seconds and get more accurate results than you would using a rafter square or speed square.

8.0.0 ◆ TRUSS CONSTRUCTION

In most cases, it is much faster and more economical to use prefabricated trusses in place of rafters and joists. Even if a truss costs more to buy than the comparable framing lumber (not always the case), it takes significantly less labor than stick framing. Another advantage is that a truss will

Give the trainees trigonometric problems using various angles and rises or runs. Have them solve for the hypotenuse to obtain the rafter length.

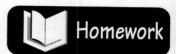

Assign reading of Sections 8.0.0–8.3.0 for the next class session.

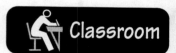

Discuss the advantages of truss construction.

Show Transparencies 39 through 41 (Figures 38 through 40).

span a greater distance without a bearing wall than stick framing. Just about any type of roof can be framed with trusses.

There are special terms used to identify the members of a truss (see *Figure 38*).

A truss is a framed or jointed structure. It is designed so that when a load is applied at any intersection, the stress in any member is in the direction of its length. *Figure 39* shows some of the many kinds of trusses. Even though some trusses look nearly identical, there is some variation in the interior (web) pattern. Each web pattern distributes weight and stress a little differently, so different web patterns are used to deal with different loads and spans. The decision of which truss to use for a particular application will be made by the architect or engineer and will be shown on the blueprints. Do not substitute or modify trusses on site, because it could affect their weight and stress-bearing capabilities. Also, be careful how you handle trusses. They are more delicate than stick lumber and cannot be thrown

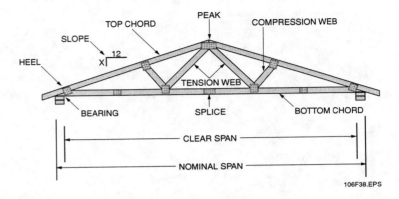

106F38.EPS

Figure 38 ◆ Components of a truss.

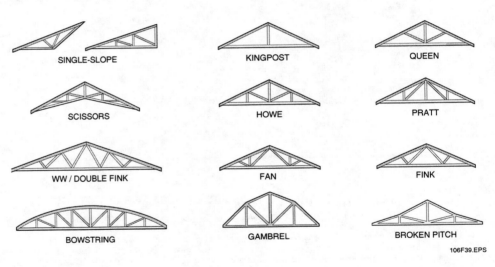

106F39.EPS

Figure 39 ◆ Types of trusses.

CARPENTRY LEVEL ONE—TRAINEE MODULE 27106

Instructor's Notes:

CARPENTRY LEVEL ONE—INSTRUCTOR'S GUIDE MODULE 27106

around or stored in a way that applies uneven stress to them.

Trusses are stored or carried on trucks either lying on their sides or upright in cradles and protected from water. It is a good idea to use a crane to lift the trusses from the truck, and unless the trusses are small and light, it is usually recommended that a crane be used to lift them into place on the building frame. *Figure 40* shows examples of erecting methods used for trusses. Note the tag lines. These lines are held by someone on the ground to stabilize the truss while it is being lifted.

A single lift line can be used with trusses with a span of less than 20'. From 20' to 40', two chokers are needed. If the span exceeds 40', a spreader bar is required.

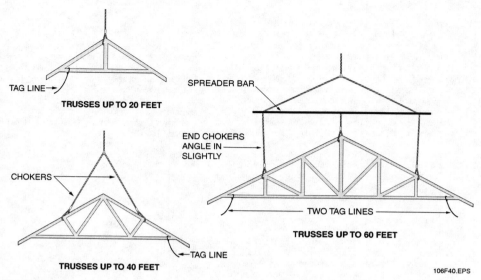

TAG LINE

TRUSSES UP TO 20 FEET

SPREADER BAR

END CHOKERS ANGLE IN SLIGHTLY

CHOKERS

TAG LINE

TRUSSES UP TO 40 FEET

TWO TAG LINES

TRUSSES UP TO 60 FEET

106F40.EPS

Figure 40 ◆ Erecting trusses.

Truss Storage

Trusses should only be stored on site for a short length of time. They are very large and take up a lot of storage area. In addition, they are more likely to get wet when stored on site. When a truss is saturated with moisture, excess shrinkage will occur, causing structural damage to the webs and chords. If on-site storage of trusses is unavoidable, make sure they are well covered with a waterproof tarp and raised off the ground on pallets. During the planning phase of a project, trusses should be figured into the schedule so that they can be delivered and erected in a timely manner.

Truss Rigging

Trusses are very large and can be difficult to position. Proper rigging technique is important to avoid seriously injuring workers and damaging equipment. The rigging setup varies depending on the type of truss being used. Enlist the help of experienced carpenters to ensure that the truss is securely fastened and balanced properly. Never stand directly below a truss that has been hoisted into the air.

Discuss the importance of proper truss storage, handling, and rigging.

 WARNING!
Installing trusses can be extremely dangerous. It is very important to follow the manufacturer's instructions for bracing and to follow all applicable safety procedures.

8.1.0 Truss Installation

Before installing roof trusses, refer to the framing plans for the proper locations. If a truss is damaged before erection, obtain a replacement or instructions from a qualified individual to repair the truss. Remember, repairs made on the ground are usually a lot better and are always easier. Never alter any part of a truss without consulting the job superintendent, the architect/engineer, or the manufacturer of the roof truss. Cutting, drilling, or notching any member without the proper approval could destroy the structural integrity of the truss and void any warranties given by the truss manufacturer.

Girders are trusses that carry other trusses or a relatively large area of roof framing. A common truss or even a double common truss will rarely serve as a girder. If there is a question about whether a girder is needed, your job superintendent should confer with the truss manufacturer. Double girders are commonplace. Triple girders are sometimes required to ensure the proper load-carrying capacity. Always be sure that multiple-member girders are properly laminated together. Spacing the trusses should be done in accordance with the truss design. In some cases, very small deviations from the proper spacing can create a big problem. Always seek the advice of the job superintendent whenever you need to alter any spacing. Make certain that the proper temporary bracing is installed as the trusses are being set.

 WARNING!
Never leave a job at night until all appropriate bracing is in place and secured well.

Light trusses (under 30' wide, for example) can be installed by having them lifted up and anchored to the top plate, then pushed into place with Y-shaped poles by crew members on the ground. Larger trusses require a crane.

When the bottom chord is in place, the truss is secured to the top plate. This can be done by toe-nailing with 10d nails. In some cases, however, metal tiedowns are required (*Figure 41*). An example is a location where high winds occur.

8.2.0 Bracing of Roof Trusses

In some circumstances, it may be necessary to add permanent bracing to the trusses. This requirement would be established by the architect and would appear on the blueprints. *Figure 42* shows an example of such a requirement.

Temporary bracing of trusses is required until the sheathing is in place. *Figure 43* shows an example of lateral bracing across the tops of the trusses. Gable ends are braced from the ground using lengths of 2 × 4 or similar lumber anchored to stakes driven into the ground.

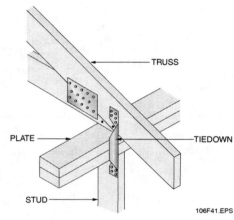

Figure 41 ◆ Use of a tiedown to secure a truss.

Instructor's Notes:

_____

_____

_____

_____

_____

_____

_____

_____

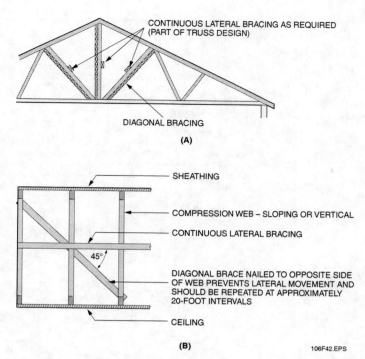

CONTINUOUS LATERAL BRACING AS REQUIRED
(PART OF TRUSS DESIGN)

DIAGONAL BRACING

(A)

SHEATHING

COMPRESSION WEB – SLOPING OR VERTICAL

CONTINUOUS LATERAL BRACING

45°

DIAGONAL BRACE NAILED TO OPPOSITE SIDE
OF WEB PREVENTS LATERAL MOVEMENT AND
SHOULD BE REPEATED AT APPROXIMATELY
20-FOOT INTERVALS

CEILING

(B)

106F42.EPS

Figure 42 ◆ Example of permanent bracing specification.

106F43.EPS

Figure 43 ◆ Example of temporary bracing of trusses.

Temporary Truss Bracing

The term *killer cleat* refers to the dangerous practice of using short 1 × 4 spacers nailed between the chords of two trusses to serve as temporary bracing. THIS IS NOT AN ACCEPTABLE METHOD OF BRACING. The recommended method is to use long lengths of 2 × 4 lumber over the top and bottom chords. The braces should be as long as possible, but no shorter than 8'. In addition to lateral bracing, diagonal bracing is also used. It is a common practice to install the bracing over the top chord, in which case while the top chord bracing is removed while the roof is being sheathed. If the bracing is run on the bottom side of the top chord, however, it does not have to be removed to apply the sheathing. The bracing installed over the bottom chord is often left in place.

The photo shows what can happen when trusses are not properly braced during installation. If people were working on or under the roof when this accident occurred, they would most likely have been seriously injured or even killed.

106P0606.EPS

Explain how to estimate the materials required for an example roof.

9.0.0 ◆ DETERMINING QUANTITIES OF MATERIALS

Estimating the material you will need for a roof depends first of all on the type of roof you are planning to construct, the size of the roof, the spacing of the framing members, and the load characteristics. Local building codes will usually dictate these factors and they will be disclosed on the building plans. Lumber for conventional framing may be from 2 × 4 to 2 × 10, depending on the span and the load. For example, 2 × 4 framing on 16" centers might support a 9' span. By comparison, 2 × 10 framing on 16" centers would support a span of more than 25'.

Instructor's Notes:

_____
_____
_____
_____
_____
_____
_____
_____
_____

9.1.0 Determine Materials Needed for a Gable Roof

Determining the rafter material:

Step 1 To determine how much lumber you will need for rafters on a gable roof, first determine the length of each common rafter (including the overhang) using the framing square or another method as described earlier.

Step 2 Figure out the number of rafters based on the spacing (16", 24", etc.). Remember that you will need one rafter for each gable end. You will also need barge rafters in the gable overhang. Note that these are usually one size smaller than the common rafters.

Step 3 Multiply the result by two to account for the two sides of the ridge.

Step 4 Convert the result to board feet.

Estimating the ridgeboard:

The ridgeboard is usually one dimension thicker than the rafters.

Step 1 Determine the length of the plate on one side of the structure and add as needed to account for gable overhang.

Step 2 Convert the result to board feet.

Estimating the sheathing:

Step 1 Multiply the length of the roof including overhangs by the length of a common rafter. This yields half the area of the roof.

Step 2 Divide the roof area by 32 (the number of square feet in a 4 × 8 sheet) to get the approximate number of sheets of sheathing you will need. Round up if you get a fractional number.

Step 3 Multiply by 2 to obtain the number of sheets needed for the full roof area.

10.0.0 ◆ DORMERS

A dormer is a framed structure that projects out from a sloped roof. A dormer provides additional space and is often used in a Cape Cod–style home, which is a single-story dwelling in which the attic is often used for sleeping rooms. A shed dormer (*Figure 44*) is a good way to obtain a large amount

106F44.EPS

Figure 44 ◆ Shed dormer.

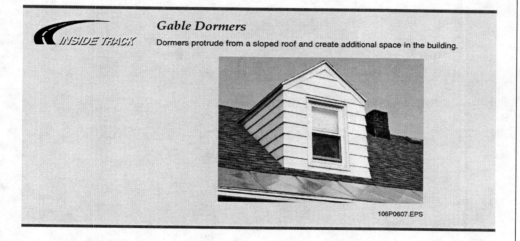

Gable Dormers

INSIDE TRACK

Dormers protrude from a sloped roof and create additional space in the building.

106P0607.EPS

Classroom

Provide a number of different roof configurations and have the trainees estimate the materials required.

Homework

Assign reading of Sections 10.0.0–11.0.0 for the next class session.

Classroom

Describe shed and gable dormers. Point out that shed dormers sometimes extend almost entirely across the length of a roof.

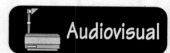

Audiovisual

Show Transparencies 45 and 46 (Figures 44 and 45).

of additional living space. If it is added to the rear of the house, it can be done without affecting the appearance of the house from the front.

A gable dormer (*Figure 45*) serves as an attractive addition to a house, in addition to providing a little extra space as well as some light and ventilation. They are sometimes used over garages to provide a small living area or studio.

11.0.0 ◆ PLANK-AND-BEAM ROOF CONSTRUCTION

Plank-and-beam framing, also known as *post-and-beam framing*, employs much sturdier framing members than the common framing previously described. It is often used in framing roofs for luxury residences, churches, and lodges, as well as other public buildings where a striking architectural effect is desired.

Because the beams used in this type of construction are very sturdy, wider spacing may be used. When plank-and-beam framing is used for a roof, the beams and planking can be finished and left exposed. The underside of the planking takes the place of an installed ceiling (*Figure 46*).

In lighter construction, solid posts or beams such as 4 × 4s are used. In heavier construction, laminated beams made of glulam, LVL, and PSL are used.

12.0.0 ◆ METAL ROOF FRAMING

When steel framing is used, the roof framing is done with metal trusses that are generally prefabricated and delivered to the site. If the trusses are fabricated on the site, the chords and webs are cut to size with a portable electric saw and placed into a jig. They are then welded and/or connected with self-tapping screws.

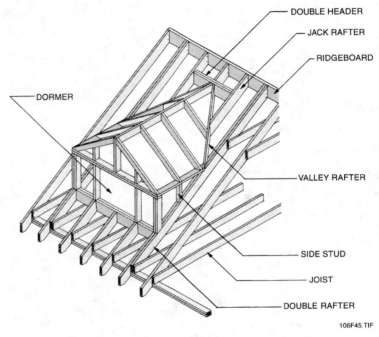

DOUBLE HEADER
JACK RAFTER
RIDGEBOARD
DORMER
VALLEY RAFTER
SIDE STUD
JOIST
DOUBLE RAFTER

106F45.TIF

Figure 45 ◆ Gable dormer framing.

Instructor's Notes:

_____

_____

_____

_____

_____

_____

_____

_____

Figure 46 ◆ Example of post-and-beam construction.

Summary

The correct layout and framing of a roof requires patience and skill. If the measurement, cutting, and installation work are not done carefully and precisely, the end result will never look right. Fortunately, there are many tools and reference tables that help to simplify the process. The important thing is to be careful and precise with the layout and cutting of the first rafter of each type. This is the rafter that is used as a pattern for the others.

106F46.EPS

Classroom

Have the trainees complete the Review Questions and go over the answers prior to administering the Module Examination.

Review Questions

1. The type of roof that has four sides running toward the center of the building is the _____ roof.
   a. gable
   b. hip
   c. gambrel
   d. gable and valley

Refer to the following illustration to answer Questions 2 through 7.

2. The letter *A* on the diagram is pointing to a _____.
   a. hip jack
   b. valley jack
   c. hip rafter
   d. cripple jack

3. The letter *B* on the diagram is pointing to the _____.
   a. main ridgeboard
   b. projection ridgeboard
   c. top plate
   d. collar beam

4. The letter *D* on the diagram is pointing to a _____.
   a. valley cripple
   b. hip jack
   c. valley rafter
   d. valley jack

5. The letter *E* on the diagram is pointing to the _____.
   a. top plate
   b. projection ridgeboard
   c. main ridgeboard
   d. gable stud

6. The letter *G* on the diagram is pointing to a _____.
   a. gable end
   b. common rafter
   c. collar beam
   d. gable stud

7. The letter *M* on the diagram is pointing to a _____ rafter.
   a. hip jack
   b. valley jack
   c. cripple jack
   d. hip

8. The roof run measurement is usually equal to _____.
   a. twice the span
   b. half the span
   c. the distance from the top plate to the ridgeboard
   d. the length of a common rafter

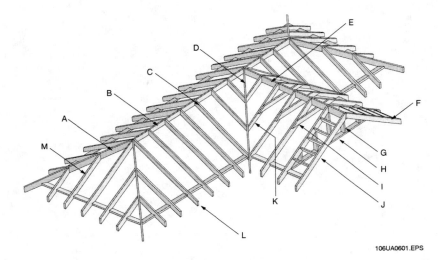

106UA0601.EPS

6.40

CARPENTRY LEVEL ONE—TRAINEE MODULE 27106

Instructor's Notes:

_____
_____
_____
_____
_____
_____
_____
_____

Refer to the following illustration to answer Questions 9 and 10.

9. The length of a common rafter for a 20' wide building with a 10" rise per foot of run is _____.
   a. 19.7'
   b. 208⅓"
   c. 13'
   d. 14'-8"

10. If the end common rafter of a hip roof is 13'-6" long and the rise per foot of run is 11", the length of the first hip jack rafter for a building framed on 16" centers is _____.
    a. 11'-8¼"
    b. 13'-6"
    c. 12'-2"
    d. 11'-10"

11. The rise per foot of run of a roof with a pitch of ½ is _____.
    a. 6"
    b. 12"
    c. 18"
    d. 24"

12. When laying out a common rafter, it is necessary to deduct _____ in order to arrive at the final measurement.
    a. half the thickness of the ridgeboard
    b. half the thickness of the rafter
    c. half the thickness of the top plate
    d. nothing

13. An L-shaped brace used to strengthen and straighten ceiling joists is known as a _____.
    a. purlin
    b. strongback
    c. collar tie
    d. ridgeboard

14. When cutting a _____ rafter, it is necessary to drop the bird's mouth cut by half the thickness of the rafter.
    a. valley
    b. common
    c. hip
    d. hip jack

15. For every foot of run of a common rafter, the hip rafter has a run of _____.
    a. 8"
    b. 12"
    c. 15"
    d. 17"

16. The _____ rafter requires a double cheek cut in order to fit into the framing.
    a. hip
    b. common
    c. hip jack
    d. valley jack

17. The major difference between a hip rafter and a valley rafter is that the valley rafter _____.
    a. is based on 12" per foot of run
    b. requires different seat and tail cuts
    c. is always shorter than the hip rafter
    d. runs at a diagonal to the ridge, while the hip rafter runs at 90°

18. When laying out jack rafters on 16" centers, the value you would read from the third line of the framing square is the _____.
    a. length of the jack rafter
    b. difference in length between one jack rafter and the next
    c. shortened length of the jack rafter
    d. distance from the tail cut to the bird's mouth plumb cut

106UA0602.EPS

Administer the Module
Examination and
Performance Test. Be sure
to record the results on
Craft Training Report Form
200 and submit the form to
the Training Program
Sponsor.

 **Performance Profile Test**

Ensure that all Performance
Profile Tests have been
completed and
Performance Profile Sheets
for each trainee are filled
out. Be sure to record the
results of the Testing on
Craft Training Report Form
200 and submit the results
to the Training Program
Sponsor. Answers can be
found on the key in the Test
Booklet.

19. Immediately after nailing the plywood sheathing on a roof frame, you should install _____.
   a. a vapor barrier
   b. a felt underlayment
   c. a drip edge
   d. plywood clips

20. When you are using a speed square, the length of the rafter is determined by_____.
   a. stepping the square
   b. reading it directly from the square
   c. using rafter tables in the instruction manual
   d. multiplying the run by the rise

Refer to the following illustration to answer Questions 21 and 22.

21. Item 6 on the illustration is pointing to the _____.
   a. center chord
   b. slope
   c. tension web
   d. compression web

22. Item 4 on the illustration is pointing to the _____.
   a. top chord
   b. clear span
   c. peak
   d. slope

23. When rigging a truss with a span of 45', you should use a _____.
   a. spreader bar
   b. single choker
   c. double choker
   d. triple choker

24. The number of common rafters needed to frame a 20' long gable roof for a house framed on 16" centers is _____.
   a. 15
   b. 32
   c. 30
   d. 16

25. The number of sheets of plywood sheathing needed for a house 30' long with a span of 20' and ¼ pitch is _____.
   a. 14
   b. 18
   c. 22
   d. 28

106UA603.EPS

Instructor's Notes:

_____

_____

_____

_____

_____

_____

_____

_____

Trade Terms Introduced in This Module

Barge rafter: A gable end roof member that extends beyond the gable to support a decorative end piece. Also known as a *fly rafter*.

False fascia (sub fascia): The board that is attached to the tails of the rafters to straighten and space the rafters and provide a nailer for the fascia. Also called *sub fascia* and *rough fascia*.

Gable: The triangular wall enclosed by the sloping ends of a ridged roof.

Lookout: A structural member used to frame an overhang.

Purlin: A horizontal roof support member parallel to the plate and installed between the plate and the ridgeboard.

Right triangle: A triangle containing a 90° angle.

Square root: A number which when multiplied by itself will yield a given number (e.g., 3 is the square root of 9).

Strongback: An L-shaped support member used to strengthen and level ceiling joists and maintain correct spacing.

ROOF FRAMING—TRAINEE MODULE 27106   6.43

Answers to Review Questions

Answer	Section
1. b	2.0.0
2. c	3.0.0/Figure 2
3. a	3.0.0/Figure 2
4. c	3.0.0/Figure 2
5. b	3.0.0/Figure 2
6. d	3.0.0/Figure 2
7. a	3.0.0/Figure 2
8. b	3.0.0
9. c	3.1.0, 3.2.2/Figures 4 and 8
10. a	3.1.0, 4.3.0/Figure 26, and 7.0.0
11. b	3.1.0/Figure 5
12. a	3.2.3
13. b	3.3.1
14. c	4.1.0
15. d	4.1.0
16. a	4.1.0
17. b	4.2.0
18. b	4.3.0
19. b	5.0.0
20. c	6.0.0
21. d	8.0.0/Figure 38
22. a	8.0.0/Figure 38
23. a	8.0.0
24. b	9.1.0
25. c	9.1.0

Instructor's Notes:

_____
_____
_____
_____
_____
_____
_____
_____

Additional Resources

This module is intended to present thorough resources for task training. The following reference works are suggested for further study. These are optional materials for continuing education rather than for task training.

Carpentry. Homewood, IL: American Technical Publishers.

Carpentry. Albany, NY: Delmar Publishers.

Full Length Roof Framing. Palo Alto, CA: AFJ Riechers.

Modern Carpentry. Tinley Park, IL: The Goodheart-Willcox Co., Inc.

MODULE 27106-01 "THINK ABOUT IT" ANSWERS

Section 3.0.0 ***Think About It—Calculating Rise and Pitch***

Find the pitch for the following roofs:

Total Rise	Span	Pitch
8'	24'	⅓
9'	36'	¼
6'	24'	¼
12'	36'	⅓
8'	32'	¼

Find the total rise for the following roofs:

Pitch	Span	Rise per Foot of Run	Total Rise
½	24'	12"	12'
⅙	32'	4"	5'-4"
⅓	24'	8"	8'
⁵⁄₁₂	30'	10"	12'-6"
¾	24'	18"	18'

Find the rise per foot of run for the following roofs:

Span	Total Rise	Rise per Foot of Run
18'	6'	8"
24'	8'	8"
28'	7'	6"
16'	4'	6"
32'	12'	9"

Section 3.2.2 ***Think About It—Calculating Common Rafter Lengths***

Find the lengths of the common rafters for the following spans:

Span	Run	Rise per Foot of Run	Rafter Length
26'	13'	6"	14.539
24'	12'	4"	12.65
28'	14'	8"	16.823
32'	16'	12"	22.262
30'	15'	7"	17.362

Section 4.3.0 ***Think About It—Calculating Jack Rafter Lengths***

Use the rafter tables to find the lengths of the following jack rafters at 16" OC when a common rafter is 14'-6" long:

Rise per Foot of Run	Jack Rafter	Length
6"	2nd	11'-6¼"
4"	4th	8'-10½"
9"	5th	6'-1⅓"
12"	3rd	8'-10"
5"	4th	8'-8⅞"

The NCCER makes every effort to keep these textbooks up-to-date and free of technical errors. We appreciate your help in this process. If you have an idea for improving this textbook, or if you find an error, a typographical mistake, or an inaccuracy in the NCCER's Craft Training textbooks, please write us, using this form or a photocopy. Be sure to include the exact module number, page number, a detailed description, and the correction, if applicable. Your input will be brought to the attention of the Technical Review Committee. Thank you for your assistance.

Instructors – If you found that additional materials were necessary in order to teach this module effectively, please let us know so that we may include them in the Equipment/Materials list in the Instructor's Guide.

Write: Curriculum Revision and Development Department
National Center for Construction Education and Research
P.O. Box 141104, Gainesville, FL 32614-1104

Fax: 352-334-0932

E-mail: curriculum@nccer.org

Craft _____ Module Name _____

Copyright Date _____ Module Number _____ Page Number(s) _____

Description _____

(Optional) Correction _____

(Optional) Your Name and Address _____

Windows and Exterior Doors

27107-01

MODULE OVERVIEW

This module introduces the carpentry trainee to methods and procedures used in the selection and installation of residential windows and exterior doors.

PREREQUISITES

Please refer to the Course Map in the Trainee Module. Prior to training with this module, it is suggested that the trainee shall have successfully completed the following modules:

Core Curriculum; Carpentry Level One, Modules 27101 through 27106

LEARNING OBJECTIVES

Upon completion of this module, the trainee will be able to:

1. Identify various types of fixed, sliding, and swinging windows.
2. Identify the parts of a window installation.
3. State the requirements for a proper window installation.
4. Install a pre-hung window.
5. Identify the common types of skylights and roof windows.
6. Describe the procedure for properly installing a skylight.
7. Identify the common types of exterior doors and explain how they are constructed.
8. Identify the parts of a door installation.
9. Identify the types of thresholds used with exterior doors.
10. Install a threshold on a concrete floor.
11. Install a pre-hung exterior door with weatherstripping.
12. Identify the various types of locksets used on exterior doors and explain how they are installed.
13. Explain the correct installation procedure for a rollup garage door.
14. Install a lockset.

PERFORMANCE OBJECTIVES

Under the supervision of the instructor, the trainee should be able to:

1. Install a pre-hung window.
2. Install a metal threshold on a concrete floor.
3. Install a pre-hung exterior door with weatherstripping.
4. Install a lockset on an entry door.

NCCER STANDARDIZED CRAFT TRAINING PROGRAM

The National Center for Construction Education and Research (NCCER) provides a standardized national program of accredited craft training. Key features of the program include instructor certification, competency-based training, and performance testing. The program provides trainees, instructors, and companies with a standard form of recognition through a National Craft Training Registry. The program is described in full in the *Guidelines for Accreditation,* published by the NCCER. For more information on standardized craft training, contact the NCCER at P.O. Box 141104, Gainesville, FL 32614-1104, 352-334-0911, visit our Web site at www.NCCER.org, or e-mail info@NCCER.org.

HOW TO USE THIS ANNOTATED INSTRUCTOR'S GUIDE

Each page presents two sections of information. The larger section displays each page exactly as it appears in the Trainee Module. The narrow column ties suggested trainee and instructor actions to each page and provides icons to call your attention to material, safety, audiovisual, or testing requirements. The bottom of each page includes space for your notes.

 **Teaching Tip** If you see the Teaching Tip icon, that means there is a teaching tip associated with this section. Also refer to any suggested teaching tips at the end of the module.

SAFETY CONSIDERATIONS

Ensure that the trainees are equipped with appropriate personal protective equipment.

PREPARATION

Before teaching this module, you should review the Module Outline, the Learning and Performance Objectives, and the Materials and Equipment List. Be sure to allow ample time to prepare your own training or lesson plan and gather all required equipment and materials.

MATERIALS AND EQUIPMENT LIST

Materials:

Transparencies
Markers/chalk
Manufacturer's catalogs and brochures on windows
Nails:
   4d finish
   6d finish
   8d finish or casing
   16d casing
Pre-hung window unit
Shims
Flashing or drip cap
Pre-hung door unit
Wood shingles for blocking shims
Fiberglass insulation or sill sealer
Lockset with manufacturer's instructions and template
   (if needed)
Weatherstripping
Screws for attaching weatherstripping
Threshold and manufacturer's installation instructions
Concrete screw anchors and screws
Copies of Job Sheets 1 through 6*
Module Examinations*
Performance Profile Sheets*

Equipment:

Overhead projector and screen
Whiteboard/chalkboard
Appropriate personal protective equipment
Miter saw
Hand levels
Handsaw
Claw hammer
Framing square
Combination square
Steel tape
30" level
Nail set
Caulking gun and sealer
Boring jig (if available)
Wood chisels
Tin snips
Utility knife
Screwdriver
Drill
Drill bits

*Packaged with this Annotated Instructor's Guide.

ADDITIONAL RESOURCES

This module is intended to present thorough resources for task training. The following reference works are suggested for both instructors and motivated trainees interested in further study. These are optional materials for continued education rather than for task training.

Carpentry, Leonard Koel. Homewood, IL: American Technical Publishers, 1997.

Carpentry, Gasper J. Lewis. Albany, NY: Delmar Publishers, 2000.

Modern Carpentry, Willis H. Wagner and Howard Bud Smith. Tinley Park, IL: The Goodheart-Willcox Company, Inc., 2000.

TEACHING TIME FOR THIS MODULE

An outline for use in developing your lesson plan is presented below. Note that each Roman numeral in the outline equates to one session of instruction. Each session has a suggested time period of 2½ hours. This includes 10 minutes at the beginning of each session for administrative tasks and one 10-minute break during the session. Approximately 12½ hours are suggested to cover *Windows and Exterior Doors*. You will need to adjust the time required for hands-on activity and testing based on your class size and resources.

Topic   **Planned Time**

Session I. Introduction; Windows

   A. Introduction   _____

   B. Windows   _____

      1. Window Construction   _____

      2. Types of Windows   _____

      3. Types of Window Glass   _____

         a. Energy-Efficient Windows   _____

         b. Safety Glass   _____

      4. Window Installation   _____

         a. Extension Jambs   _____

      5. Glass Blocks   _____

Session II. Laboratory

   A. Laboratory

   Hand out Job Sheet 27107-1. Under your supervision, have the trainees perform the tasks on the Job Sheet. Note the proficiency of each trainee on his or her Job Sheet and Skill Test Record.

Session III. Exterior Doors; Laboratory

   A. Exterior Doors

      1. Exterior Door Sizes   _____

      2. Thresholds   _____

      3. Weatherstripping   _____

   B. Laboratory   _____

   Hand out Job Sheets 27107-2 and 27107-3. Under your supervision, have the trainees perform the tasks on the Job Sheets. Note the proficiency of each trainee on his or her Job Sheets and Skill Test Record.

Session IV. Installing an Exterior Pre-Hung Door; Installing a Garage Door; Laboratory

   A. Installing an Exterior Pre-Hung Door   _____

      1. Locksets

   B. Installing a Garage Door   _____

   C. Laboratory   _____

   Hand out Job Sheets 27107-4 and 27107-5. Under your supervision, have the trainees perform the tasks on the Job Sheets. Note the proficiency of each trainee on his or her Job Sheets and Skill Test Record.

Session V. Laboratory; Module Examination and Performance Testing

   A. Laboratory   _____

   Hand out Job Sheet 27107-6. Under your supervision, have the trainees perform the tasks on the Job Sheet. Note the proficiency of each trainee on his or her Job Sheet and Skill Test Record.

B. Summary   _____

   1. Summarize module.

   2. Answer questions.

C. Module Examination   _____

   1. Trainees must score 70% or higher to receive recognition from the NCCER.

   2. Record the testing results on Craft Training Report Form 200 and submit the results to the Training Program Sponsor.

D. Performance Testing   _____

   1. Trainees must perform each task to the satisfaction of the instructor to receive recognition from the NCCER.

   2. Record the testing results on Craft Training Report Form 200 and submit the results to the Training Program Sponsor.

Module 27107-01

Windows and Exterior Doors

Instructor's Notes:

_____
_____
_____
_____
_____
_____
_____
_____

Course Map

This course map shows all of the modules in the first level of the Carpentry curriculum. The suggested training order begins at the bottom and proceeds up. Skill levels increase as a trainee advances on the course map. The training order may be adjusted by the local Training Program Sponsor.

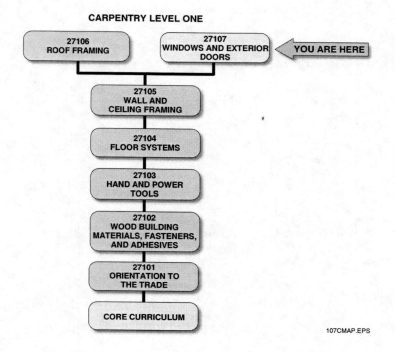

CARPENTRY LEVEL ONE

27106
ROOF FRAMING

27107
WINDOWS AND EXTERIOR DOORS

YOU ARE HERE

27105
WALL AND CEILING FRAMING

27104
FLOOR SYSTEMS

27103
HAND AND POWER TOOLS

27102
WOOD BUILDING MATERIALS, FASTENERS, AND ADHESIVES

27101
ORIENTATION TO THE TRADE

CORE CURRICULUM

107CMAP.EPS

Assign reading of Module 27107.

1.0.0   INTRODUCTION 7.1

2.0.0   WINDOWS 7.1

    2.1.0   Window Construction 7.2

    2.2.0   Types of Windows 7.3

    2.3.0   Types of Window Glass 7.9

    2.3.1   Energy-Efficient Windows 7.9

    2.3.2   Safety Glass 7.10

    2.4.0   Window Installation 7.10

    2.4.1   Extension Jambs 7.13

    2.5.0   Glass Blocks 7.13

3.0.0   EXTERIOR DOORS 7.14

    3.1.0   Exterior Door Sizes 7.17

    3.2.0   Thresholds 7.17

    3.3.0   Weatherstripping 7.17

    3.4.0   Installing an Exterior Pre-Hung Door 7.17

    3.4.1   Locksets 7.19

    3.5.0   Installing a Garage Door 7.22

SUMMARY 7.24

REVIEW QUESTIONS 7.25

GLOSSARY 7.26

ANSWERS TO REVIEW QUESTIONS 7.27

ADDITIONAL RESOURCES 7.28

Figures

Figure 1   Parts of a sash 7.3

Figure 2   Parts of a double-hung wood window 7.4

Figure 3   Examples of fixed window applications 7.5

Figure 4   Examples of windows 7.5

Figure 5   Bay window 7.5

Figure 6   Examples of bay windows 7.6

Figure 7   Supported bay window 7.7

Figure 8   Bow windows 7.7

Figure 9   Sliding (gliding) window 7.8

Figure 10 Roof windows and skylights 7.8

Figure 11 Skylight curb 7.8

Figure 12 Cooling and heating load factors 7.9

Figure 13   Double- and triple-pane windows 7.10
Figure 14   Cutaway of a vinyl-clad window 7.12
Figure 15   Leveling a window with shims 7.12
Figure 16   Plumb the side jambs 7.12
Figure 17   Insulating between the jambs and trimmer studs 7.12
Figure 18   Jamb extension 7.13
Figure 19   Glass block installation 7.13
Figure 20   Entry doors 7.14
Figure 21   Parts of a typical exterior door installation 7.15
Figure 22   Types of entry doors 7.17
Figure 23   Thresholds 7.18
Figure 24   Fixed bottom sweep 7.18
Figure 25   Weatherstripping 7.19
Figure 26   Door swing 7.19
Figure 27   Examples of exterior door locksets and security locks 7.20
Figure 28   Exploded view of a mortise lock 7.20
Figure 29   Disassembled view of a tubular lockset 7.21
Figure 30   Heavy-duty cylindrical lockset 7.21
Figure 31   Using an installation template 7.22
Figure 32   Boring jig and mortise marker 7.22
Figure 33   Installing a door strike 7.22
Figure 34   Rollup garage door and hardware 7.23
Figure 35   Electric garage door openers 7.23

Instructor's Notes:

_____

_____

_____

_____

_____

_____

_____

_____

Windows and Exterior Doors

Materials

Ensure you have everything required to teach the course. Check the Materials and Equipment List at the front of this Instructor's Guide.

Objectives

Upon completion of this module, the trainee will be able to:

1. Identify various types of fixed, sliding, and swinging windows.
2. Identify the parts of a window installation.
3. State the requirements for a proper window installation.
4. Install a pre-hung window.
5. Identify the common types of skylights and roof windows.
6. Describe the procedure for properly installing a skylight.
7. Identify the common types of exterior doors and explain how they are constructed.
8. Identify the parts of a door installation.
9. Identify the types of thresholds used with exterior doors.
10. Install a threshold on a concrete floor.
11. Install a pre-hung exterior door with weather-stripping.
12. Identify the various types of locksets used on exterior doors and explain how they are installed.
13. Explain the correct installation procedure for a rollup garage door.
14. Install a lockset.

Prerequisites

Successful completion of the following Task Modules is required before beginning study of this Task Module: Core Curriculum; Carpentry Level One, Modules 27101 through 27105.

Required Trainee Materials

1. Trainee Task Module
2. Appropriate personal protective equipment

1.0.0 ◆ INTRODUCTION

The final step in drying-in a house is closing off the structure with windows and exterior doors. There are so many types and styles of windows and exterior doors, made by so many manufacturers, that all the possibilities cannot be covered in this module. The purpose of this module is to introduce you to the special terms you will need to know and to give you an overview of the various kinds of windows and exterior doors and the important installation practices related to them.

2.0.0 ◆ WINDOWS

In this section, we will introduce the many kinds of windows used in residential construction and will provide an overview of important installation practices that apply to windows. We will start by defining what a window is. When we use the term *window* in this program, it will refer to the entire window assembly, including the glass, sash, and frame. These terms are explained in the next paragraph. In this section, we will focus on pre-hung windows; that is, windows that are delivered complete with sash, frame, and hardware.

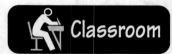

Audiovisual

Show Transparency 1, Course Objectives.

Show Transparency 2, Performance Profile Tasks.

Classroom

Discuss some of the basic terms related to windows. Explain that terms shown in bold (blue) are defined in the Glossary at the back of this module.

Gather photos of local buildings with various window arrangements. Discuss window aesthetics.

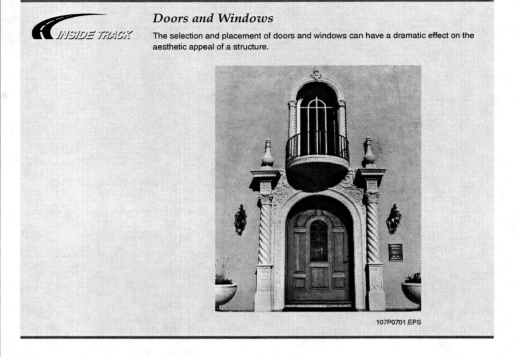

INSIDE TRACK

Doors and Windows

The selection and placement of doors and windows can have a dramatic effect on the aesthetic appeal of a structure.

107P0701.EPS

Identify the components of a sash.

Show Transparency 3 (Figure 1).

2.1.0 Window Construction

Window sashes can be made from wood, metal, or vinyl. Aluminum and steel are used in metal window construction. Wood windows, which are subject to decay, must be protected with wood preservatives and paint. Ponderosa pine, carefully selected and kiln-dried to a moisture content of 6% to 12%, is commonly used in making wooden windows. Wood is often preferred over aluminum because it does not conduct heat as readily. Metal window frames are usually filled with insulating material to make them more energy efficient. Although many windows are made of wood, it has become more common to find wooden windows clad on the outside with aluminum or vinyl.

Aluminum windows are much lighter, easier to handle, and more durable than wooden windows. Aluminum windows are generally coated to prevent corrosion. However, they are not recommended for coastal areas where the corrosive effect of salt air can be extreme.

Steel windows are stronger than aluminum and wood, but are far more expensive. They are more common in commercial buildings than residential construction.

Vinyl window frames are made of impact-resistant polyvinyl chloride (PVC). They are resistant to heat loss and condensation. Vinyl windows cannot be painted, so they must be purchased in a color that is compatible with the building. Vinyl windows may distort when exposed to temperature extremes. This can cause difficulty in operating the window, as well as increased air infiltration.

The basic component of a window assembly is the sash, which is the framework around the glass. *Figure 1* identifies the component parts of the sash. Note that the sections of glass are called lights. The sash may contain several lights or just one light with a false muntin known as a *grille*. The grille does not support the glass; it is simply installed on the face of the glass for decorative purposes.

The sash fits into, or is attached to, a window frame consisting of stiles and rails. The entire window assembly, including the frame and sash

7.2

CARPENTRY LEVEL ONE—TRAINEE MODULE 27107

Instructor's Notes:

_____

_____

_____

_____

_____

_____

_____

_____

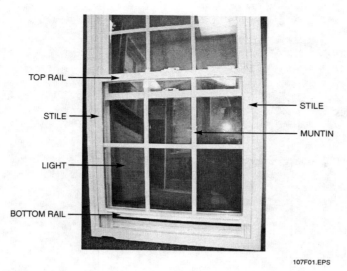

TOP RAIL

STILE

STILE

MUNTIN

LIGHT

BOTTOM RAIL

107F01.EPS

Figure 1 ◆ Parts of a sash.

(or sashes), is shipped from the factory. The window frame consists of the head jamb, side jambs, and sill, as shown in *Figure 2*.

2.2.0 Types of Windows

There are several types of windows commonly used in residential construction. This section provides an overview of each type.

- *Fixed windows*—In a fixed window, the glass cannot be opened. Fixed windows come in a huge variety of styles and shapes. They are often used to create an attractive, distinctive architectural appearance (*Figure 3*). Decorative glazing known as *art glass* (similar to stained glass) is sometimes used to achieve dramatic effects in these windows.
- *Single- and double-hung windows*—(See *Figure 2*.) These windows contain two sashes. In the double-hung window, the upper sash can be lowered and the lower sash can be raised. In a single-hung window, the upper sash is stationary and the lower sash can be raised. Many double-hung windows have a tilt-and-wash feature that allows the outside of the window to be washed from the inside. In one application of this feature, the jamb can be pushed in enough to allow the top of the sash to pivot inward.

- *Casement windows*—Casement windows are hinged on the left or right of the frame so that they can swing open like a door (*Figure 4*). They are usually operated by a hand crank. A swivel arm prevents the window from swinging wide open. Casement windows can be equipped with a limited ventilation control that limits the opening to a few inches. While this feature provides a certain amount of security from outside entry and accidental falls by occupants, it can also prevent people on the inside from getting out during an emergency. It is therefore not recommended for windows that may be used as emergency exits. Casement windows are often installed side by side and may be used in combination with decorative fixed windows.
- *Awning and hopper windows*—Awning windows are hinged on the top and pushed outward to open (*Figure 4*). They are often operated by a hand crank. Awning windows are commonly used in combination with fixed windows. The hopper window is hinged on the bottom and opens inward from the top. It is commonly used for basement windows. The hopper window is equipped with a locking mechanism. It also has pivot arms on the sides to keep the window from falling.

Identify different types of residential windows, and explain the applications for which they are used.

Show Transparencies 4 through 12 (Figures 2 through 4 and 6 through 11).

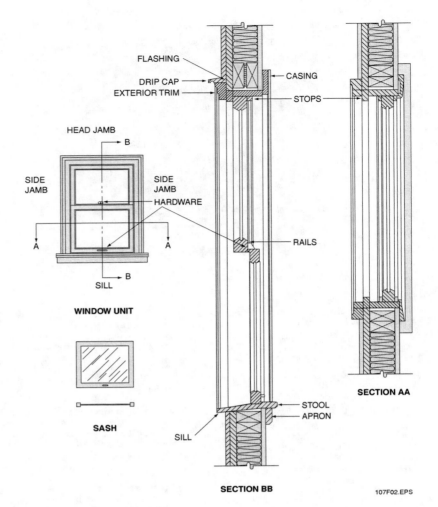

Figure 2 ◆ Parts of a double-hung wood window.

- *Jalousie windows*—A jalousie window consists of a series of horizontal glass slats, each in a pivoting metal frame (*Figure 4*). It can provide security while at the same time providing ventilation. The frames of a jalousie window are joined by pivoting arms so that they operate in unison. A hand crank is used to open and close the slats.

- *Bay windows*—A bay window is a three-walled window that projects outward from the structure (*Figure 5*). Bay windows may be constructed from casement, fixed, or double-hung windows (*Figure 6*). Combinations are often used, such as a fixed window in the center and casement or double-hung windows on the sides. The angles of the bay window may be 30°, 45°, or 90°. The latter is known as a *box bay window*.

Instructor's Notes:

_____

_____

_____

_____

_____

_____

_____

_____

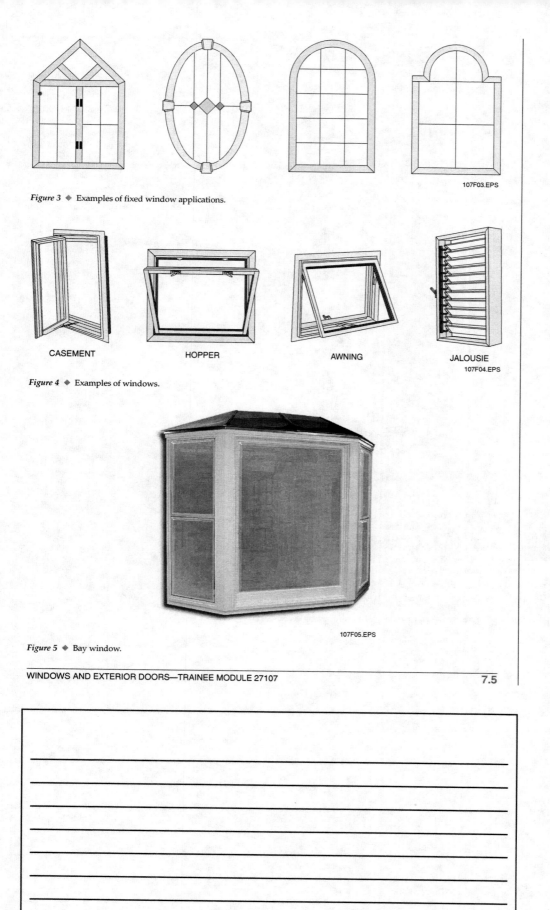

Figure 3 ◆ Examples of fixed window applications.

CASEMENT   HOPPER   AWNING   JALOUSIE

107F04.EPS

Figure 4 ◆ Examples of windows.

107F05.EPS

Figure 5 ◆ Bay window.

DOUBLE-HUNG BAY

CASEMENT BAY

BOX OR GARDEN BAY

107F06.EPS

Figure 6 ◆ Examples of bay windows.

Under certain load conditions, bay windows must be supported by cables that are secured to the building structure (*Figure 7*). This method is usually recommended for projecting units that do not have a support wall beneath them.

• *Bow windows*—A bow window projects out from the structure in a curved radius (*Figure 8*). A bow window is normally made of several narrow, flat planes set at slight angles to each other. Like the bay window, a bow window is commonly made up of casement windows or a combination of casement and fixed windows.

Transom Windows

Transom windows were common in the high-ceiling buildings erected in the early part of the 20th century. Today, transom windows are becoming popular again because of the growing number of new high-ceiling homes. These decorative windows are usually shaped as a semicircle or rectangle, and may be either fixed windows or awning types. They are often used as decorative "toppers" for double-hung windows, as shown here.

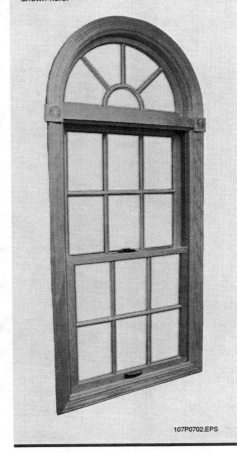

107P0702.EPS

Instructor's Notes:

_____

_____

_____

_____

_____

_____

_____

_____

_____

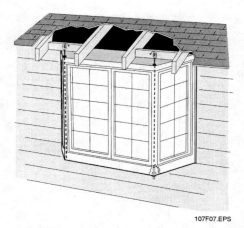

Figure 7 ◆ Supported bay window.

- *Horizontal sliding windows*—A horizontal sliding (gliding) window may be constructed from a number of sashes (*Figure 9*). The most common designs have either two sashes, with one or both sashes movable, or three sashes with the middle sash fixed and the other two movable. Frames are either wood or metal. A locking handle is normally added at installation.
- *Skylights and roof windows*—The difference between a skylight and a roof window is that the roof window can be opened (*Figure 10*). There are many designs and shapes of skylights. Domed and hipped skylights are the most common because they more readily shed rain and snow. Skylights and roof windows are often installed using a built-up curb (*Figure 11*). Metal flashing is installed around the curb to prevent leakage. Another style of skylight is *self-flashing*. This type has an integral metal flange that is screwed into the roof and sealed on both sides with roofing cement during installation.

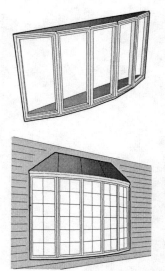

INSIDE TRACK

Flashing

Metal flashing is essential to preventing moisture from entering the roof around the seams of roof windows, skylights, and chimneys.

107P0703.EPS

Figure 8 ◆ Bow windows.

107F09.EPS

Figure 9 ◆ Sliding (gliding) window.

107F10.EPS

Figure 10 ◆ Roof windows and skylights.

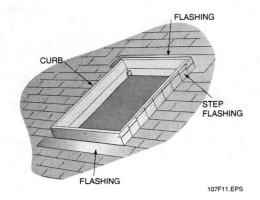

FLASHING

CURB

STEP FLASHING

FLASHING

107F11.EPS

Figure 11 ◆ Skylight curb.

Instructor's Notes:

_____

_____

_____

_____

_____

_____

_____

_____

2.3.0 Types of Window Glass

There are several kinds of sheet glass made for use in window glazing. Single-strength (SS) glass is about ½" thick. It is used only for small lights of glass. Double-strength (DS) glass, which is about ⅛" thick, can be used for larger lights. Heavy-duty glass ranges in thickness from 5/16" to ⅞".

Glass is a good conductor of heat. In any building, glass accounts for the majority of the heat loss in cold months and heat gain in warm months. When an engineer is sizing the heating/cooling system for a building, window glass normally accounts for 38% of the cooling load and 20% of the heating load (*Figure 12*). The term *load* refers to the amount of heat that must be added (heating mode) or removed (cooling mode) to keep the building comfortable for occupants.

Money spent on energy-efficient windows is well spent. For example, the use of high-efficiency reflective glass instead of standard single-pane glass could reduce the cooling load by one-third. The load directly affects the size of the heating/cooling system, and therefore directly affects the

original cost of the system, as well as the energy cost involved in operating the system.

2.3.1 Energy-Efficient Windows

A single pane of glass provides very little insulation. It has an *R-value* (insulating value) of less than 1. (The greater the R-value, the greater the insulating value.) Adding another pane with ½" of air space more than doubles the R-value. The air space between the panes of glass acts as an insulator. The larger the air space, the more insulation it provides. Windows are commonly designed with two or three layers of glass separated by 3/16" to 1" of air space in order to improve insulation quality (*Figure 13*). To obtain even more insulating value, the space between panes in some windows is filled with argon gas, which conducts heat at a lower rate than air. Where single-pane glass is used, it is common to add storm windows.

A special type of glass known as *low-e* (low-emissivity) provides even greater insulating properties. Emissivity is the ability of a material to

Summarize the physical properties of glass, and identify the impact of glass types on energy use.

Show Transparencies 13 and 14 (Figures 12 and 13).

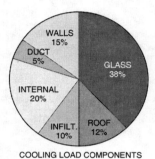

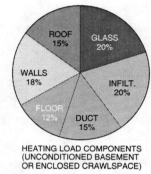

TYPICAL LOAD PERCENTAGES

107F12.EPS

Figure 12 ◆ Cooling and heating load factors.

Reflective Glass

Scientists working in the space program developed the first reflective glass. Spacecraft are subjected to high levels of friction when flying through the Earth's atmosphere at great speeds. Reflective glass was developed in an attempt to reflect, transfer, and dissipate high temperatures. Today, reflective glass is used on the Space Shuttle. This gives astronauts a clear field of vision to navigate the shuttle, and also provides heat protection upon re-entry into the atmosphere.

Provide an overview of basic window installation.

Show Transparencies 15 through 18 (Figures 14 through 17).

Demonstrate how to install a double-hung window.

Energy-Efficient Glass

Low-e glass is coated with a thin metallic substance and is an effective way to control radiant heat transfer. Special heat-absorbing glass is also available. It contains tints that can absorb approximately 45% of incoming solar energy. This energy (heat) is then transferred from the window to the building structure. If these new types of glass aren't available, there are other alternatives that will increase energy efficiency in existing windows. For example, reflective film can be applied to windows to help contain heat loss in the winter and reflect sunlight in the summer. Caulking, weatherstripping, and storm windows, which are discussed in detail later in this module, are also simple approaches to energy conservation.

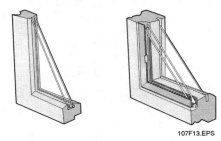

107F13.EPS

Figure 13 ◆ Double- and triple-pane windows.

absorb or radiate heat. Low-e glass is coated with a very thin metallic substance on the inside of the inner pane of a double-pane window. In cold weather, radiated heat from walls, floors, furniture, etc., is reflected back into the room by the low-e coating instead of escaping through the windows. This reduces the heat loss, which in turn saves heating costs. In summer months, radiated heat from outdoor sources such as the sun, roads, parking lots, etc., is reflected away from the building by the low-e coating. Although windows with low-e glass are considerably more expensive than standard windows, they usually pay for themselves in reduced heating and cooling costs within three or four years.

2.3.2 Safety Glass

Some local and state codes require the use of safety glass in windows with very low sills, and in those located in or near doors. Skylights and roof windows also require safety glass. There are several types of safety glass. Laminated glass contains two or more layers of glass bonded to transparent plastic. Tempered glass is treated with heat or chemicals. When broken, it disintegrates into tiny, harmless pieces. Wired glass has a layer of mesh sandwiched between the panes. The mesh keeps the glass from shattering. Transparent plastic (plexiglass) is also used for safety glazing applications, especially in skylights and doors.

2.4.0 Window Installation

This section contains the basic procedure for installing a pre-hung window. As you know, there are many types of windows and many types of installations. The best approach is to follow the manufacturer's recommended procedure, using the recommended tools, materials, and fasteners.

The basic procedure for installing a pre-hung window is as follows:

Step 1 Make sure the window is shut before you start the installation. Also, ensure that the opening is plumb, level, and square, and is large enough to accommodate the window. You will want to leave ¼" to ½" between the rough header and the window head jamb to account for settling.

Check Opening for Squareness

To check the window opening for squareness, use a framing square. If the opening is not square, make the necessary adjustments and check again. You shouldn't proceed until the opening is square and level.

Instructor's Notes:

_____

_____

_____

_____

_____

_____

_____

_____

House Wrap

When house wrap is installed over a rough opening, don't cut the wrap diagonally across the entire opening. Instead, cut diagonally from the bottom corners to the center, then cut straight up from the center to the top of the opening, as shown below. Wrap the material over the sill and sides of the opening and secure it. Cut the wrap flush with the bottom of the header, but do not secure it. This will allow the window fin to slide under the wrap for a more secure installation.

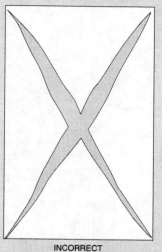

INCORRECT

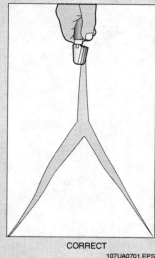

CORRECT

107UA0701.EPS

Step 2 Remove protection blocks used to protect the window during shipping. Also remove any *horns,* which are side jamb extensions that are also used for protection during shipping. If there are diagonal braces on the window, leave them in place temporarily.

Step 3 Install the window in the opening. Make sure you have enough help to avoid injuring yourself or damaging the window.

Step 4 If necessary, insert wedges, wooden shingle tips, or shims under the window to raise it to the correct height. Temporarily nail one corner of the window through the flange or wooden frame.

Note

Vinyl and metal windows, as well as clad wooden windows, are equipped with nailing flanges (fins) around the outside (*Figure 14*). Use screws or roofing nails to secure the window.

Step 5 Check the level of the sill. If it is not level, use shims to correct it. On windows with long sills, use shims at intermediate points to ensure that the sill is level with no sag (*Figure 15*). As each side is leveled, secure it with a nail or screw as recommended by the manufacturer.

Emphasize the proper methods for securing window units to the building structure.

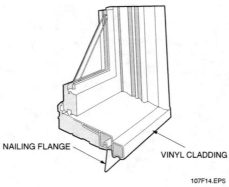

NAILING FLANGE

VINYL CLADDING

107F14.EPS

Figure 14 ◆ Cutaway of a vinyl-clad window.

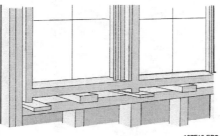

107F15.EPS

Figure 15 ◆ Leveling a window with shims.

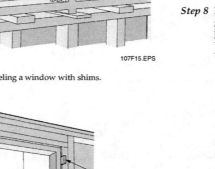

SHIM

107F16.EPS

Figure 16 ◆ Plumb the side jambs.

Step 6 Plumb the side jambs using shims (*Figure 16*), but be sure not to shim too tight. When each jamb is plumb, secure that side.

Step 7 Check to see that the unit is plumb, level, and square. Operate the sash to make sure it works smoothly and does not bind. When it is satisfactory, finish securing the unit. If casing nails are used, use a nail set to drive the nail in order to avoid denting the frame. Check the manufacturer's instructions to see if there are restrictions on nailing into the header.

> **Note**
> Casing nails should be used to secure the unit to the building structure. Don't drive the nails all the way in yet. Use a nail set to finish the job and prevent denting of the frame. You can keep your shims permanently in place by nailing through them during this step. Once the basic unit is installed, finish nails should be used on the trim.

Step 8 Pack insulation or expanding foam in the gaps between the trimmer studs and the jambs (*Figure 17*). Check the window manufacturer's recommendations before using expanding foam.

107F17.EPS

Figure 17 ◆ Insulating between the jambs and trimmer studs.

7.12

Instructor's Notes:

2.4.1 Extension Jambs

Modern windows are made for thinner walls than those found in older buildings. When replacing windows in older buildings, it may be necessary to build out the jamb so that it is flush with the wall (*Figure 18*). Jamb extensions are available as optional accessories from the window manufacturer or they can be ripped from a length of 1 × lumber that is wide enough to fill the gap. Also, for 2 × 6 wall construction, most manufacturers supply extension jambs at 4⅞", 5⅛", 6⅞", etc., as options.

2.5.0 Glass Blocks

Glass block panels are sometimes used in place of standard windows to provide light while at the same time providing privacy. Glass blocks are generally 3⅞" thick and come in three common nominal sizes: 6" × 6", 8" × 8", and 12" × 12". Actual dimensions are ¾" less than nominal size to account for mortar joints. They are made of two formed pieces of glass fused together in a way that leaves an air space in between. The air space provides excellent insulating qualities. They are available in pre-assembled panels in a range of sizes.

Glass blocks are installed using masonry tools and white mortar or silicone. Prefabricated glass block windows are available in a variety of sizes for direct installation. Some are available with a built-in vent for air circulation. Glass block windows can also be constructed from individual blocks (panels), as shown in *Figure 19*. The illustration shows a glass block window system consisting of spacers, panel anchors, expansion strips, and glass blocks. If the window is larger than 25 square feet, panel reinforcing is also required.

The following considerations are important when framing-in glass block windows. To determine the size of the opening required, multiply the nominal block size by the number of blocks and add ⅜" for expansion strips at the header and jambs. When the size of the opening is less than 25 square feet, the height should be no greater than 7'

and the width no greater than 5'. When the opening is greater than 25 square feet, the panels should never be more than 10' wide or 10' high. A built-up sill should be used to protect the bottom row of blocks from damage.

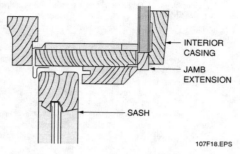

Figure 18 ◆ Jamb extension.

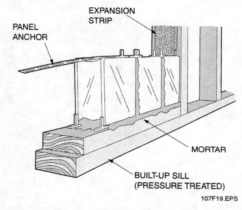

Figure 19 ◆ Glass block installation.

Discuss extension jambs.

Show Transparency 19 (Figure 18).

Explain the procedure for installing glass blocks.

Show Transparency 20 (Figure 19).

Have each trainee complete to your satisfaction Performance Profile Test 1 on installing a pre-hung window. Fill out Performance Profile Sheets for each trainee.

Assign reading of Sections 3.0.0–3.3.0 for the next class session.

3.0.0 ◆ EXTERIOR DOORS

Exterior doors are designed to provide security as well as insulation. Exterior doors, especially main entrance doors, are often designed to give a building an attractive appearance (*Figure 20*). Most residential exterior doors you encounter will be pre-hung in frames with hinges and exterior casings applied. Once the door opening is framed in, all you have to do is install and level the door. The first part of this discussion covers residential doors.

Figure 21 shows the component parts of an exterior door installation. Like windows, doors have headers, side jambs, and sills. There are also several different kinds of exterior doors, as shown in *Figure 22*. You can go to any building supply store and get color brochures containing a wide variety of styles and decorative designs for pre-hung exterior doors.

The panel door is the most common factory-built exterior door. It is made up of vertical members called *stiles,* cross members called *rails,* and filler panels. The panels are usually thinner than the stiles and rails. Panel doors are more decorative than flush doors. The doors shown in *Figure 20* are examples of panel doors. Panel doors can be made of wood or metal.

An exterior flush door has a smooth surface made of wood veneer or metal. It usually has a solid core of wood, composition board, or solidly-packed foam. Hollow-core doors consist of a framework of wood strips, metal honeycombs, or other materials that give the door rigidity. Hollow-core doors are not recommended for exterior use because they provide very little insulation and limited security. In addition, local building codes may prohibit their use in exterior entries. A flush door may contain a glass inset or it may come with decorative moldings attached to the surface.

Sash doors may have a fixed or movable window. The window may be divided into lights. Various types of glass are used, including insulated, reinforced, and leaded. Louvered doors are popular as entry doors in warmer climates.

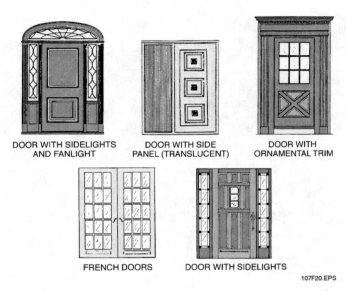

DOOR WITH SIDELIGHTS AND FANLIGHT

DOOR WITH SIDE PANEL (TRANSLUCENT)

DOOR WITH ORNAMENTAL TRIM

FRENCH DOORS

DOOR WITH SIDELIGHTS

107F20.EPS

Figure 20 ◆ Entry doors.

Instructor's Notes:

_____
_____
_____
_____
_____
_____
_____
_____

Trimming Door Length

If you need to trim a veneer door to length, lay a straightedge on your marks and score through the layer of veneer with a utility knife. This will prevent the veneer from splintering when you finish the cut with a saw. Before you make the cut, place a piece of tape on top of the veneer so that the saw baseplate doesn't scratch the surface.

Identify the components of an exterior door and frame.

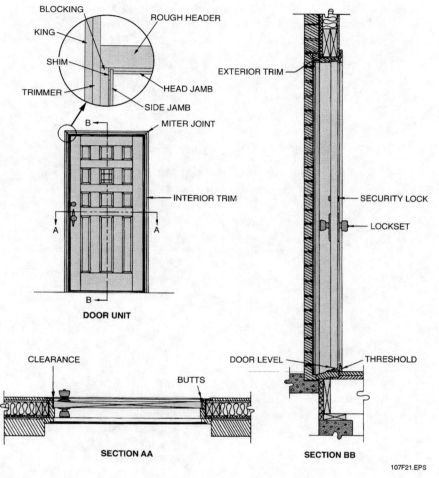

Figure 21 ◆ Parts of a typical exterior door installation.

Sliding Patio Doors

INSIDE TRACK

This type of exterior door is normally used for patios and porches. The door shown is solid oak with a painted aluminum exterior and a transom for use in high-ceiling rooms.

107P0704.EPS

Weatherstripping

Weatherstripping systems, such as thresholds, prevent moisture from entering under a door and reduce heat loss by creating a seal between a door and the floor.

PVC TUBE WEATHER-STRIP

DOOR SWEEP WEATHER-STRIP

107P0705.EPS

Instructor's Notes:

_____

_____

_____

_____

_____

_____

_____

_____

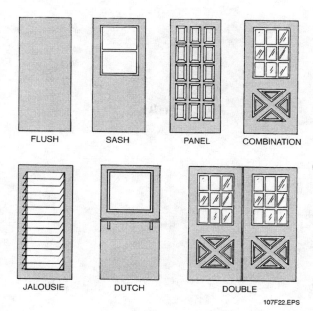

FLUSH SASH PANEL COMBINATION

JALOUSIE DUTCH DOUBLE

107F22.EPS

Figure 22 ◆ Types of entry doors.

3.1.0 Exterior Door Sizes

Residential exterior doors are generally 1¾" thick and come in a variety of widths and heights. A standard width for a main entry door is 36", although the available range is 18" to 40". Double-width doors are available in 60", 64", and 72" widths. A typical height for an exterior door is 80". Other available heights are 78", 79", 84", 90", and 96". Keep in mind that local building codes sometimes specify the types and sizes of entry doors, as well as the location and minimum number of such doors.

3.2.0 Thresholds

A threshold is a wood or metal piece used to close the gap between the entry floor or sill and the door. It is beveled on both sides. *Figure 23* shows examples of thresholds. Weatherseal thresholds have a rubber or plastic strip in the center that is compressed by the closed door to keep out drafts. An installation detail for a threshold is shown in *Figure 23*.

3.3.0 Weatherstripping

Weatherstripping material is added to the bottom of a door to prevent heat from escaping and moisture from entering. One weatherstripping technique uses a rubber or vinyl sweep attached to the bottom of the door (*Figure 24*). Other methods are shown in *Figure 25*. A wide variety of self-stick and tack-on weatherstripping materials are available to seal off door jambs.

3.4.0 Installing an Exterior Pre-Hung Door

A pre-hung door comes equipped with the frame, threshold, and exterior casing. Often, the lockset is included. Doors can be either left-hand or right-hand swing (*Figure 26*). Different manufacturers use different methods of describing the door. The direction of swing is determined when facing the door from the outside of the building.

Discuss the common sizes of exterior doors.

Demonstration

Demonstrate the procedures for installing thresholds and weatherstripping.

Audiovisual

Show Transparencies 24 through 26 (Figures 23 through 25).

Laboratory

Hand out Job Sheets 27107-2 and 27107-3. Under your supervision, have the trainees perform the tasks on the Job Sheets. Note the proficiency of each trainee on his or her Job Sheets and Skill Test Record.

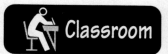

Classroom

Provide an overview of the procedure for installing exterior pre-hung doors.

Performance Profile Test

Have each trainee complete to your satisfaction Performance Profile Tests 2 and 3. Fill out Performance Profile Sheets for each trainee.

Homework

Assign reading of Sections 3.4.0–3.5.0 for the next class session.

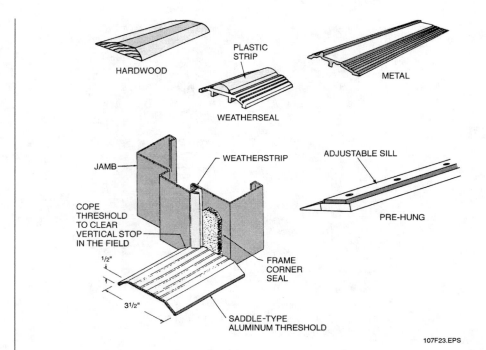

HARDWOOD

PLASTIC STRIP

METAL

WEATHERSEAL

JAMB

WEATHERSTRIP

ADJUSTABLE SILL

COPE
THRESHOLD
TO CLEAR
VERTICAL STOP
IN THE FIELD

PRE-HUNG

1/2"

FRAME
CORNER
SEAL

31/2"

SADDLE-TYPE
ALUMINUM THRESHOLD

107F23.EPS

Figure 23 ◆ Thresholds.

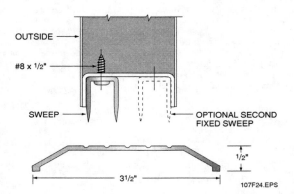

OUTSIDE

#8 x 1/2"

SWEEP

OPTIONAL SECOND
FIXED SWEEP

1/2"

31/2"

107F24.EPS

Figure 24 ◆ Fixed bottom sweep.

Instructor's Notes:

_____

_____

_____

_____

_____

_____

_____

_____

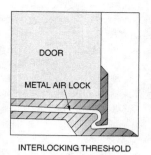
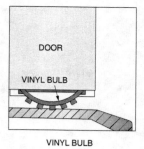
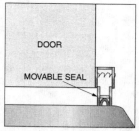

INTERLOCKING THRESHOLD   VINYL BULB   AUTOMATIC DOOR BOTTOM

107F25.EPS

Figure 25 ◆ Weatherstripping.

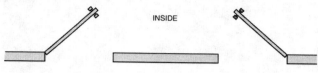

INSIDE

LEFT-HAND DOOR   RIGHT-HAND DOOR

107F26.EPS

Figure 26 ◆ Door swing.

The following is a typical procedure for installing a pre-hung exterior door:

Step 1 As soon as the door is removed from the packaging, inspect it to make sure there is no damage. Remove the nails if the door was nailed shut for shipping.

Step 2 Make sure the rough opening is the right size.

Step 3 Put the door unit in place and mark the inside of the threshold. Remove the door unit and apply a double bead of caulking to the bottom of the opening inside the line you marked.

Step 4 Center the door unit in the opening. Leave the factory-installed spacer shims between the door and the frame in place until the frame is securely attached to the rough opening.

Step 5 Ensure that the sill is level and the hinge jamb is plumb. Use shims to correct, if necessary.

Step 6 Adjust for correct clearance by inserting shims between the side jambs and rough openings at the bottom, top, and middle. Additional shimming may be required on larger doors.

Step 7 Nail the jamb through the shims using 16d finishing nails. Drive the nails home with a nail set to avoid marring the finish.

Step 8 Adjust the threshold so that it makes smooth contact with the door without binding or leaving a space.

Step 9 Remove the top inner screws on the hinge and install 2½" to 3" screws through the door jamb to the trim.

Step 10 Cut off all shims flush with the edge of the door casing and pack any gaps between the trimmer stud and the door casing with insulation.

3.4.1 Locksets

Figure 27 shows examples of locksets that might be used on residential exterior doors. Mortise locksets (*Figure 28*) are more secure and are therefore considerably more expensive than the other types shown. For that reason, they are more common in commercial buildings than in residential construction. Tubular locksets (*Figure 29*) are less secure than cylindrical locksets (*Figure 30*). Neither is an excellent choice where security is a major concern.

Show Transparency 27 (Figure 26).

Demonstrate how to install a pre-hung door.

Hand out Job Sheet 27107-4. Under your supervision, have the trainees perform the tasks on the Job Sheet. Note the proficiency of each trainee on his or her Job Sheet and Skill Test Record.

Discuss the types of locksets used on residential exterior doors and provide an overview of their installation requirements.

Show Transparencies 28 through 34 (Figures 27 through 33).

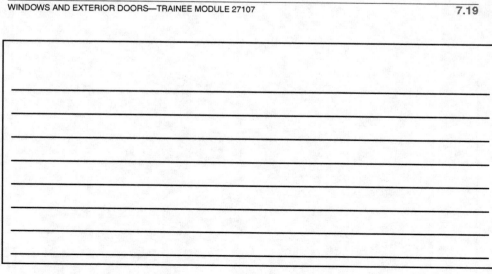

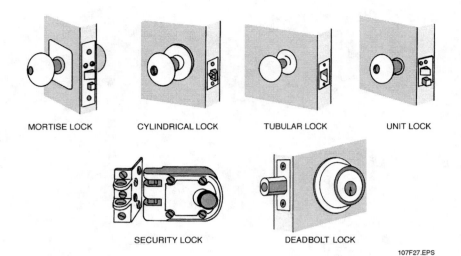

MORTISE LOCK   CYLINDRICAL LOCK   TUBULAR LOCK   UNIT LOCK

SECURITY LOCK   DEADBOLT LOCK

107F27.EPS

Figure 27 ◆ Examples of exterior door locksets and security locks.

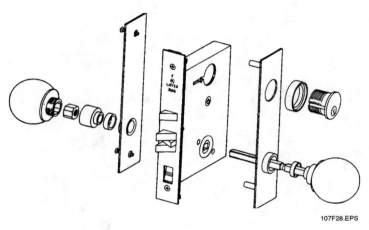

107F28.EPS

Figure 28 ◆ Exploded view of a mortise lock.

Installing Door Strikes

When mortising for a doorplate strike or hinge, score around the plate or hinge, then score every ¼" down the mortise before applying the chisel. This will allow you to chisel out small distances and produce a clean, accurate mounting area on the jamb for the strike or hinge.

Instructor's Notes:

_____

_____

_____

_____

_____

_____

_____

_____

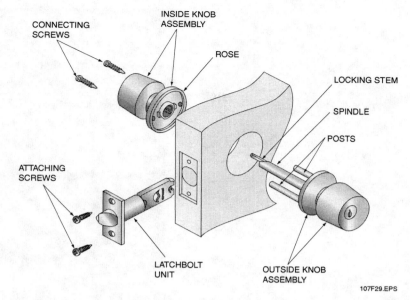

Figure 29 ◆ Disassembled view of a tubular lockset.

CONNECTING SCREWS

INSIDE KNOB ASSEMBLY

ROSE

LOCKING STEM

SPINDLE

POSTS

ATTACHING SCREWS

LATCHBOLT UNIT

OUTSIDE KNOB ASSEMBLY

107F29.EPS

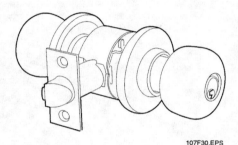

107F30.EPS

Figure 30 ◆ Heavy-duty cylindrical lockset.

In many cases, homeowners will want a security lock or deadbolt lock in addition to the locking mechanism built into the doorknob assembly. Deadbolts are either single- or double-cylinder types. The double-cylinder type requires a key on both sides. It would be used in cases where there is glass close to the lock. If the door is solid, a security lock or single-cylinder deadbolt (requiring a key on the outside only) will suffice.

Lock manufacturers include installation instructions and drilling templates (*Figure 31*) in the packaging for locksets and other locking devices. Professionals generally prefer to use a boring jig and boring bit, along with a mortise marker (*Figure 32*) instead of a template, because this method is faster and more accurate.

Locksets are usually installed at a height of 36" to center from the floor. Correct measuring is extremely important because the bolt on the lockset must fit into a strike plate installed on the door jamb (*Figure 33*). Check the manufacturer's instructions to determine if the backset is 2⅜" or 2¾".

Demonstrate how to install a lockset.

Hand out Job Sheet 27107-5. Under your supervision, have the trainees perform the tasks on the Job Sheet. Note the proficiency of each trainee on his or her Job Sheet and Skill Test Record.

Emphasize the importance of checking the door opening size before attempting to position a door.

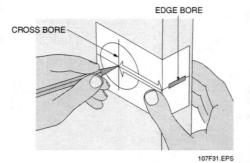

Figure 31 ◆ Using an installation template.

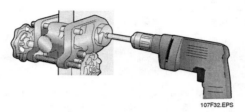

Figure 32 ◆ Boring jig and mortise marker.

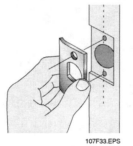

Figure 33 ◆ Installing a door strike.

INSIDE TRACK

Measuring the Rough Opening

Doors can be unwieldy to carry and position, so always check the rough opening size before attempting to move the door into place.

TAPE MEASURE 107P0706.EPS

Shims

Always keep plenty of shims on hand when installing windows and doors. Shims are easy to cut by simply scoring them with a utility knife, then snapping them off along the score.

SHIM 107P0707.EPS

7.22 CARPENTRY LEVEL ONE—TRAINEE MODULE 27107

Instructor's Notes:

_____

_____

_____

_____

_____

_____

_____

_____

3.5.0 Installing a Garage Door

There are three basic types of garage door assemblies: hinged or swinging, rollup, and swingup. *Figure 34* shows a rollup type, which is the most common type for residential use. These doors come in standard widths of 8', 9', 10', 16', and 18'. Standard heights are 6'-6", 7'-0", and 8'-0".

The following is the basic procedure for framing and installing a rollup garage door (this procedure is generally done by a specialist):

Step 1 Make sure the rough opening is level, square, and plumb.

Step 2 Cut the frame members to size. For most residential doors, the header jamb should be the width of the door plus twice the thickness of the side jamb.

Step 3 Install and level the finish header jamb, then nail it in place.

Step 4 Install the finish side jambs. Plumb and square them, then nail them in place.

Step 5 Check the finish opening to be sure it is the size of the garage door.

Step 6 Install the inside casing and track-mount supports. The casing runs full length between the floor and ceiling. It is sometimes mounted to studs.

Step 7 Stack the door sections in the opening and attach the hinges and rollers. Make sure the bottom section is level. Use shims if necessary.

Step 8 Install the vertical sections of the track over the rollers. Level the tracks, then temporarily secure the tracks to the wall.

Step 9 Install the horizontal track sections and square them with the door opening.

Step 10 Raise the door, prop it open, and install the counterbalance springs.

Step 11 Install the automatic garage door opener, if applicable (*Figure 35*).

Step 12 Install compression stops on the outside around the door.

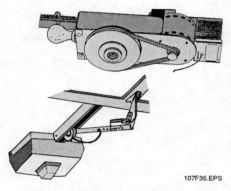

Figure 35 ◆ Electric garage door openers.

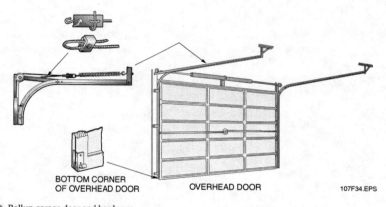

BOTTOM CORNER
OF OVERHEAD DOOR

OVERHEAD DOOR

107F34.EPS

Figure 34 ◆ Rollup garage door and hardware.

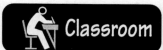

Explain the procedure for installing a garage door.

Show Transparencies 35 and 36 (Figures 34 and 35).

Demonstrate how to install a standard door frame, casing, door, and lock.

Hand out Job Sheet 27107-6. Under your supervision, have the trainees perform the tasks on the Job Sheet. Note the proficiency of each trainee on his or her Job Sheet and Skill Test Record.

Have each trainee complete to your satisfaction Performance Profile Tests 4, 5, and 6. Fill out Performance Profile Sheets for each trainee.

Summary

There are many different styles and types of windows and exterior doors, made by many different manufacturers. The overwhelming majority of the windows and exterior doors encountered by carpenters are the pre-hung type. While the basic designs and installation requirements are essentially the same, the specifics will vary from one manufacturer to another. Selection and installation may also depend to some extent on local building codes. The important thing to remember is to follow the instructions provided by the manufacturer for each window and become familiar with local building codes.

Instructor's Notes:

_____

_____

_____

_____

_____

_____

_____

_____

Review Questions

1. The framework around the glass in a window is known as the _____.
   a. muntin
   b. sash
   c. casing
   d. header

2. A double-hung window contains _____.
   a. a single, fixed sash
   b. two fixed sashes
   c. one fixed sash and one movable sash
   d. two movable sashes

3. A *self-flashing* skylight is called that because it _____.
   a. has a built-in metal flange
   b. comes with a roll of aluminum tape
   c. comes with a roof curb
   d. reflects sunlight

4. Of the types of glazing listed below, the one that has the greatest insulating value is _____.
   a. DS
   b. low-e
   c. double-pane
   d. argon-filled

5. Extension jambs are most likely to be needed when _____.
   a. the window opening is too high
   b. the new window is not as wide as the old window
   c. a modern replacement window is installed in an older building
   d. casement windows are used

6. When a glass block window opening is greater than 25 square feet, the maximum height permitted is _____.
   a. 7'
   b. 5'
   c. 25'
   d. 10'

7. A flush exterior door usually has _____.
   a. a hollow core
   b. a solid core
   c. decorative work carved into its face
   d. stiles, rails, and filler panels

8. The most common thickness of a residential exterior door is _____.
   a. 1¼"
   b. 1½"
   c. 1¾"
   d. 2"

9. Which of the following types of locksets is the most secure?
   a. Cylindrical
   b. Mortise
   c. Deadbolt
   d. Tubular

10. The header jamb on a garage door should be _____.
    a. the same width as the door
    b. twice the width of the door
    c. the width of the door plus the combined thickness of the two side jambs
    d. the same thickness as the door

Classroom

Have the trainees complete the Review Questions and go over the answers prior to administering the Module Examination.

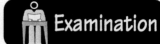

Examination

Administer the Module Examination and Performance Test. Be sure to record the results on Craft Training Report Form 200 and submit the results to the Training Program Sponsor.

Performance Profile Test

Ensure that all Performance Profile Tests have been completed and Performance Profile Sheets for each trainee are filled out. Be sure to record the results of Testing on Craft Training Report Form 200 and submit the results to the Training Program Sponsor. Answers can be found on the key in the Test Booklet.

Trade Terms Introduced in This Module

Casing: The trim around a window or door.

Curb: A framework on which a skylight is mounted.

Flashing: Sheet metal strips used to seal a roof or wall against leakage.

Glazing: Material such as glass or plastic used in windows, skylights, and doors.

Hipped: The external angle formed by the meeting of two adjacent sloping sides.

Jamb: The top and sides of a door or window frame that are in contact with the door or sash.

Lights: The glass inserts in a door.

Muntin: A thin framework used to secure panes of glass in a door.

Rail: The horizontal member of a window sash or panel door.

Sash: The part of a window that holds the glass.

Sill: The lowest member of a window or exterior door frame.

Stile: The vertical member of a window sash or panel door.

Instructor's Notes:

_____

_____

_____

_____

_____

_____

_____

_____

Answers to Review Questions

Answer	Section
1. b	2.1.0
2. d	2.2.0
3. a	2.2.0
4. b	2.3.1
5. c	2.4.1
6. d	2.5.0
7. b	3.0.0
8. c	3.1.0
9. b	3.4.1
10. c	3.5.0

Additional Resources

This module is intended to present thorough resources for task training. The following reference works are suggested for further study. These are optional materials for continuing education rather than for task training.

Carpentry. Homewood, IL: American Technical Publishers.

Carpentry. Albany, NY: Delmar Publishers.

Modern Carpentry. Tinley Park, IL: The Goodheart-Willcox Company, Inc.

Instructor's Notes:

_____

_____

_____

_____

_____

_____

_____

_____

The NCCER makes every effort to keep these textbooks up-to-date and free of technical errors. We appreciate your help in this process. If you have an idea for improving this textbook, or if you find an error, a typographical mistake, or an inaccuracy in the NCCER's Craft Training textbooks, please write us, using this form or a photocopy. Be sure to include the exact module number, page number, a detailed description, and the correction, if applicable. Your input will be brought to the attention of the Technical Review Committee. Thank you for your assistance.

Instructors – If you found that additional materials were necessary in order to teach this module effectively, please let us know so that we may include them in the Equipment/Materials list in the Instructor's Guide.

Write: Curriculum Revision and Development Department
National Center for Construction Education and Research
P.O. Box 141104, Gainesville, FL 32614-1104

Fax: 352-334-0932

E-mail: curriculum@nccer.org

Craft _____ Module Name _____

Copyright Date _____ Module Number _____ Page Number(s) _____

Description _____

_____

_____

_____

_____

(Optional) Correction _____

_____

_____

(Optional) Your Name and Address _____

_____

_____

Photo Credits

Module 27101

Arcways, 101F05.TIF
Thomas Burke, 101F04.TIF
Steve Metz, 101F03.TIF
Photodisk, Inc., © 2000, 101P0102.TIF
Gary Wilson, 101P0101.TIF

Module 27102

Michael Anderson, 102F04.EPS, 102P0203.EPS, 102P0204.EPS
Thomas Burke, 102P0201.EPS, 102F11.EPS, 102F12.EPS, 102P0205.EPS
Gerald Shannon, 102P0202.EPS, 102F21.EPS
Trus Joist, 102F09A.EPS, 102F09B.EPS, 102F09C.EPS, 102F10.EPS

Module 27103

Daniele Dixon, 103P0306.EPS, 103P0308.EPS, 103P0310.EPS
Donald Dixon, 103F09.EPS
Jonathan Liston, 103F21.EPS
Gerald Shannon, 103F02.EPS, 103F03.EPS, 103P0301.EPS, 103P0302.EPS, 103P0303.EPS, 103P0304.EPS, 103P0305.EPS, 103F10.EPS, 103P0307.EPS, 103F17.EPS, 103F19.EPS, 103F20.EPS, 103F24A.EPS, 103F24B.EPS, 103F26.EPS, 103P0309.EPS, 103F29.EPS, 103F30.EPS, 103F33.EPS, 103F34.EPS, 103F35.EPS

Module 27104

Thomas Burke, 104P0401.EPS, 104P0404.EPS, 104P0406.EPS
Gerald Shannon, 104P0403.EPS, 104P0405.EPS
Trus Joist, 104P0402.EPS, 104F20.EPS

Module 27105

Thomas Burke, 105P0501.EPS, 105P0502.EPS, 105P0503.EPS

Module 27106

Thomas Burke, 106P0602.EPS, 106P0603.EPS, 106P0604.EPS
Gerald Shannon, 106P0601.EPS, 106F29.EPS, 106P0605.EPS, 106F43.EPS, 106F44.EPS, 106P0607.EPS
Southern Forest Products Association, 106F46.EPS
Texas Instruments, 106P0604.EPS
Wood Truss Council of America (www.woodtruss.com), 106P0606.EPS

Module 27107

Erin Kellem, 107P0701.EPS
Gerald Shannon, 107F01.EPS, 107F05.EPS, 107P0702.EPS, 107P0703.EPS, 107P0704.EPS, 107P0705.EPS, 107P0706.EPS, 107P0707.EPS

PC.1